Withdrawn
University of Waterloo

Principles of
String Theory

Series of the Centro de Estudios Científicos de Santiago

Series Editor: Claudio Teitelboim
Centro de Estudios Científicos de Santiago
Santiago, Chile
and University of Texas at Austin
Austin, Texas, USA

IONIC CHANNELS IN CELLS AND MODEL SYSTEMS
Edited by Ramon Latorre

PHYSICAL PROPERTIES OF BIOLOGICAL MEMBRANES AND
THEIR FUNCTIONAL IMPLICATIONS
Edited by Cecilia Hidalgo

PRINCIPLES OF STRING THEORY
Lars Brink and Marc Henneaux

QUANTUM MECHANICS OF FUNDAMENTAL SYSTEMS
Edited by Claudio Teitelboim

Principles of String Theory

Lars Brink
*Chalmers University of Technology
and University of Göteborg
Göteborg, Sweden*

and
Marc Henneaux
*Université Libre de Bruxelles
Brussels, Belgium
and Centro de Estudios Científicos de Santiago
Santiago, Chile*

PLENUM PRESS • NEW YORK AND LONDON

Library of Congress Cataloging in Publication Data

Brink, Lars.
  Principles of string theory / Lars Brink and Marc Henneaux.
      p.   cm. — (Series of the Centro de Estudios Científicos de Santiago)
  Bibliography: p.
  Includes index.
  ISBN 0-306-42657-9
  1. String models. I. Henneaux, Marc. II. Title. III. Series.
  QC794.6.S85B75   1988
  530.1—dc19   87-29815
                                                                                 CIP

© 1988 Plenum Press, New York
A Division of Plenum Publishing Corporation
233 Spring Street, New York, N.Y. 10013

All rights reserved

No part of this book may be reproduced, stored in a retrieval system, or transmitted in any form or by any means, electronic, mechanical, photocopying, microfilming, recording, or otherwise, without written permission from the Publisher

Printed in the United States of America

Preface

The almost irresistible beauty of string theory has seduced many theoretical physicists in recent years. Even hardened men have been swept away by what they can already see and by the promise of even more.

It would appear fair to say that it is not yet clear what form the theory will finally take and in what precise way it will relate to the physical world. However, it would seem equally fair to state that, most likely, strings are here to stay and will play a profound and central role in our conception of the universe.

There is therefore a pressing need to provide both practicing physicists and advanced students with ways to master quickly, but soundly, the basic principles of the theory. The present volume is a step in that direction. It contains a lucid presentation of the basic principles of string theory in forms which may survive future developments.

The book is an outgrowth of lectures given by Lars Brink and Marc Henneaux at the Centro de Estudios Científicos de Santiago. The lectures covered in a self-contained manner different but complementary aspects of the foundations of string theory.

The reader will find in Brink's contribution a masterly and careful review of the light-cone gauge quantization of string models, which goes from the bosonic and the spinning string to superstrings and the heterotic string, with precious information about the light-cone gauge supersymmetry algebra. Building on that analysis, he then proceeds to discuss interactions and the field theory of superstrings.

Highlights of Henneaux's presentation include a detailed gauge-invariant Hamiltonian analysis, a thorough and incisive treatment of the Becchi-Rouet-Stora-Tyutin quantization of both the bosonic and the spinning string, and a covariant formulation of superstrings.

For those of us who had the privilege of attending them, the lectures were, to use a dry but eloquent adjective, extremely useful. We are happy to share them with colleagues and students in what we believe should be a volume of long-lasting interest and utility.

Professors Brink and Henneaux were among the first visitors to our center and we are extremely grateful to them. Thanks are also due to Jorge Zanelli, María Luisa Valdovinos, and Yolanda Flores for their help, and especially to Rafael Rosende for his assistance in preparing the manuscript for publication.

The lectures were conducted under a program sponsored by the Tinker Foundation and the visits were also made possible by support from the United Nations Development Program and the Fonds National Belge de la Recherche Scientifique, respectively.

Claudio Teitelboim

Quebrada de las Terneras, Chile

Contents

Comparison of Conventions Used in Parts I and II . . . xiii

Part I. Lectures on Superstrings
Lars Brink

1. Introduction 3

2. Bosonic Strings 9

3. Spinning Strings 21

4. Superstrings 31

5. The Heterotic String 41

6. The Operator Formalism 49

7. Field Theory for Free Superstrings 61

8. Interaction Field Theory for Type IIB Superstrings . . 69

9. Other String Interactions and the Possible Occurrence of Anomalies 77

10. Outlook 83

Appendix. Some Notations and Conventions 87

References 89

Part II. Lectures on String Theory, with Emphasis on Hamiltonian and BRST Methods

Marc Henneaux

11. Introduction 95

    General References 96

12. The Nambu–Goto String: Classical Analysis 97
    12.1. Action Principle 97
        12.1.1. Nambu–Goto Action 97
        12.1.2. Quadratic Form of the Action 100
        12.1.3. σ-Model Interpretation of the Action . . 103
        12.1.4. Gauge Invariances 103
        12.1.5. Global Symmetries 104
        12.1.6. Conformal Symmetry 105
        12.1.7. Boundary Conditions 110
    12.2. Hamiltonian Formalism 114
        12.2.1. Constraints 114
        12.2.2. Meaning of the Constraints—Simplification of the Formalism 118
        12.2.3. Hamiltonian Form of the Boundary Conditions (Open Case) 122
        12.2.4. Hamiltonian Expression for the Poincaré Charges 124
    12.3. A Closer Look at the Constraint Algebra . . . 124
        12.3.1. Explicit Computation 124
        12.3.2. Virasoro Conditions 127
    12.4. Fourier Modes 129

Contents

  12.4.1. Open String 129
  12.4.2. Closed String 131
 12.5. Light-Cone Gauge 133
  12.5.1. Conformal Gauges 133
  12.5.2. Light-Cone Gauge 136
  12.5.3. General Solution of the String Classical Equations 139
  12.5.4. Independent Degrees of Freedom—Dirac Brackets 143
  12.5.5. Light-Cone Gauge Action—Light-Cone Gauge Hamiltonian 145
  12.5.6. Poincaré Generators 146
  12.5.7. Peculiarities of the Closed String . . . 147

13. Quantization of the Nambu-Goto String . . . 151
 13.1. General Considerations—Virasoro Algebra . . . 151
  13.1.1. Introduction 151
  13.1.2. Fock Representation—Virasoro Operators . 152
  13.1.3. Virasoro Algebra 154
  13.1.4. Virasoro Constraints versus the Wheeler-De Witt Equation 155
  13.1.5. Virasoro Algebra and Kac-Moody Algebras . 159
  13.1.6. Virasoro Algebra in Curved Backgrounds . 161
 13.2. Becchi-Rouet-Stora-Tyutin (BRST) Quantization of the String 162
  13.2.1. BRST Quantization—A Rapid Survey . . 162
  13.2.2. Classical Expression of the BRST Charge . 165
  13.2.3. Ghost Fock Space 166
  13.2.4. Nilpotency of the Quantum BRST Operator . 168
  13.2.5. Critical Dimension in Curved Backgrounds . 169
  13.2.6. Physical Subspace 170
  13.2.7. Remarks on the Doubling 176
  13.2.8. Miscellanea 178
 13.3. Light-Cone Gauge Quantization 181
  13.3.1. Poincaré Invariance of the Quantum Theory . 181
  13.3.2. Description of the Spectrum 183
  13.3.3. Closed String—Poincaré Invariance . . . 185
  13.3.4. Spectrum (Closed String) 186
 13.4. Covariant Quantization 187
  13.4.1. Elimination of Ghosts as the Central Issue in the Covariant Approach 187
  13.4.2. Vertex Operators 191
  13.4.3. DDF States 193
  13.4.4. No-Ghost Theorem for $d = 26$, $\alpha_0 = 1$ . . 196
  13.4.5. Quantum Gauge Invariance 199

14. The Fermionic String: Classical Analysis 201
  14.1. Local Supersymmetry in Two Dimensions . . . 201
  14.2. Superconformal Algebra 202
    14.2.1. Square Root of the Bosonic Constraints and
           Fermionic Constraints 202
    14.2.2. Boundary Conditions 207
    14.2.3. Supergauge Transformations—Light-Cone Gauge
           Conditions 209
    14.2.4. Poincaré Generators 210
  14.3. Fourier Modes (Open String) 210
    14.3.1. Fourier Expansion of the Fields . . . 210
    14.3.2. Super-Virasoro Generators 211
    14.3.3. Poincaré Generators 213
    14.3.4. Remarks on the Closed String 213
    14.3.5. Super-Virasoro Algebra 214

15. The Fermionic String: Quantum Analysis 217
  15.1. Becchi-Rouet-Stora-Tyutin (BRST) Quantization of the
       Neveu-Schwarz Model 217
    15.1.1. Ghost Fock Space 217
    15.1.2. BRST Operator 218
    15.1.3. Critical Dimension 219
    15.1.4. Structure of the Physical Subspace . . . 219
  15.2. Becchi-Rouet-Stora-Tyutin (BRST) Quantization of the
       Ramond Model 222
    15.2.1. Ghost Fock Space 222
    15.2.2. BRST Operator 223
    15.2.3. Critical Dimension 224
    15.2.4. Structure of the Physical Subspace . . . 226
    15.2.5. Remarks on the Closed String 228
  15.3. Light-Cone Gauge Quantization of the Neveu-Schwarz
       Model 229
    15.3.1. Poincaré Invariance 229
    15.3.2. Neveu-Schwarz Spectrum 230
    15.3.3. The Closed Neveu-Schwarz Spectrum . . 231
  15.4. Light-Cone Gauge Quantization of the Ramond Model 232
    15.4.1. Poincaré Invariance 232
    15.4.2. Ramond Spectrum 232
    15.4.3. Closed String 234
  15.5. Supersymmetry in Ten Dimensions 235
    15.5.1. Open String 235
    15.5.2. Closed String 236

Contents

16. The Superstring 237
    16.1. Covariant Action 237
        16.1.1. SUSY(N)/SO($d-1,1$) as Target Space . . 237
        16.1.2. Invariant Actions 240
        16.1.3. Local Supersymmetry 243
        16.1.4. Equations of Motion and Boundary Conditions 246
        16.1.5. Structure of Gauge Symmetries 247
        16.1.6. Super-Poincaré Charges 249
        16.1.7. Hamiltonian Formalism 250
        16.1.8. Light-Cone Gauge 251
    16.2. Quantum Theory 255
    16.3. The Superparticle 256
        16.3.1. Action–Gauge Symmetries 256
        16.3.2. Super-Poincaré Charges 258
        16.3.3. Hamiltonian Formalism 258
        16.3.4. Meaning of the Constraints 260
        16.3.5. Siegel's Model 261
        16.3.6. Light-Cone Gauge 266

17. The Heterotic String 273

Appendix A. BRST-Based Demonstration of the No-Ghost Theorem for the Bosonic String 275

Appendix B. γ Matrices in Ten Dimensions 279
    B.1. Symmetry Properties 279
    B.2. Fierz Rearrangements 281

Appendix C. Light-Cone Gauge Decomposition of Ten-Dimensional Spinors 283

References 289

Index 293

Comparison of Conventions Used in Parts I and II

There is little difference between the notations used in the two parts of this volume. The following table should enable the reader to pass back and forth between both parts without difficulty.

	Part I	Part II
Space-time index	$\mu, \nu, \cdots = 0, 1, \cdots, d-1$ ($d = 26$ or 10)	$A, B, \cdots = 0, 1, \cdots, d-1$ ($d = 26$ or 10)
Range of spatial coordinates for closed strings	$0 \leq \sigma \leq \pi$	$0 \leq \sigma \leq 2\pi$
	$\sigma_{\text{Part I}} = \frac{1}{2}\sigma_{\text{Part II}}$	
Fermi oscillator for Ramond model	d^μ_m	$(1/\sqrt{2})\Gamma^A_m$
SO(8) light cone gauge spinors for superstring	S^a	$\sqrt{p^+}\,\theta^a$

Part I

Lectures on Superstrings

Lars Brink

Chapter 1

Introduction

In the 1960s the four basic interactions, the strong, weak, electromagnetic, and gravitational interactions, were treated in quite different ways. While quantum field theory had been successful for quantum electrodynamics and successful and mysterious for the weak interactions, for the strong interactions very different methods were attempted. At this time gravity was hardly part of elementary particle physics.

One of the most important discoveries in elementary particle physics around this time was the existence of an abundance of hadronic resonances. Plotting them as spin versus mass^2 they fall on (nearly) linear trajectories. These so-called Regge trajectories played a crucial role in Regge theory [1], a theory based on the analyticity of the S-matrix. In this theory high-energy behavior is governed by "Regge poles," which are the Regge trajectories in the cross channel. The fact that there seemed to be a dual description of S-matrix amplitudes either in terms of intermediate resonances or asymptotically by Regge poles [2] spurred huge efforts to construct models incorporating both aspects. These models necessarily needed infinitely many states. The culmination came with the Veneziano (dual) model [3], which proved to be quite a successful model for the scattering of meson. An even greater effort was now made to understand the details of the model. It was soon realized that the version of the model which best fitted data turned out to have theoretical shortcomings, such as negative-norm states as intermediary states. However, Virasoro [4] found that by choosing a specific nonphysical (from a hadronic point of view) intercept for the leading trajectory, the model possesses a huge symmetry, corresponding to an infinite algebra, the Virasoro algebra. For this case it was finally proven by Brower, and Goddard and Thorn [5] that there are no negative-norm states in the model (although the model has tachyonic states).

Shortly after the discovery of the Veneziano model a remarkable observation was made by Nambu, Nielsen, and Susskind [6]. They realized that the model describes the scattering of one-dimensional objects, strings. Attempts had certainly been made before to construct scattering models for extended objects, but all had failed. The main obstacles had been to implement causality. For the Veneziano model this problem was indeed solved. The string picture that emerged was quite appealing for mesons. Somehow it was believed by many that the missing parts in the Veneziano model were the quarks at the ends.

Another shortcoming of the model was the absence of fermions. In order to allow for fermions Ramond [7] generalized the Virasoro algebra to also include anticommuting generators, in fact laying the ground for supersymmetry! Soon after Neveu and Schwarz [8] found another bosonic model realizing such a symmetry. Eventually it was shown that the Ramond fermions fitted into the Neveu-Schwarz model, building up one model with both bosons and fermions [9].

For both these models the very striking fact was found that the model could seemingly be consistent (apart from the tachyons) only in specific "critical dimensions" of space time, $d = 26$ for the Veneziano model and $d = 10$ for the Ramond-Neveu-Schwarz model. These facts found a natural explanation when Goddard *et al.* [10] considered the quantization of a relativistic string. The action for such an object, taken to be proportional to the area of the world sheet swept out by the string, had been proposed by Nambu [11] and later by Hara and Goto [12]. The string was quantized in two ways. In one way it was quantized covariantly reproducing the description of physical states in the Veneziano model, and in another way in the light-cone gauge, where all states have explicitly positive norm but where the Lorentz algebra is realized nonlinearly. The constraint $d = 26$ appears, since it is only for this case that the algebra closes appropriately. An easy way to understand the critical dimensions was found [13] by considering the string as an infinite set of harmonic oscillators. Each oscillator will contribute a zero-point fluctuation to the energy upon quantization. Summing this infinite sum and renormalizing the velocity of light led to the critical dimensions. It was the first indication, to my knowledge, that quantum mechanics demands a finite velocity of light. The quantization in arbitrary dimensions was for a long time obscured, but eventually, after some ten years, it was illuminated by Polyakov [14], who showed one such possible scheme, which in fact leads to the possibility of having string theories in four dimensions. This still seems quite likely to be a good limiting case of QCD, at least in the multicolor limit.

The generalization of the Nambu action to a "spinning string" corresponding to the Ramond-Neveu-Schwarz model turned out to be not quite straightforward. Such strings demand (classically) Grassmann degrees of freedom and there does not seem to be a generalization of the area concept above to a world surface endowed with anticommuting degrees of freedom. The solution to this problem was to consider the theories as general relativity theories in two space time dimensions (the world sheet) [15]. With the advent of supergravity the spinning string theory could be constructed as a supergravity theory on the world sheet. Another way of interpreting these actions is in terms of σ-models. Long before the general action for the spinning string was written, the basic structure became known [16]. It showed clearly the need for two-dimensional supersymmetry and the uniqueness of the critical dimension $d = 10$.

The spinning-string model is evidently not a theory for hadrons and it was natural to search for new string models. The fact that the model is based on supersymmetry on the world sheet made it natural to ask if this symmetry could be extended. In fact, it can be extended to couple to an $SO(N)$ symmetry [17]. Unfortunately, it turned out that only the $SO(2)$ model is ghost-free, but it has a critical dimension of $d = 2$ and thus has no chance of being an interesting model in particle physics [18].

The Veneziano model originally incorporated S-matrix Born terms. When loop corrections were attempted, it was found that certain loop graphs contain a new set of poles [19]. In fact, this is true only for $d = 26$ (in other dimensions one has unitarity-violating cuts) and was the first hint of the critical dimension. The new trajectory carries the quantum numbers of the vacuum and has double the intercept and half the slope of the leading "ρ trajectory," which are the signs of the Pomeron. This trajectory had been missing in the model and its appearance aroused a lot of interest. In the string picture the new set of states was easy to understand. While the trajectories in the tree amplitudes correspond to strings with open ends, the new ones correspond to closed strings and in a string interaction it is natural that the two open ends of a string can join to make up a closed string.

At this stage the question seemed to be how to lower the ρ and Pomeron trajectories to their physical values while keeping the ratio of intercepts, and still have a consistent model. The Ramond-Neveu-Schwarz model did not solve this problem and, in fact, it has not yet been solved. The failure to solve it caused interest in string models for hadronic physics to fade away some 13 years ago. Many of the concepts developed in dual models have, however, influenced modern thinking of gauge field theories and hadron physics in terms of QCD, for example, the Nielsen-Olesen vortex lines [20]. These were developed as an example of stringlike solutions in ordinary field theory, but led to major developments in nonperturbative aspects of gauge field theories.

The consistent Veneziano model demanded a "ρ trajectory" of intercept one, i.e., the spin-1 particle must be massless. Neveu and Scherk [21] then checked the scattering amplitudes for this state in the limit when the slope of the trajectory goes to zero (pushing all massive states to infinite mass) and found that the S-matrix is just that of Yang–Mills theory. Later Yoneya [22] and Scherk and Schwarz [23] showed that the massless spin-2 particle on the "Pomeron trajectory" in the zero-slope limit interacts appropriately to be identified as a graviton. Scherk and Schwarz [23] then made the bold suggestion that string models should be regarded as unified models including gravity! Strings are characterized by a fundamental length scale L $[\sim(\alpha')^{1/2} \sim (T)^{-1/2}$, where α' is the Regge slope and T the string tension]. In the case of hadron physics L is taken to be of order 10^{-13} cm. If we instead regard the string quanta as fundamental quanta including gravity and Yang–Mills fields, the natural length scale is the Planck length, 10^{-33} cm.

The Ramond–Neveu–Schwarz model considered in this new light is still not a consistent model. It contains tachyons and a detailed study of some of the couplings, such as those of the closed-string state of spin $\frac{3}{2}$, would reveal inconsistencies. Gliozzi, Scherk, and Olive [24], however, pointed out that the tachyons could all be eliminated by considering only bosons with even "G-parity." This fact was certainly known, but we had been reluctant to impose this constraint since the "pion" in the model (which is tachyonic) has odd G-parity. However, once we give up aspirations to construct hadronic amplitudes, this step is most natural. Furthermore, they pointed out, by making the fermions satisfy both the Weyl and the Majorana constraints, which is possible if $d = 10$, the number of bosonic and fermionic degrees of freedom are equal, indicating a supersymmetric spectrum. The supersymmetry of this new model was finally proven by Green and Schwarz [25]. These "superstring theories" were subsequently constructed by Green and Schwarz partly in collaboration with me [26]. These models are, in fact, free from all the problems that plague the old models. There are two classes: The type I model, which corresponds to the Ramond–Neveu–Schwarz model with the above constraints, has both open and closed strings and its zero-slope limit (infinite-tension limit) corresponds to $N = 1$ supergravity coupled to $N = 1$ Yang–Mills field theory (for $d = 10$). The type II models describe only closed strings and correspond to $N = 2$ supergravity for $d = 10$.

The type II models were proven to be one-loop finite while type I models seemingly all were renormalizable to this order. These facts looked quite promising for the quantum theory. However, both the type I model and one of the type II models (IIb) are constructed with fermions of one kind of chirality and this should lead to anomalies. However, the type IIb model was found to be anomaly-free due to impressive cancellations among the contributions coming from the chiral fermions and a self-dual fourth-rank antisymmetric tensor field [27]. Such a cancellation cannot be expected for the type I strings. However, an explicit calculation by Green and Schwarz [28] revealed that the type I string with gauge group SO(32) is anomaly-free (to the one-loop order)! Since anomalies are due to massless chiral fields, they could understand the cancellation in the underlying supergravity theory. They also realized that the same mechanism would work for the gauge group $E_8 \times E_8$, too. However, no type I string theory with this gauge group can be constructed [29]. It was then realized by Gross *et al.* [30] that a new string model (the heterotic-string model) could be constructed by combining the superstring with the bosonic string. This construction gave the unique choice of gauge groups as $E_8 \times E_8$ or SO(32)!

This long journey has taken us from the world of hadronic physics to a (so far) almost unique unified theory for all interactions. In this process quite a few new and important concepts have been discovered, such as supersymmetry, to mention just one. In retrospect a detailed study of any of these concepts would lead to a string theory. Einstein's gravity, as well as supergravity extensions, has one dimensionful coupling constant. Such theories are nonrenormalizable. The only way they can be consistent theories, if they are quantized straightforwardly, is if all divergencies cancel. For Einstein's gravity one has indeed found divergencies at the two-loop order [31], while supergravity theories have been shown to be finite to this order at the S-matrix level, but there is really no chance that finiteness can persist to all orders [32]. One way out of this dilemma is to introduce another dimensionful constant. The string tension is such a parameter and one is then naturally led into string theories.

A string theory can be described by an infinite set of harmonic modes. When quantized such a set will provide infinite vacuum fluctuations. For a free string one can renormalize away such infinities. However, an interaction between two extended objects must be local in order to ensure causality and the infinite vacuum fluctuations will destroy such locality [32]. Supersymmetry, though, demands an equal number of bosonic and fermionic modes leading to a cancellation of the infinities. In string theory supersymmetry is really mandatory.

Subsequent chapters will not follow the historic developments but will instead start with free strings and lead the way to interactions. All the basic ingredients will be described but recent developments will only be sketched. For other approaches and for a summary of all modern aspects the reader is referred to the literature [34].

Chapter 2

Bosonic Strings

In the description of pointlike particles, it is appropriate to start with a free spinless particle. It will follow a one-parameter trajectory $x^\mu(\tau)$. The classical action describing this particle must be independent of how the trajectory is parametrized and is taken to be proportional to the arc-length traveled by the particle (a spacelike metric is used)

$$S = m \int_{s_i}^{s_f} ds = m \int_{\tau_i}^{\tau_f} d\tau \sqrt{-\dot{x}^2(\tau)} \qquad (2.1)$$

where m is identified with the mass of the particle and $\dot{x}^\mu = dx^\mu/d\tau$. The fact that the action is reparametrization invariant has a ring of gravity to it. By regarding $x^\mu(\tau)$ as a set of scalar fields in one dimension we can couple them to a metric $g \equiv g_{\tau\tau}$ and write an action just the way scalar fields are coupled to gravity in four dimensions,

$$S = -\tfrac{1}{2} \int d\tau \sqrt{g} (g^{-1}\dot{x}^2 - m^2) \qquad (2.2)$$

(We note that the mass term is like a cosmological constant.)

Eliminating the metric field g and inserting its solution back into equation (2.2) gives relation (2.1), so actions (2.1) and (2.2) are equivalent, at least classically. In fact, equation (2.2) is more general since it allows us to take the limit $m = 0$.

To construct an action for a freely falling string we follow the same lines. We describe the simplest string by just its d-dimensional Minkowski coordinate $x^\mu(\sigma, \tau)$. The parameters σ and τ span the world sheet traced out by the string when it propagates; σ is spacelike and τ timelike. To write an action we can generalize either equation (2.1) or equation (2.2). We choose the second way and introduce a metric $g_{\alpha\beta}$ and its inverse $g^{\alpha\beta}$ on the world surface. A reparametrization-invariant action is then [15]

$$S = -\tfrac{1}{2} T \int_0^\pi d\sigma \int_{\tau_i}^{\tau_f} d\tau \sqrt{-g} \, g^{\alpha\beta} \partial_\alpha x^\mu \partial_\beta x_\mu \tag{2.3}$$

where $g = \det g_{\alpha\beta}$ and T is a proportionality factor (ensuring that x^μ is a length) which will turn out to be the string tension.

The symmetries of equation (2.3) are global Poincaré invariance:

$$\delta x^\mu = l^\mu{}_\nu x^\nu + a^\mu, \qquad \delta g_{\alpha\beta} = 0 \tag{2.4}$$

and local reparametrization invariance:

$$\delta x^\mu = \xi^\alpha \partial_\alpha x^\mu$$
$$\delta g_{\alpha\beta} = \xi^\gamma \partial_\gamma g_{\alpha\beta} + \partial_\alpha \xi^\gamma g_{\gamma\beta} + \partial_\beta \xi^\gamma g_{\alpha\gamma} \tag{2.5}$$

together with local Weyl invariance:

$$\delta g_{\alpha\beta} = \Lambda g_{\alpha\beta}; \qquad \delta x^\mu = 0 \tag{2.6}$$

Classically, one can eliminate $g_{\alpha\beta}$ from equation (2.3) by solving its field equations algebraically and substituting this solution back in the action. The resulting expression is the Nambu–Hara–Goto action [11, 12], namely, the area of the world sheet (definitely a reparametrization-invariant expression). Quantum mechanically the elimination of g involves performing a path integral. In general an extra "Liouville mode" is left over (the Weyl invariance is broken) except in the special case of $d = 26$ [14]. Whether or not it is possible to make sense of the bosonic string theory for $d < 26$ is still not completely settled. For the rest of Part I we will only consider the string theories in their critical dimensions.

We note that we are not allowed to introduce a cosmological constant this time, if we insist on the Weyl invariance (2.6).

In order to quantize the system properly we will perform a Hamiltonian treatment of the system. We first compute the canonically conjugate momenta:

$$p^{\alpha\beta} = \frac{\delta\mathcal{L}}{\delta\dot{g}_{\alpha\beta}} = 0 \tag{2.7}$$

and

$$p_\mu = \frac{\delta\mathcal{L}}{\delta\dot{x}^\mu} = -T\sqrt{-g}\,g^{\alpha 0}\partial_\alpha x^\mu \tag{2.8}$$

Equation (2.7) defines a set of primary constraints

$$\psi^{\alpha\beta} = p^{\alpha\beta} = 0 \tag{2.9}$$

On computing the Hamiltonian density we find

$$\mathcal{H} = -\frac{1}{2g^{00}}\left[(-g)^{-1/2}\frac{1}{T}p^2 + 2g^{01}\acute{x}\cdot p + T(-g)^{-1/2}\acute{x}^2\right] \tag{2.10}$$

where $\acute{x} = \partial x/\partial\sigma$.

Following the analysis of Dirac [35] for constrained systems we define the total Hamiltonian as

$$H = \int_0^\pi d\sigma(\mathcal{H} + A_{\alpha\beta}p^{\alpha\beta}) \tag{2.11}$$

where $A_{\alpha\beta}$ are arbitrary coefficients. We introduce the Poisson brackets

$$\{g_{\alpha\beta}(\sigma), p^{\gamma\delta}(\sigma')\}_\tau = \delta^{\gamma\delta}_{\alpha\beta}\delta(\sigma - \sigma') \tag{2.12}$$

and

$$\{x^\nu(\sigma), p_\mu(\sigma')\}_\tau = \delta^\nu_\mu\delta(\sigma - \sigma') \tag{2.13}$$

If we now check the time-dependence of the constraints by checking the Poisson bracket of H with $p^{\alpha\beta}$, we find two secondary constraints:

$$\phi_1 = \frac{1}{2}\left(\frac{1}{T^2}p^2 + \acute{x}^2\right) = 0 \tag{2.14}$$

and

$$\phi_2 = \frac{1}{T}\acute{x}\cdot p = 0 \tag{2.15}$$

The total Hamiltonian is now the sum of all constraints. The next step is then to check the algebra of constraints:

$$\{\psi^{\alpha\beta}, \phi_i\} = 0 \tag{2.16}$$

$$\{\phi_1(\sigma), \phi_1(\sigma')\} = \frac{1}{T}[\phi_2(\sigma) + \phi_2(\sigma')]\partial_{\sigma'}\delta(\sigma - \sigma') \tag{2.17}$$

$$\{\phi_1(\sigma), \phi_2(\sigma')\} = -\frac{1}{T}[\phi_1(\sigma) + \phi_1(\sigma')]\partial_{\sigma}\delta(\sigma - \sigma') \tag{2.18}$$

$$\{\phi_2(\sigma), \phi_2(\sigma')\} = \frac{1}{T}[\phi_2(\sigma) + \phi_2(\sigma')]\partial_{\sigma}\delta(\sigma - \sigma') \tag{2.19}$$

The algebra closes and hence all constraints are first class. It is now straightforward to quantize by letting $\{A, B\} \to -(i/\hbar)[A, B]$.

At this point there are two alternative approaches to studying the quantum theory. The first is covariant quantization, where we choose an orthonormal gauge by

$$g_{\alpha\beta} = \eta_{\alpha\beta} \tag{2.20}$$

This only fixes the gauge such that it allows us to eliminate the $\psi^{\alpha\beta}$ constraints. The remaining ones are retained. The resulting Hamiltonian is (we now choose $T = 1/\pi$ for simplicity)

$$H = \frac{1}{2\pi}\int_0^\pi d\sigma[\pi^2 p^2 + \acute{x}^2] \tag{2.21}$$

and

$$p^\mu(\sigma) = \frac{1}{\pi}\dot{x}^\mu(\sigma) \tag{2.22}$$

The equations of motion are

$$\ddot{x}^\mu - \acute{\acute{x}}^\mu = 0 \tag{2.23}$$

In deriving them we have assumed boundary conditions such that no surface terms arise. There are then two choices:

1. Open string, $\acute{x}^\mu(\sigma = 0) = \acute{x}^\mu(\sigma = \pi) = 0$.
2. Closed strings, x^μ periodic in σ.

2 • Bosonic Strings

The covariant quantization entails using an indefinite-metric Hilbert space obtained from the commutator

$$[x^\mu(\sigma, \tau), p_\nu(\sigma', \tau)] = i\delta^\mu_\nu \delta(\sigma - \sigma') \tag{2.24}$$

A solution to the equations of motion in the open-string case is given by

$$x^\mu(\sigma, \tau) = x^\mu + p^\mu \tau + i \sum_{n \neq 0} \frac{\alpha^\mu_n}{n} \cos n\sigma \, e^{-in\tau} \tag{2.25}$$

The canonical commutator (2.24) amounts to

$$[\alpha^\mu_m, \alpha^\nu_n] = m\delta_{m+n,0} \eta^{\mu\nu} \tag{2.26}$$

i.e., we have an infinite set of harmonic oscillators, where the time components generate negative-norm states. The remaining constraints now have to be imposed on the states of the Hilbert space. Let us write the constraints as

$$\phi_\pm = \phi_1 \pm \phi_2 \tag{2.27}$$

On introducing solution (2.25) we find that

$$\phi_\pm = \phi(\tau \pm \sigma) \tag{2.28}$$

and

$$\phi(\sigma) = \frac{1}{2} \left(\sum_{n=-\infty}^{\infty} \alpha^\mu_n e^{-in\sigma} \right)^2 \tag{2.29}$$

with ($\alpha^\mu_0 \equiv p^\mu$).

By imposing the constraints at $\tau = 0$, $\phi(\sigma)$, $-\pi \leq \sigma \leq \pi$, is the full constraint function. We consider its Fourier modes

$$L_n = \int_{-\pi}^{\pi} d\sigma \, e^{in\sigma} \phi(\sigma) = \frac{1}{2} \sum_{m=-\infty}^{\infty} \alpha_{n-m} \cdot \alpha_m \tag{2.30}$$

and

$$L_0 = \frac{1}{2}\alpha_0^2 + \sum_{n=1}^{\infty} \alpha_{-n} \cdot \alpha_n \tag{2.31}$$

In the quantum case they become operators and, to normal order, one must obtain well-defined expressions. Then these operators satisfy the Virasoro algebra (with a central charge found by Weis [36])

$$[L_n, L_m] = (n - m)L_{n+m} + \frac{d}{12}(n^3 - n)\delta_{n+m,0} \qquad (2.32)$$

The Hilbert space of states is defined by

$$\alpha_n^\mu|0\rangle = 0, \qquad n > 0$$

The constraints are imposed as in the Gupta-Blueler quantization of QED. We demand

$$\langle\text{phys}|L_n|\text{phys}'\rangle = 0, \qquad n \neq 0 \qquad (2.33)$$

i.e., $L_n|\text{phys}\rangle = 0$, $n > 0$.

The zeroth component has to satisfy

$$(L_0 - 1)|\text{phys}\rangle = 0 \qquad (2.34)$$

Later, we will discuss this condition at length.

These conditions on physical states are identical to the conditions that physical states satisfy in the Veneziano model as found by Virasoro [4].

The alternative method used is to quantize in the *light-cone gauge* [10]. In this case the gauge is specified completely. We introduce light-cone coordinates

$$x^\pm = \frac{1}{\sqrt{2}}(x^0 \pm x^{d-1}), \qquad x^i, \qquad i = 1, \ldots, d - 2 \qquad (2.35)$$

and impose the gauge choices (2.20) together with the choices

$$x^+(\sigma, \tau) = x^+ + p^+\tau \qquad (2.36)$$

and

$$\int_0^\sigma d\sigma' \, p^+(\sigma', \tau) = \frac{1}{\pi}p^+\sigma \qquad (2.37)$$

The latter gauge choices show that, in this gauge, each point along the string carries the same "light-cone time" x^+ and the same p^+.

2 • Bosonic Strings

The constraints (2.14) and (2.15) can now be completely solved:

$$\dot{x}^- = \frac{1}{2p^+}(\dot{x}^{i2} + \acute{x}^{i2}) \tag{2.38a}$$

and

$$\acute{x}^- = \frac{1}{p^+}\dot{x}^i\acute{x}^i \tag{2.38b}$$

which can be integrated to give $x^-(\sigma, \tau)$ in terms of the transverse coordinates $x^i(\sigma, \tau)$, p^+ and a single integration constant x^-. The equations of motion are

$$\ddot{x}^i - \acute{\acute{x}}^i = 0 \tag{2.39}$$

The dynamics of the string in the light-cone gauge can be completely described by the action

$$S = -\frac{1}{2\pi}\int_0^\pi d\sigma \int d\tau\, \eta^{\alpha\beta}\partial_\alpha x^i \partial_\beta x^i \tag{2.40}$$

This action, although not Lorentz covariant, must be Poincaré invariant. The transformations are realized *nonlinearly*. Before giving these transformations, let us check the dynamical content of the action. The canonically conjugate momentum density is

$$p^i(\sigma, \tau) = \frac{1}{\pi}\dot{x}^i(\sigma, \tau) \tag{2.41}$$

As in the previous case we have two choices of boundary condition:

1. $\acute{x}^i(\sigma = 0, \pi) = 0$, open strings, with solution given by

$$x^i(\sigma, \tau) = x^i + p^i\tau + i\sum_{n\neq 0}\frac{1}{n}\alpha_n^i \cos n\sigma\, e^{-in\tau} \tag{2.42}$$

2. x^i periodic, closed strings, with solution given by

$$x^i(\sigma, \tau) = x^i + p^i\tau + \frac{i}{2}\sum_{n\neq 0}\frac{1}{n}(\alpha_n^i\, e^{-2in(\tau-\sigma)} + \tilde{\alpha}_n^i\, e^{-2in(\tau+\sigma)}) \tag{2.43}$$

It is straightforward to quantize the theory,

$$[x^i(\sigma, \tau), p^j(\sigma', \tau)] = i\, \delta^{ij}\delta(\sigma - \sigma') \qquad (2.44)$$

The classically free parameters now become operators satisfying

$$[\alpha_n^i, \alpha_m^j] = n\delta_{n+m,0}\delta^{ij} \qquad (2.45)$$

$$[\tilde{\alpha}_n^i, \tilde{\alpha}_m^j] = n\delta_{n+m,0}\delta^{ij} \qquad (2.46)$$

Again we find an infinity of harmonic oscillators. The Hilbert space of states is constructed by introducing a vacuum $|0\rangle$ and demanding

$$\alpha_n^i |0\rangle = 0, \qquad n > 0 \qquad (2.47a)$$

and

$$\tilde{\alpha}_n^i |0\rangle = 0, \qquad n > 0 \qquad (2.47b)$$

It is evident that the Hilbert space only consists of positive-norm states.

In order to construct the Poincaré generators, one starts with the covariant (classical) ones. Then one adds a gauge transformation to stay in the gauge and introduces finally the solution (2.38). The resulting representation is then [10]

$$p^+ = p^+ \qquad (2.48a)$$

$$p^i = \int_0^\pi p^i(\sigma)\, d\sigma \qquad (2.48b)$$

$$p^- = \frac{1}{\pi}\int_0^\pi d\sigma\, \dot{x}^- = \frac{1}{2\pi p^+}\int_0^\pi d\sigma[\pi^2 p^{i^2} + \dot{x}^{i^2}] \qquad (2.48c)$$

$$j^{ij} = \int_0^\pi d\sigma(x^i p^j - x^j p^i) \qquad (2.49a)$$

$$j^{+i} = \int_0^\pi d\sigma(x^+ p^i - x^i p^+) \qquad (2.49b)$$

$$j^{+-} = x^+ p^- - x^- p^+ \qquad (2.49c)$$

$$j^{-i} = \int_0^\pi d\sigma[x^-(\sigma)p^i(\sigma) - x^i(\sigma)p^-(\sigma)] \qquad (2.49d)$$

where $x^-(\sigma)$ is solved from expression (2.38b) and $p^-(\sigma)$ is the integrand of p^- in relation (2.48c).

In the quantum case we symmetrize the generators in order to retain hermiticity. This affects the checking of the commutator $[j^{i-}, j^{j-}] = 0$, since j^{i-} is cubic. *A detailed computation shows that the commutator is satisfied only if $d = 26$* [10] (the critical dimension).

Substitution of solution (2.42) into p^- given by relation (2.48c) yields

$$p^- = \frac{p^{i2}}{2p^+} + \frac{1}{p^+} \sum_{n=1}^{\infty} \alpha^i_{-n} \alpha^i_n \qquad (2.50)$$

i.e.,

$$p^2 = -2 \sum_{n=1}^{\infty} \alpha^i_{-n} \alpha^i_n = -m^2 \qquad (2.51)$$

We reintroduce the dimensionful constant as $\alpha' = 1/2\pi T$, the Regge slope. Then the correct expression is

$$\alpha' m^2 = \sum_{n=1}^{\infty} \alpha^i_{-n} \alpha^i_n \qquad (2.52)$$

The mass-squared is hence built up by an infinite set of "harmonic oscillator energies" [15]. For the quantum case each oscillator will contribute an energy due to zero-point fluctuations (due to the above symmetrization). The lowest mass level will then be

$$\alpha' m_0^2 = \frac{d-2}{2} \sum_{n=1}^{\infty} n \qquad (2.53)$$

This is clearly a divergent sum which must be regularized. We do so by comparing the sum to the Riemann ζ-function [37]

$$\zeta(s) = \sum_{n=1}^{\infty} n^{-s}, \quad \text{Re } s > 1 \qquad (2.54)$$

This is a function that can be continued analytically to $s = -1$ and

$$\zeta(-1) = -\tfrac{1}{12} \qquad (2.55)$$

In this way the infinite series has been regularized into

$$\alpha' m_0^2 = -\frac{d-2}{24} \qquad (2.56)$$

Hence this term should be included in p^-. Since the next state on the trajectory is a vector state with $d-2$ degrees of freedom, and hence must be massless, we again find $d=26$. This fact is also the reason for the -1 in condition (2.34).

There is also a more standard way [15] of obtaining this result by adding counterterms to the action (2.40), which amounts to renormalizing the speed of light (which, of course, is another parameter of the theory). It is quite attractive that it is the quantum theory that demands a finite speed of light.

The really important consequence of equation (2.56) is that the *lowest state is a tachyon*. This really means that an interacting theory based on this bosonic string will not make sense. Also, since the tachyon is the scalar state with no excitations of the higher modes, I find it hard to believe that there exists a consistent truncation in which the tachyon is left out.

Let us now revise the last analysis for closed strings. On inserting the solution (2.43) we find

$$\frac{\alpha'}{2} m^2 = \sum_{n=1}^{\infty} (\alpha^i_{-n}\alpha^i_n + \tilde{\alpha}^i_{-n}\tilde{\alpha}^i_n) \equiv N + \tilde{N} \qquad (2.57)$$

Also, in this sector we find a tachyon. For closed strings we obtain a further constraint. We consider again equation (2.38b). Integrating it between 0 and σ yields $x^-(\sigma)$. Although $x^-(\sigma)$ depends on the x^i we must demand that it represents a component of $x^\mu(\sigma)$ and hence must be periodic. Then

$$\int_0^\pi d\sigma \, \dot{x}^- = 0 = \frac{1}{p^+} \int_0^\pi d\sigma \, \dot{x}^i \dot{x}^i = \frac{\pi}{p^+}(N - \tilde{N}) \qquad (2.58)$$

i.e., classically $N = \tilde{N}$ and quantum mechanically we impose this condition on the physical states.

The Poincaré algebra spanned by the generators (2.48)-(2.49) with the constraint (2.38b) contains all the information about the strings. This is typical for the light-cone gauge. By finding the nonlinear representation we know the complete dynamics of the system, since the Hamiltonian, p^-, is one of the generators. The regrettable thing is that, so far, we have not found a deductive way to determine the generators directly without deriving them from a covariant theory.

By describing the dynamics of free bosonic strings we have found that such a theory is much more constrained than a corresponding theory for point particles. This is certainly a most desirable property, since one lesson we have learned from modern gauge field theories for point particles is that seemingly whole classes of theories are theoretically consistent and only experiments can tell which theories Nature is using. In string theories we can entertain the hope that only one model is consistent and this then should be the theory of Nature!

Chapter 3

Spinning Strings

The representation of the Poincaré algebra presented in the previous chapter was found to lead to an inconsistent theory. In order to obtain a consistent theory, we need to change expression (2.50) for p^- so as to avoid tachyons. The most natural approach is to introduce a set of anticommuting harmonic oscillators such that their zero-point fluctuations compensate those from the commuting oscillators.

The first problem is to determine in which representation of the transverse symmetry group $SO(d-2)$ the new oscillators should be chosen. The x coordinates belong to the vector representation. We can always try this representation also for the new set, which we shall do first, but we should keep in mind that for certain values of $d-2$ there are other representations with the same dimension as the vector representation.

We therefore consider a set of anticommuting harmonic oscillators d_n^i satisfying

$$\{d_m^i, d_n^j\} = \delta_{n+m,0}\delta^{ij} \tag{3.1}$$

If the relevant mass formula for open strings is

$$\alpha' p^2 = -\left(\sum_{n=1}^{\infty} \alpha_{-n}^i \alpha_n^i + \sum_{n=1}^{\infty} n d_{-n}^i d_n^i\right) \tag{3.2}$$

we can deduce p^- from this expression. It can be written in a coordinate basis if we introduce two normal-mode expansions,

$$\lambda^{1i} = \sum_{n=-\infty}^{\infty} d_n^i e^{-in(\tau-\sigma)} \qquad (3.3)$$

and

$$\lambda^{2i} = \sum_{n=-\infty}^{\infty} d_n^i e^{-in(\tau+\sigma)} \qquad (3.4)$$

such that

$$\{\lambda^{Ai}(\sigma, \tau), \lambda^{Bj}(\sigma', \tau)\} = \pi \delta^{AB} \delta^{ij} \delta(\sigma - \sigma') \qquad (3.5)$$

The generator p^- can then be expressed (classically) in the form

$$p^- = \frac{1}{2\pi p^+} \int_0^\pi d\sigma (\pi^2 p^{i2} + \acute{x}^{i2} - i\lambda^{1i}\acute{\lambda}^{1i} + i\lambda^{2i}\acute{\lambda}^{2i}) \qquad (3.6)$$

Before trying to construct the remaining generators we examine the dynamics following from expression (3.6). Since

$$p^- \sim i\frac{\partial}{\partial x^+} = \frac{1}{p^+}\left(i\frac{\partial}{\partial \tau}\right) \sim \frac{1}{p^+} H$$

the corresponding action is

$$S = -\frac{1}{2\pi} \int d\tau \int_0^\pi d\sigma [\eta^{\alpha\beta} \partial_\alpha x^i \partial_\beta x^i + i\bar{\lambda}^i \rho^\alpha \partial_\alpha \lambda^i] \qquad (3.7)$$

where we combine the two λ^i into two-dimensional two-component spinors and use the Majorana representation for the 2×2 Dirac matrices, here called ρ^α. To get a consistent theory the surface terms obtained upon variation of equation (3.7) must be zero. In the case of open strings the boundary conditions for the λ^i are

$$\lambda^{1i}(0, \tau) = \lambda^{2i}(0, \tau) \qquad (3.8)$$

and

$$\lambda^{1i}(\pi, \tau) = \begin{cases} \lambda^{2i}(\pi, \tau) & (3.9a) \\ -\lambda^{2i}(\pi, \tau) & (3.9b) \end{cases}$$

In fact we have two choices. The first choice together with equations of motion give the solutions (3.3) and (3.4). In this sector we know that there are no tachyons. The other choice results in expansions

$$\lambda^{1i} = \sum_r b_r^i e^{-ir(\tau-\sigma)} \tag{3.10}$$

and

$$\lambda^{2i} = \sum_r b_r^i e^{-ir(\tau+\sigma)} \tag{3.11}$$

where the index r takes all half-integer values. Quantities b_r^i and b_s^j satisfy the anticommutators

$$\{b_r^i, b_s^j\} = \delta_{r+s,0}\delta^{ij} \tag{3.12}$$

The (classical) mass-shell condition now reads

$$\alpha' m^2 = \sum_{n=1}^{\infty} \alpha_{-n}^i \alpha_n^i + \sum_{r=1/2}^{\infty} r b_{-r}^i b_r^i \tag{3.13}$$

Computation of the contributions from the zero-point fluctuations yields

$$\alpha' m_0^2 = \frac{d-2}{2}\left[\sum_{n=1}^{\infty} n - \sum_{n=1}^{\infty}(n-\tfrac{1}{2})\right]$$

$$= \frac{d-2}{2} \sum_{n=1}^{\infty} n(1 - \tfrac{1}{2} + 1)$$

$$\to -\frac{d-2}{16} \tag{3.14}$$

when the sum is renormalized. Again we find tachyons! Hence this sector of the model is unphysical and basing an interacting theory upon it would lead to inconsistencies.

The states we have discovered spanned by the d and b oscillators together with the αs are in fact the spectrum of the Ramond–Neveu–Schwarz [7, 8] model. The states constructed from d oscillators must all transform as fermions and constitute the states of the Ramond sector, while those constructed from b modes, which are bosonic, constitute the Neveu–Schwarz sector. Both sectors are needed in order to obtain a model with both fermions and bosons.

There is, in fact, a way to truncate the spectrum to avoid tachyons; this approach can be proven to be consistent with interactions [24]. We consider the projector

$$P = \tfrac{1}{2}[1 + (-1)^{\sum b_{-r} b_r}] \tag{3.15}$$

By demanding that it is zero on physical states,

$$P|\text{phys}\rangle = 0 \tag{3.16}$$

we obtain a tachyon-free spectrum. A consistent interaction can be set up if also the spinors are chosen to be of Majorana–Weyl type (the critical dimension is 10 and such a choice is then possible). This leads to the superstrings, which we will describe in the next chapter.

The Poincaré generators can be constructed and, in the quantum case, the algebra only works in $d = 10$. We will not give them here but will discuss them in the next chapter.

In the case of closed strings there are also two sectors depending on which boundary conditions are chosen. One sector is obtained if the λs are periodic in σ. This leads to two sets of integer-moded oscillators d_n^i and $\tilde{d}_n^i$. The other sector is obtained by choosing the λs antiperiodic, which leads to two sets of half-integer-moded oscillators b_r^i and $\tilde{b}_r^i$. This sector has a tachyon. One can also have sectors with d_n^i and $\tilde{b}_r^i$ or $\tilde{d}_n^i$ and b_r^i.

We have discussed the spinning string completely in the light-cone gauge. This is certainly enough, as we have seen, but it is often advantageous to have a covariant formalism and, in fact, there is an action generalizing (2.3). It was done mimicking supergravity [15]. One considers two-dimensional supergravity (on the world sheet σ, τ) with a "zweibein" $V_\alpha{}^a$, related to the metric $g_{\alpha\beta}$ in the usual way, and a Majorana Rarita–Schwinger field ψ_α. Then the action is

$$S = -\frac{T}{2} \int d\sigma \, d\tau \, \eta_{\mu\nu} V \{ g^{\alpha\beta} \partial_\alpha x^\mu \partial_\beta x^\nu + i V_a{}^\alpha \bar{\lambda}^\mu \rho^a \partial_\alpha \lambda^\nu$$

$$+ 2 V_a{}^\alpha V_b{}^\beta \bar{\psi}_\alpha \rho^b \rho^a \lambda^\mu (\partial_\beta x^\nu + \tfrac{1}{2} \bar{\lambda}^\nu \psi_\beta) \} \tag{3.17}$$

This is an action with a lot of symmetry. Its physics is just the physics of free field theory, although it contains interaction terms. It is not unlikely that this action could be used in other fields of physics if properly interpreted.

The action (3.17) can be quantized covariantly as in the case of the bosonic string. The local symmetries are reparametrization invariance, Weyl invariance, and local supersymmetry. Going through the Dirac analysis and

3 • Spinning Strings

fixing a gauge

$$g^{\alpha\beta} = \eta^{\alpha\beta} \qquad (3.18)$$

$$\psi^{\alpha\beta} = 0 \qquad (3.19)$$

we find the equations of motion

$$\ddot{x}^\mu - \ddot{x}^\mu = 0 \qquad (3.20)$$

$$\rho^\alpha \partial_\alpha \lambda^\mu = 0 \qquad (3.21)$$

together with the constraints

$$\partial_\alpha x^\mu \partial_\beta x_\mu - \tfrac{1}{2}\eta_{\alpha\beta}\partial^\gamma x^\mu \partial_\gamma x_\mu + \tfrac{1}{4}\bar\lambda^\mu(\rho_\alpha\partial_\beta + \rho_\beta\partial_\alpha)\lambda_\mu = 0 \qquad (3.22)$$

$$\rho^\alpha \rho_\beta \partial_\alpha x \cdot \lambda = 0 \qquad (3.23)$$

where we have reintroduced $\dot{x}^\mu = (1/\pi)p^\mu$. The light-cone gauge can now be reached by further specifying the gauge as in relations (2.36) and (2.37) together with $\lambda^+ = 0$.

The equations of motion (3.20) and (3.21) can be comfortably described by an action [38]

$$S = -\frac{1}{2\pi}\int d\sigma\, d\tau[\eta^{\alpha\beta}\partial_\alpha x^\mu \partial_\beta x_\mu + i\bar\lambda^\mu \rho^\alpha \partial_\alpha \lambda_\mu - F^\mu F_\mu] \qquad (3.24)$$

which differs from action (3.7) in that the fields run over d values. (The constraints will bring it down to $d - 2$.) Furthermore, we have introduced an auxiliary field $F^\mu(\sigma, \tau)$ on the world sheet to exhibit a supersymmetry of the action under the transformations

$$\delta x^\mu = i\bar\alpha \lambda^\mu$$

$$\delta \lambda^\mu = \rho^\alpha \partial_\alpha x^\mu \alpha + \alpha F^\mu$$

$$\delta F^\mu = i\bar\alpha \rho^\alpha \partial_\alpha \lambda^\mu \qquad (3.25)$$

By using this fact we may describe the string in terms of a "supercoordinate" [17, 18]

$$\phi^\mu(\sigma, \tau, \theta^A) = x^\mu(\sigma, \tau) + i\bar\theta\lambda^\mu(\sigma, \tau) + \tfrac{1}{2}i\bar\theta\theta F^\mu(\sigma, \tau) \qquad (3.26)$$

This is quite a remarkable result. We build in the ten-dimensional spin density $\lambda^\mu(\sigma, \tau)$ on the world sheet by considering instead a superworld sheet spanned by σ, τ and a two-component spinor θ^A. By introducing a covariant derivative

$$D^A = \frac{\partial}{\partial \bar{\theta}^A} + i(\rho^\alpha \theta)^A \partial_\alpha \tag{3.27}$$

the action (3.24) can be written as

$$S = -\frac{1}{8\pi} \int d\tau \, d\sigma \, d^2\theta \, \bar{D}\phi^\mu D\phi_\mu \tag{3.28}$$

The supersymmetry transformations are implemented by

$$\delta\sigma^\alpha = -i\bar{\alpha}\rho^\alpha\theta, \qquad \sigma^\alpha = (\tau, \sigma) \tag{3.29}$$

$$\delta\theta^A = \alpha^A \tag{3.30}$$

The action (3.28) has, in fact, an infinite-dimensional symmetry, a superconformal symmetry. Its "supergauge" transformation is given by relations (3.29) and (3.30), where $\alpha^A = \alpha^A(\sigma^\alpha)$ satisfies the Dirac equation.

Since σ has fixed boundaries 0 and π, boundary conditions have to be introduced in the θs:

at $\sigma = 0$ $\qquad\qquad \theta_1 + \theta_2 = 0$ \qquad (3.31)

at $\sigma = \pi$ $\qquad\qquad \theta_1 \pm \theta_2 = 0$ \qquad (3.32)

and similarly for α^A, where the upper sign refers to the Ramond sector and the lower one to the Neveu–Schwarz sector.

The constraints (3.22) and (3.23) can also be written in a superspace formulation as

$$J_\beta = \rho^\alpha \rho_\beta \partial_\alpha \phi \cdot D\phi = 0 \tag{3.33}$$

This is the supercurrent of the theory generating the superconformal invariance. One way of constructing a string theory is hence to construct a superconformal action and demand that the supercurrent be zero.

If we solve the equations of motion following from equations (3.24), or equivalently equations (3.20) and (3.21), x^μ and λ^μ can be written as x^i in equation (2.25) and λ^i_A in expressions (3.3), (3.4) or (3.10), and (3.11) with $\alpha^i \to \alpha^\mu$, $b^i \to b^\mu$, or $d^i \to d^\mu$ in the case of an open string and similarly for closed strings. These solutions allow the constraints to be written as [39]

$$\Pi(\tau) \cdot D\Pi(\tau) = 0 \tag{3.34}$$

where

$$\Pi^\mu(\tau) = iD\phi^\mu(0, \tau, \theta = \theta_1 = -\theta_2) \tag{3.35}$$

and

$$D = -i\theta \frac{\partial}{\partial \tau} + \frac{\partial}{\partial \theta} \tag{3.36}$$

Here we have again extended the σ interval to $-\pi \leq \sigma \leq \pi$ in order that the constraint can be expressed as a function only of $\sigma + \tau$. Taking the constraint at $\tau = 0$, we call the remaining variable τ so that the identification (3.35) can be made.

The constraint (3.34) can now be Fourier decomposed with respect to both τ and θ. In the Neveu–Schwarz sector we then obtain (in the quantum case)

$$L_n = -\frac{1}{4\pi} \int_{-\pi}^{\pi} d\tau \int d\theta \, e^{in\tau} : \Pi \cdot D\Pi : \tag{3.37}$$

and

$$G_r = -\frac{1}{2\pi} \int_{-\pi}^{\pi} d\tau \int d\theta \, \theta \, e^{ir\tau} : \Pi \cdot D\Pi : \tag{3.38}$$

with the algebra

$$[L_n, L_m] = (n - m)L_{n+m} + \frac{d}{8} n(n^2 - 1)\delta_{n+m,0} \tag{3.39}$$

$$[L_n, G_r] = \left(\frac{n}{2} - r\right) G_{n+r} \tag{3.40}$$

$$\{G_r, G_s\} = 2L_{r+s} + \frac{d}{2}(r^2 - \tfrac{1}{4})\delta_{r+s,0} \tag{3.41}$$

and in the Ramond sector

$$L_n = -\frac{1}{4\pi} \int_{-\pi}^{\pi} d\tau \int d\theta \, e^{in\tau} : \Pi \cdot D\Pi : + \tfrac{5}{8}\delta_{n,0} \tag{3.42}$$

and

$$F_n = -\frac{1}{2\pi} \int_{-\pi}^{\pi} d\tau \int d\theta\, \theta\, e^{in\tau} : \Pi \cdot D\Pi : \quad (3.43)$$

where we have added a term to L_0 to have a Möbius (SU(1, 1)) subalgebra. The full algebra is

$$[L_n, L_m] = (n - m)L_{n+m} + \frac{d}{8} n(n^2 - 1)\delta_{n+m,0} \quad (3.44)$$

$$[L_n, F_m] = \left(\frac{n}{2} - m\right) F_{n+m} \quad (3.45)$$

$$\{F_n, F_m\} = 2L_{n+m} + \frac{d}{2}(n^2 - \tfrac{1}{4})\delta_{n+m,0} \quad (3.46)$$

The quantization of the theory is carried out in a straightforward manner by introducing the canonical commutators following from action (3.28),

$$[x^\mu(\sigma, \tau), p^\nu(\sigma', \tau)] = -i\eta^{\mu\nu}\delta(\sigma - \sigma') \quad (3.47)$$

$$\{\lambda^{A\mu}(\sigma, \tau), \lambda^{B\nu}(\sigma', \tau)\} = -\pi \delta^{AB} \eta^{\mu\nu}\delta(\sigma - \sigma') \quad (3.48)$$

The constraints are again imposed on the states as

$$L_n|\text{phys}\rangle = 0, \quad n > 0 \quad (3.49)$$

$$G_r|\text{phys}\rangle = 0, \quad r > 0 \quad (3.50)$$

$$F_n|\text{phys}\rangle = 0, \quad n > 0 \quad (3.51)$$

The mass-shell conditions are

$$(L_0 - \tfrac{1}{2})|\text{phys}\rangle = 0 \quad \text{(NS)} \quad (3.52)$$

$$F_0|\text{phys}\rangle = 0 \quad \text{(R)} \quad (3.53)$$

In fact, this was the method used in one attempt to find new string models [17]. One can construct extended superconformal symmetries, where the supergauge generators G_r, F_n transform as vectors under an SO(N) algebra. In this way one finds an infinite set of extensions of the Virasoro algebra. The algebras are quite interesting in their own right. They involve, for example, Virasoro generators, supergauge generators, and Kac–Moody generators. However, only one of the algebras corresponds to a ghost-free string model, the SO(2) one [18]. Sadly enough the critical dimension is 2, and it seems difficult to find a place in physics for this model.

Chapter 4

Superstrings

In the previous chapter we added anticommuting degrees of freedom to cancel zero-point fluctuation. They transform as the vector representation of $SO(d-2)$. For $d = 3, 4, 6$, and 10 we could also choose the lowest spinor representation, since it has the same dimension as the vector one. Since $SO(d-2)$ is a compact group, the scalar product of two spinors is just the contracted sum as for vectors (which is the only product used in Chapter 3). We can then carry over all the work in Chapter 3. We only make the substitution

$$\lambda^{Ai} \to S^{Aa} \tag{4.1}$$

where A is still a two-component spinor index and a is a $(d-2)$-component spinor index. This leads to the superstring theory, which was discussed in another formulation in the previous chapter.

The action for the superstring theory is then given by [40]

$$S = -\frac{1}{2\pi} \int d\tau \int_0^\pi d\sigma [\eta^{\alpha\beta} \partial_\alpha x^i \partial_\beta x^i + i \bar{S}^a \rho^\alpha \partial_\alpha S^a] \tag{4.2}$$

We know that this action leads to a sector (for open strings) which starts with massless particles as the lowest-lying states. In this sector the solutions to the equations of motion are

$$S_a^1 = \sum_{n=-\infty}^{\infty} S_n^a \, e^{-in(\tau-\sigma)} \tag{4.3}$$

$$S_a^2 = \sum_{n=-\infty}^{\infty} S_n^a \, e^{-in(\tau+\sigma)} \tag{4.4}$$

with the anticommutation rule

$$\{S_a^A(\sigma,\tau), S_b^B(\sigma',\tau)\} = \pi \delta_{ab} \delta^{AB} \delta(\sigma-\sigma') \tag{4.5}$$

$$\{S_n^a, S_m^b\} = \delta_{n+m,0} \delta^{ab} \tag{4.6}$$

The operators S_{-n}^a with n positive are creation operators. They will act on a bosonic state and take it to a fermionic state. Hence this section will contain both bosons and fermions, in fact equally many of each kind at each mass level, building up supermultiplets at each level (as we will soon prove).

We know that the other sector, which follows by using the other set of boundary conditions corresponding to relations (3.8) and (3.9b), has tachyons. Furthermore, the fermionic oscillators will be half-integer moded and there will not be an equal number of bosons and fermions at each mass level, thus ruining the possibility of having a supersymmetry. Since the first sector contains all we require, we simply decree that we only use the boundary conditions (3.8) and (3.9a).

Similarly for closed strings we decree that we only use periodic boundary conditions. This yields the following solutions to the equations of motion:

$$S_a^1 = \sum_{n=-\infty}^{\infty} S_n^{1a} \, e^{-2in(\tau-\sigma)} \tag{4.7}$$

$$S_a^2 = \sum_{n=-\infty}^{\infty} S_n^{2a} \, e^{-2in(\tau+\sigma)} \tag{4.8}$$

The arduous task now is to check whether there is a representation of the Poincaré algebra spanned on this string theory. In fact there is [41], and the marvelous fact is that it can also be extended to a super-Poincaré algebra! Since light-cone supersymmetry [42] might not be too familiar, we shall first review it. The supersymmetry charge Q in ten dimensions decomposes into two SO(8) light-cone spinors Q_+^a and $Q_-^{\dot{a}}$, where the indices

$a, \dot{a} = 1, \ldots, 8$ denote the two inequivalent 8-component spinors of SO(8). The algebra is

$$\{Q_+^a, Q_+^b\} = 2p^+ \delta^{ab} \tag{4.9a}$$

$$\{Q_-^{\dot{a}}, Q_-^{\dot{b}}\} = 2p^- \delta^{\dot{a}\dot{b}} \tag{4.9b}$$

$$\{Q_+^a, Q_-^{\dot{b}}\} = \sqrt{2}(\gamma_i)^{a\dot{b}} p^i \tag{4.9c}$$

For further notation, see the Appendix to Part I.

We therefore write the representation of the super-Poincaré algebra. Again it is stressed that there is some guesswork behind its construction. For the case of closed strings the algebra turns out to be an $N = 2$ super-Poincaré algebra. The most general algebra is

$$p^+ = p^+ \tag{4.10a}$$

$$p^i = \int_0^\pi d\sigma\, p^i(\sigma, \tau) \tag{4.10b}$$

$$p^- = \frac{1}{2\pi p^+} \int_0^\pi d\sigma\, [\pi^2 p^{i^2} + \dot{x}^{i^2} - i(S^1 \dot{S}^1 - S^2 \dot{S}^2)] \tag{4.10c}$$

$$q_1^{+a} = \sqrt{\frac{2p^+}{\pi}} \int_0^\pi d\sigma\, S_1^a \tag{4.11a}$$

$$q_2^{+a} = \sqrt{\frac{2p^+}{\pi}} \int_0^\pi d\sigma\, S_2^a \tag{4.11b}$$

$$q_1^{-\dot{a}} = \frac{1}{\pi\sqrt{p^+}} \int_0^\pi d\sigma (\gamma^i S_1)^{\dot{a}} (\pi p^i - \dot{x}^i) \tag{4.11c}$$

$$q_2^{-\dot{a}} = \frac{1}{\pi\sqrt{p^+}} \int_0^\pi d\sigma (\gamma^i S_2)^{\dot{a}} (\pi p^i + \dot{x}^i) \tag{4.11d}$$

$$j^{ij} = \int_0^\pi d\sigma \left[x^i p^j - x^j p^i + \frac{1}{4\pi} (S^1 \gamma^{ij} S^1 + S^2 \gamma^{ij} S^2) \right] \tag{4.12a}$$

$$j^{+i} = \int_0^\pi d\sigma (x^+ p^i - x^i p^+) \tag{4.12b}$$

$$j^{+-} = x^+p^- - x^-p^+ \tag{4.12c}$$

$$j^{-i} = \tfrac{1}{2}\int_0^\pi d\sigma \left[\{x^-(\sigma), p^i\} - \{x^i, p^-(\sigma)\}\right. \tag{4.12d}$$

$$\left. -\frac{i}{4\pi\sqrt{\pi}p^+}(S^1\gamma^{ij}S^1(\pi p^j - \acute{x}^j) + S^2\gamma^{ij}S^2(\pi p^j + \acute{x}^j)) + 4\frac{p^i}{p^+}\right] \tag{4.12e}$$

where

$$\acute{x}(\sigma) = \frac{\pi}{p^+}p^i\acute{x}^i + \frac{i}{2p^+}(S^1\acute{S}^1 + S^2\acute{S}^2) \tag{4.13}$$

In fact this algebra is enough to cover all known string models (apart from the $d = 2$ model [18]):

1. *Type IIb superstrings.* This is the full algebra (4.10)–(4.12) with periodic boundary conditions for the coordinates. We note that this is a chiral model, since the creation operators S^{1a}_{-n} and S^{2a}_{-n} create spinors of only one chirality.
2. *Type IIa superstrings.* The anticommuting coordinate S^a_2 can instead be chosen to transform as the other spinor representation, $S^{\dot{a}}_2$. Nothing is affected in the algebra since S_1 and S_2 are never contracted with each other. S_1 and S_2 cannot be put together as a two-component spinor, but who cares? This model is not chiral since the spinor states can be combined to Majorana states.
3. *Type I superstrings.* For open strings we must use the boundary conditions corresponding to relations (2.8), and (3.8) and (3.9a). Then $q_1^+ = q_2^+$ and $q_1^- = q_2^-$ and the supersymmetry is reduced to an $N = 1$ one. This truncation can also be performed for closed strings.
4. *The bosonic strings.* Put $S^1 = S^2 = 0$. No supersymmetry, of course.
5. *The heterotic string.* This string model will be discussed in the next chapter.
6. *The spinning string.* By performing triality transformations back to vectors λ^i, such as $S^{1a}S^{1a} \to \lambda^{1i}\lambda^{1i}$ and $S^1\gamma^{ij}S^1 \to \lambda^{1i}\lambda^{1j}$, and similarly for S^2, one can easily read off the representation for the spinning string. No supersymmetry survives, of course.

Considering the superstring theories I, IIa, and IIb we know we have a quantum theory of free strings in which the lowest-lying states are massless. Next we wish to know what is the spin content of these massless states.

For the bosonic string one can choose a scalar vacuum state $|0\rangle$ as the ground state. If all fermionic oscillators in equations (4.12) are set equal to zero, we find this state to be indeed a scalar one. However, in the superstring case there is an extra piece in the zero-mode part of j^{ij} in equations (4.12),

$$s_0^{ij} = \tfrac{1}{4}(S_0^1 \gamma^{ij} S_0^1 + S_0^2 \gamma^{ij} S_0^2) \tag{4.14}$$

Its effect on a vacuum state will be nonzero in general.

We should also realize that the massless level must be a supermultiplet. Such a one can be constructed from the anticommutator (4.9a) combined with the knowledge of expression (4.14), taking into account that q_+ is linearly realized. The generator q_+^a is real. In the four-dimensional case it is customary to go from an SO(2) to a U(1) description forming complex generators (with no Lorentz index), which then build up a Clifford algebra, and we can define creation and annihilation operators from which we construct the supermultiplet. For $d = 10$ we have so far used an SO(8) covariant notation. To decompose the generators into creation and annihilation operators we must break the covariance into SU(4) × U(1). This is the formalism to be used for the field theories governing open strings. Here, however, we describe an alternative formalism, which uses the full SO(8) covariance. This method can equally well be used in $d = 4$. Consider the zero-mode part of j^{ij} given by equation (4.12a), which is the relevant part on vacuum states, and start with open strings

$$j_0^{ij} = l^{ij} + \tfrac{1}{2} S_0 \gamma^{ij} S_0 \equiv l^{ij} + s_0^{ij} \tag{4.15}$$

The last term in this relation is the spin contribution. If we try to start with a scalar vacuum $|0\rangle$ we need a constraint

$$s_0^{ij}|0\rangle = 0 \tag{4.16}$$

On multiplying (4.16) by another S_0 and using constraint (4.16) and Fierz rearrangements, we find that $S_0^a|0\rangle = 0$ and hence $|0\rangle = 0$, unless $d = 4$.

For $d = 10$ we must try the next simplest thing. We take the vacuum to be a vector $|i\rangle$ with $\langle i|j\rangle = \delta^{ij}$. Then we insist on a constraint

$$s_0^{ij}|k\rangle = \delta_{ik}|j\rangle - \delta_{jk}|i\rangle \tag{4.17}$$

This time, when we multiply equation (4.17) by S_0, we find for the general state

$$\psi^{ia} \equiv q_+^a|i\rangle \tag{4.18}$$

with the decomposition

$$\psi^{ia} = \tilde{\psi}^{ia} + \tfrac{1}{8}\gamma^i\gamma^j\psi^j \qquad (4.19)$$

where $\gamma^i\tilde{\psi}^{ia} = 0$, that

$$\tilde{\psi}^{ia} = 0 \qquad (4.20)$$

but that there is no constraint on

$$|\dot{a}\rangle \equiv \tfrac{1}{8}(\gamma^i q_+)^{\dot{a}}|i\rangle \qquad (4.21)$$

with $\langle \dot{b}|\dot{a}\rangle = p^+\delta_{\dot{a}\dot{b}}$. On checking the Lorentz properties of this state we find that

$$s_0^{ij}|\dot{a}\rangle = -\tfrac{1}{2}(\gamma^{ij})^{\dot{a}\dot{b}}|\dot{b}\rangle \qquad (4.22)$$

Hence it transforms properly as a spinor. From the Fierz property (A.7)

$$q_+^a q_+^b = p^+\delta^{ab} + \tfrac{1}{16}(\gamma^{ij})^{ab} q_+ \gamma^{ij} q_+ \qquad (4.23)$$

we find that no other massless states can be constructed and the massless sector then contains one vector state and one spinor state which we recognize as the Yang–Mills multiplet in ten dimensions. We can, of course, let both states transform according to some representation (such as the adjoint one) of some internal group. For type I theories, there is a standard way for including internal symmetry quantum numbers in scattering amplitudes introduced by Chan and Paton [43] a long time ago. What one does is to associate a matrix $(\lambda_i)_{ab}$ with the ith external string state and multiply the N-point scattering amplitude by a group theory factor $\text{tr}(\lambda_1 \cdots \lambda_N)$. When checking that such factors factorize properly so as not to destroy the factorization properties of the scattering amplitudes, one finds that the possible internal symmetry groups are SO(n), U(n), or Sp($2n$) [29]. We note that the exceptional groups are not possible.

It may also be instructive to check the first excited level. Here one can form 128 bosonic states $\alpha^i_{-1}|j\rangle$ and $S^a_{-1}|\dot{b}\rangle$ and 128 fermionic states $\alpha^i_{-1}|\dot{a}\rangle$ and $S^a_{-1}|i\rangle$. These form various reducible SO(8) multiplets, which can be reassembled into SO(9) representation (since they are massive) using the Lorentz generators. Doing this, one finds for the bosons the SO(9) representations

☐☐ and ☐/☐

of dimensions 44 and 84, respectively. The fermions form a single 128-dimensional Rarita–Schwinger SO(9) multiplet [like the $\tilde{\psi}^{ia}$ in equation (4.19)].

In the closed-string case the massless spectrum is generated by two supersymmetry generators q_+^1 and q_+^2. If both belong to the same SO(8) representation [this means that they have the same chirality in ten dimensions (type IIb)], we can form complex generators

$$q_+^a = \frac{1}{\sqrt{2}}(q_+^1 + iq_+^2)^a \qquad (4.24)$$

and hence use q_+^a and q_+^{*a} as creation and annihilation operators. For this case we can introduce a scalar vacuum (which has to be complex, as can be seen from the supersymmetry transformations). Checking j_0^{ij} with the s_0^{ij} term as in relation (4.14), we see that the vacuum is indeed a scalar state. If we act with q_+^a, then the following massless supermultiplet can be formed:

$$|0\rangle \sim \phi$$

$$q_+^a|0\rangle \sim \phi^a$$

$$q_+^a q_+^b|0\rangle \sim \phi^{ab}$$

$$\vdots$$

$$q_+^{a_1} \cdots q_+^{a_8}|0\rangle \sim \phi^{a_1 \cdots a_8}$$

It is easily seen that this is a reducible multiplet. To obtain an irreducible representation we impose the conditions

$$(\phi^{a_1 a_2 \cdots a_{2N}})^* = \frac{1}{(8-2N)!} \varepsilon^{a_1 a_2 \cdots a_8} \phi^{a_{2N+1} \cdots a_8} \qquad (4.25)$$

and

$$(\psi^{a_1 a_2 \cdots a_{2N+1}})^* = \frac{1}{(7-2N)!} \varepsilon^{a_1 a_2 \cdots a_8} \phi^{a_{2N+2} \cdots a_8} \qquad (4.26)$$

Owing to the triality properties of the representations of SO(8) we can rewrite the states with vector indices. We now find the bosonic spectrum:

- 2 scalars ϕ
- 2 antisymmetric tensors $A^{ij} \sim (\gamma^{ij})^{ab} \phi^{ab}$

- 1 self-dual antisymmetric tensor A^{ijkl}
- 1 graviton g^{ij}

where the last two sets of states follow from $\phi^{a_1\cdots a_4}$. The fermionic spectrum is

- 2 spinors ψ^a
- 2 Rarita–Schwinger states $\tilde{\psi}^{i\dot{a}}$

If q_+^1 and q_+^2 have opposite Weyl properties (type IIa) we cannot form creation operators in an SO(8) covariant way. We then have to follow the procedure of the $N = 1$ case. We can start with a tensor state

$$|ij\rangle \sim |i\rangle \otimes |j\rangle \sim \phi^{ij}$$

and generate the states by stuttering the $N = 1$ procedure,

$$|a\rangle \otimes |i\rangle \sim \psi^{ai}$$

$$|i\rangle \otimes |\dot{a}\rangle \sim \chi^{\dot{a}i}$$

$$|a\rangle \otimes |\dot{a}\rangle \sim \phi^{a\dot{a}}$$

The bosonic spectrum is then

- 1 graviton $g^{ij} = \phi^{(ij)} - \frac{1}{8}\phi^{kk}\delta^{ij}$
- 1 antisymmetric tensor $A^{ij} = \phi^{[ij]}$
- 1 scalar $\phi = \phi^{ii}$
- 1 vector $A^i = (\gamma^i)^{a\dot{a}}\phi^{a\dot{a}}$
- 1 antisymmetric tensor $A^{ijk} = (\gamma^{ijk})^{a\dot{a}}\phi^{a\dot{a}}$

The fermionic spectrum is

- 2 spinors $\psi^{\dot{a}}$, χ^a
- Rarita–Schwinger states $\tilde{\psi}^{ai}$, $\tilde{\chi}^{\dot{a}i}$

This technique could, of course, also have been used in the other case above.

We can impose constraints on the $N = 2$ spectrum to obtain an $N = 1$ spectrum. These closed strings (type I) are important, since they will be able to couple to the open strings. To find this spectrum we linearly combine $Q^1 + Q^2 = Q$ and use the $N = 1$ technique. The spectrum is then clearly

$$|i\rangle \otimes |j\rangle$$

$$|i\rangle \otimes |a\rangle$$

i.e.,

- 1 graviton $\qquad g^{ij}$
- 1 antisymmetric tensor $\qquad A^{ij}$
- 1 scalar $\qquad \phi$
- 1 spinor $\qquad \psi^a$
- 1 Rarita–Schwinger state $\qquad \tilde{\psi}^{ai}$

So far we have only discussed free-string states. We could anticipate that all closed-string states also belong to some representation of some internal group. However, it will turn out that interaction is only possible if this representation is the trivial one.

Since the strings are extended objects in one dimension there is a possibility that they can carry an intrinsic orientation, like an "arrow" pointing in one direction along its length. When strings are oriented there are two distinct classical states for a single spatial configuration, corresponding to the two possible orientations. An open string is oriented, loosely speaking, if the end points are different, while a closed string is oriented if one can distinguish a mode running one way around the string from a mode running the other way.

Hence the basic question is whether a string described by $x^\mu(\sigma)$, $S^{Aa}(\sigma)$ is the same as one described by $x^\mu(\pi - \sigma)$, $S^{Aa}(\pi - \sigma)$. We consider the closed-string solution (2.43), (4.7), and (4.8). The replacement $\sigma \to \pi - \sigma$ corresponds to the interchanges $\alpha_n \leftrightarrow \tilde{\alpha}_n$, $S_n^1 \leftrightarrow S_n^2$. In the type II case the two strings connected by the interchanges are different, while for type I the constraints are just such as to make the two strings the same state. Hence we conclude that type II strings are oriented while type I closed strings are nonoriented. Heterotic strings, which have different left-going and right-going modes, are evidently oriented. The analysis here is classical but can be taken over to the quantum case by considering matrix elements of the operators $x^\mu(\sigma)$ and $S^{Aa}(\sigma)$. The same conclusions are reached then.

For open strings we argued above that one can allow for an internal (global) symmetry (the remnant of the gauge symmetry in a covariant formalism). The interaction allows for letting the states transform as ϕ_b^a, where indices a and b run over the fundamental representation and its complex conjugate, respectively. Hence ϕ_b^a transforms as either the adjoint representation or the singlet one. The intuitive way to interpret this is to say that each end carries a "quark" and an "antiquark." Hence if the "quarks" are different from the "antiquarks" the string is oriented, otherwise not. Now these strings can join their ends to form closed strings, if they are singlets. Then the $U(n)$ strings will form oriented closed strings, while the $SO(n)$ and $Sp(2n)$ strings will form nonoriented closed strings. But these strings must be of type I, since the open strings are, and hence should be nonoriented. Thus we conclude by this nonrigorous argument that only the gauge groups $SO(n)$ or $Sp(2n)$ are possible [44].

The notation of orientability will be important for perturbation expansions. For oriented strings the interaction must be such that the orientations match. This will make the perturbation expansion of type I strings different from that of type II strings.

We have so far discussed superstrings in the light-cone gauge. This is sufficient to understand the structure of the theory and to build an interacting theory. However, it would be advantageous to have a covariant formalism. Such an action has been found by Green and Schwarz [45] but it is not yet clear if that action can be quantized covariantly [46].

Chapter 5

The Heterotic String

We saw in the previous chapter several representations embodied in the algebra (4.10)–(4.12). The most economical one is obtained in the bosonic string by writing $x^\mu(\tau, \sigma) = x^\mu(\tau - \sigma) + \tilde{x}^\mu(\tau + \sigma)$, and similarly for p^μ. The algebra then splits into one piece from the right-moving coordinates $x^\mu(\tau - \sigma)$, $p^\mu(\tau - \sigma)$ and one from the left-moving part. The two parts do not mix and we can freely set one of them to zero. Similarly for the full algebra (4.10)–(4.12) it is straightforward to see that for the terms where S^A couple to x and p, S^1 couple to the right-moving and S^2 to the left-moving parts. Since they are right-moving and left-moving, respectively, again a consistent truncation can be made by setting, say, the left-moving parts equal to zero. This reduces the algebra to an $N = 1$ supersymmetry.

The heterotic string [30] is now constructed by putting together the algebra from one right-moving superstring constructed as above with the algebra from a 26-dimensional left-moving bosonic string. There is an obvious mismatch in dimensions! The solution to this dilemma is that we only add in the $SO(1, 9)$ subalgebra of the full bosonic algebra. Eventually we must interpret the extra coordinates $x^I(\tau, \sigma)$, $I = 1, \ldots, 16$ and check what has happened to the $SO(16)$ symmetry left out.

We now check the algebra closer. Consider the bosonic algebra. For a left-moving string

$$p^- = \frac{1}{4\pi p^+} \int_0^\pi d\sigma (\pi p^I + \hat{x}^I)^2 \qquad (5.1)$$

The combination $\pi p^I + \acute{x}^I$ is only left-moving. If we now *insist* that p^I and x^I both *separately* are left-moving, we see in the mode expansion that they are the same (up to a π) and we have to change the canonical commutator to

$$[x^I(\sigma, \tau), p^J(\sigma', \tau)] = \frac{i}{2} \delta(\sigma - \sigma') \delta^{IJ} \qquad (5.2)$$

and impose

$$[x^I(\sigma, \tau), x^J(\sigma, \tau)] = -\frac{i}{2} \pi \frac{1}{\partial_\sigma} \delta(\sigma - \sigma') \delta^{IJ} \qquad (5.3)$$

where

$$\frac{1}{\partial_\sigma} \delta(\sigma - \sigma') = \varepsilon(\sigma - \sigma')$$

This change does not alter the closure of the rest of the algebra. Hence in the full algebra of the heterotic string

$$p^- = \frac{1}{2\pi p^+} \int_0^\pi d\sigma [\pi^2 p^{i^2} + \acute{x}^{i^2} - i S^a \acute{S}^a + 2\pi p^I \acute{x}^I] \qquad (5.4)$$

and the constraint on $x^-(\sigma)$ is

$$\acute{x}^-(\sigma) = \frac{\pi}{p^+} p^i \acute{x}^i + \frac{i}{2p^+} S^a \acute{S}^a + \frac{\pi}{p^+} p^I \acute{x}^I \qquad (5.5)$$

with $S \equiv S^1$. The remaining generators of the $N = 1$ super-Poincaré algebra are obtained by putting $S^2 = 0$ in algebras (4.10)–(4.12).

The action corresponding to the Hamiltonian (5.4) is

$$S = -\frac{T}{2} \int_0^\pi d\sigma \int d\tau \left[\eta^{\alpha\beta} (\partial_\alpha x^i \partial_\beta x^i + \partial_\alpha x^I \partial_\beta x^I) + i S^a \left(\frac{\partial}{\partial \tau} + \frac{\partial}{\partial \sigma} \right) S^a \right] \qquad (5.6)$$

together with constraints

$$\Phi^I = \left(\frac{\partial}{\partial \tau} - \frac{\partial}{\partial \sigma} \right) x^I = 0 \qquad (5.7)$$

5 • The Heterotic String

Alternatively a term $\lambda[(\partial/\partial\tau - \partial/\partial\sigma)x^I]^2$ can be added to the action. Siegel [47] has shown that the resulting action possesses a local gauge symmetry, which allows the Lagrange multiplier λ to be gauged away, leaving us with the constraints (5.7).

The canonical structure of the action can be analyzed using Dirac's analysis. Taking into account the fact that the constraints (5.7) are second class, the canonical commutators are found to be (2.44), (4.5), and (5.2).

We therefore consider a solution to the equation of motion for the x^I given by

$$x^I(\tau + \sigma) = x^I + p^I(\tau + \sigma) + \frac{i}{2} \sum_{n \neq 0} \frac{\tilde{\alpha}_n^I}{n} e^{-2in(\tau+\sigma)} \quad (5.8)$$

However, it is not consistent with periodic boundary conditions in σ! To solve this dilemma we are forced to allow for more general boundary conditions. If the coordinates x^I lie on a hypertorus with radii R, then the function $x^I(\sigma, \tau)$ maps the circle $0 \leq \sigma \leq \pi$ onto the circle $0 \leq x^I \leq 2\pi R$, and such maps fall into homotopy classes characterized by a winding number L^I that counts how many times x^I wraps around the circle [48]. Then expression (5.8) is an allowed solution with $p^I = 2L^I R$. We are hence forced to consider the extra 16 coordinates x^I to span a hypertorus. This is physically very appealing, since such a hypertorus can be small and we do not need to worry about the coordinates x^I on a macroscopic scale.

Before performing this compactification in detail, we consider its effect on equations (5.4) and (5.5). If the solutions for x^i, S^a, and x^I are substituted into equations (5.4) and (5.5), we obtain in analogy to relations (2.57) and (2.58) ($\alpha' = \frac{1}{2}$)

$$\tfrac{1}{4}m^2 = N + \tilde{N} - 1 + \tfrac{1}{2}\sum_{I=1}^{16}(p^I)^2 \quad (5.9)$$

and

$$N = \tilde{N} - 1 + \tfrac{1}{2}\sum_{I=1}^{16}(p^I)^2 \quad (5.10)$$

where

$$N = \sum_{n=1}^{\infty}(\alpha^i_{-n}\alpha^i_n + nS^a_{-n}S^a_n) \quad (5.11)$$

and

$$\tilde{N} = \sum_{n=1}^{\infty}(\tilde{\alpha}^i_{-n}\tilde{\alpha}^i_n + \tilde{\alpha}^I_{-n}\tilde{\alpha}^I_n) \quad (5.12)$$

In expressions (5.9) and (5.10) the subtraction of -1 is due to the regularization of the contribution from the zero-point fluctuations. It is noteworthy that although this contribution is not canceled, there is no tachyon! It is enough to have the vacuum contributions cancel among the right-moving modes. Furthermore, we find from relation (5.10) that p^{I^2} must be an even integer.

In further study, the 16-dimensional hypertorus spanned by the x^I may be thought of as R^{16} modulo a lattice Γ^{16}, generated by 16 independent basis vectors e_i^I ($i = 1, \ldots, 16$). We choose an *even* lattice (to be justified shortly) with the length of each e_i equal to $\sqrt{2}$:

$$x^I = x^I + \sqrt{2}\pi \sum_{i=1}^{16} n_i R_i e_i^I \qquad (5.13)$$

with n_i integers and R_i the radii.

For a torus R^{16}/Γ^{16}, the allowed momenta p^I lie on the dual lattice $\tilde{\Gamma}^{16}$ generated by

$$\sum_{I=1}^{16} e_i^I e_j^{*I} = \delta_{ij} \qquad (5.14)$$

From equation (5.2) $2p^I$ generates translations, i.e.,

$$p^I = \frac{1}{\sqrt{2}} \sum_{i=1}^{16} \frac{m_i}{R_i} e_i^{*I}, \qquad m_i \text{ integers} \qquad (5.15)$$

However, we have also seen that $p^I = L^I$, a winding number, where

$$L^I = \frac{1}{\sqrt{2}} \sum_{i=1}^{16} R_i n_i e_i^I \qquad (5.16)$$

From this we conclude that the allowed states must have momenta, which lie in the intersection of Γ^{16} and $\tilde{\Gamma}^{16}$. Furthermore, reintroducing the Regge slope parameter α' and comparing equations (5.15) and (5.16), we find

$$R_i \sim \sqrt{\alpha'} \qquad (5.17)$$

This shows that the internal coordinates are small. (The parameter α' is so far a free parameter, which we might hope the theory determines eventually. It is certainly small, say $\sqrt{\alpha'} < 10^{-17}$ m, since we have not seen any Regge recurrencies, states with $N = 1$ or higher, experimentally.) Finally, since $\tilde{\Gamma} < \Gamma$, which is even, also $\tilde{\Gamma}$ is even and p^{I^2} will be even integers.

5 • The Heterotic String

When the interacting case is investigated a further condition is derived. The (Euclidean) two-dimensional world sheet, which describes a one-loop closed-string amplitude, is a torus. This torus should be symmetric subject to the interchange of σ and τ, owing to the global reparametrization invariance still left in the theory (duality). To achieve this symmetry the lattice has to be *self-dual*, $\Gamma = \tilde{\Gamma}$.

Even self-dual lattices are extremely rare. In fact they only exist in $8n$ dimensions. In a pioneering work Goddard and Olive [49] showed that in 16 dimensions there are only two such lattices, $\Gamma_8 \times \Gamma_8$ and Γ_{16}, which they proposed should be important in physics. Γ_8 is the root lattice of E_8. The construction of Γ_{16} is more complicated. The basis vectors for the SO(32) root lattice can be written in terms of a set of orthonormal vectors $\{u_i\}$, $i = 1, \ldots, 16$ as $e_i = u_i - u_{i+1}, i = 1, \ldots, 15, e_{16} = u_{15} + u_{16}$. This root lattice is not self-dual. However, a self-dual lattice can be constructed from the root lattice by adding points that are multiplets of one of the spinor weights of Spin(32). We can choose the spinor weight to be $s_1 = \frac{1}{2}(u_1 - u_2 + u_3 - u_4 + \cdots + u_{15} - u_{16})$, $s_1^2 = 4$ and take the basis vectors for Γ_{16} to be s_1 and e_i, $i = 2, \ldots, 16$. Since e_1 is an integer linear combination of the basis vectors, Γ_{16} contains all the points of the root lattice of SO(32) plus additional points that correspond to spinor weights of Spin(32)/Z_2. This is not the same as SO(32). The center of Spin(32) is $Z_2 \times Z_2$. Removing the diagonal combination of the two Z_2 factors eliminates all spinor representations leaving SO(32). Removing one Z_2 factor eliminates one of the two spinor representations of Spin(32) and all representations in the same Z_2 conjugacy class as the spinor (including the vector).

Each of the two 16-dimensional self-dual lattice has 480 vectors of the minimal length^2 = 2.

We now check the lowest-lying states using relations (5.9) and (5.10). In the superstring sector the lowest-lying states are, in analogy to the analysis in Chapter 4,

$$|i\rangle_R$$

$$|\dot{a}\rangle_R$$

We must satisfy the condition (5.10). Hence the lowest-lying states are

$$|i\rangle_R \otimes \tilde{\alpha}^j_{-1}|0\rangle_L \qquad g^{ij}, A^{ij}, \phi$$

$$|i\rangle_R \otimes \tilde{\alpha}^I_{-1}|0\rangle_L \qquad \text{16 vectors}$$

$$|i\rangle_R \otimes |p^I\rangle_L \qquad \text{480 vectors}$$

$$|\dot{a}\rangle_R \otimes \tilde{\alpha}^i_{-1}|0\rangle_L \quad \tilde{\psi}^{\dot{a}i}, \psi^a$$

$$|\dot{a}\rangle_R \otimes \tilde{\alpha}^I_{-1}|0\rangle_L \quad \text{16 spinors}$$

$$|\dot{a}\rangle_R \otimes |p^I\rangle_L \quad \text{480 spinors}$$

in the notation of Chapter 4.

This is the spectrum of $N = 1$ supergravity coupled to a Yang–Mills theory with a gauge group comprising 496 generators.

Finally, we need to investigate what has happened to the symmetry in the 16 compact dimensions. In fact, it is already known from the work of Frenkel and Kac, Segal, and Goddard and Olive [50] that, connected to the lattices $\Gamma_8 \times \Gamma_8$ and Γ_{16}, one can construct representations of the groups $E_8 \times E_8$ and $\mathrm{Spin}(32)/Z_2$, respectively. To this end we construct an operator $E(K^I)$, which acts on left-moving states; $E(K^I)$ represents the generator of G that translates states on the weight lattice by a root vector K^I. The construction is as follows. We consider

$$E(K) = \oint_0 \frac{dz}{2\pi i z} : \exp[2iK^I x^I(z)] : C(K) \tag{5.18}$$

with $K^{I2} = 2$, $z = e^{2i(\tau+\sigma)}$.

These operators contain the usual translation operator $\exp(2iK^I x^I)$, which translates the internal momenta (= winding numbers) by K^I. The additional term $C(K^I)$ can be viewed as an operator 1-cocycle, which will be chosen such that the 480 $E(K^I)$ (5.18) together with the 16 operators p^I satisfy the Lie algebra of $E_8 \times E_8$ or $\mathrm{Spin}(32)/Z_2$.

The properties of harmonic-oscillator coherent states can be employed to straightforwardly verify that

$$: \exp[2iK^I x^I(z)] :: \exp[2iL^I x^I(w)] :$$
$$= : \exp\{2i[K^I x^I(z) + L^I x^I(w)]\} : (z-w)^{K^I L^I} \quad \text{for } |w| < |z| \tag{5.19}$$

We consider now the commutator

$$[E(K), E(L)]$$

The contour in z may be moved such that we only pick up contributions from the possible singularities at $z = w$. In order for the commutator to close properly we demand that

$$C(K)C(L) = (-1)^{K \cdot L} C(L) C(K) \tag{5.20}$$

5 • The Heterotic String

and

$$C(K)C(L) = \varepsilon(K,L)C(K+L) \qquad (5.21)$$

The commutator will be nonzero if $K \cdot L = -1$. Then $K + L$ is another root vector since $(K+L)^2 = 2$. The commutator is also nonzero if $K \cdot L = -2$ due to the normal ordering of the zero modes. In this case $(K+L)^2 = 0$. Hence $K = -L$. These are the only cases for which the commutator is nonzero. Therefore

$$[E(K), E(L)] = \begin{cases} \varepsilon(K,L)E(K+L) & K+L \text{ root} \\ K^I \cdot p^I & K = -L \\ 0 & \text{otherwise} \end{cases} \qquad (5.22)$$

and

$$[p^I, E(K)] = K^I \qquad (5.23)$$

These are precisely the commutators of the generators of $E_8 \times E_8$ or $\text{Spin}(32)/Z_2$ as long as $\varepsilon(K, L)$ are chosen to equal the structure constants (± 1 in this basis). It has been shown that one can construct $C(K)$ and $\varepsilon(K, L)$ and hence we have the two explicit representations on the Fock space of the left-movers. It is remarkable that the SO(16) internal symmetry group with which we started under the compactification turns into a 16-rank group. The extension of the string plays a crucial role here!

Chapter 6

The Operator Formalism

When we now turn to interactions, we again want to follow the methods developed for point particles. Three different approaches have been used: operator formalism, path integral formalism, and field theory. While the last one (to be discussed in the following chapters) can treat both perturbative and non-perturbative aspects, the first two are essentially perturbative.

Consider a $\lambda\phi^3$ theory. The perturbation expansion for an S-matrix element can be thought of as a summation of "branch polymers," where a particle has a certain probability of splitting up, and where we consider freely propagating particles in between. Such amplitudes can be obtained schematically from a path integral

$$A \sim \lambda^{N-2} \int \mathrm{D}x^\mu(\tau)\psi_1(x_1)\cdots\psi_N(x_N)\exp(iS[x^\mu]) \qquad (6.1)$$

where $\psi_K(x_K)$ is the wave function for the Kth external state. The action in integral (6.1) is the free action (2.1) together with a gauge choice and a Feynman–Fadde'ev–Popov ghost action.

This method is really streamlined for theories with only a three-point coupling and then especially for the first type of diagram in Figure 6.1. Such diagrams can be thought of as one-particle propagating, chopping off particles at times τ_k. The functional integral can then be divided into integrals from τ_1 to τ_2, from τ_2 to τ_3, and so on. In these intervals the particle is

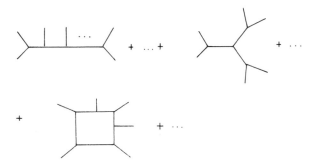

Figure 6.1. S-matrix elements for a $\lambda\phi^3$ theory.

propagating freely. The other types of diagrams in Figure 6.1 are most easily obtained using unitarity and the results of the first class of diagrams.

There is an alternative approach, which is also quite useful, especially for the first class of diagrams, namely an operator formalism. The wave functions in expression (6.1) are $\exp(ik_n \cdot x_n)$ for external particles of momenta k_n. We are then using a representation, where the x_n are taken to be c numbers. We now consider an operator wave function $\exp[ik \cdot x^\mu(\tau)]$ with $x^\mu(\tau) = x^\mu + p^\mu \tau$, the solution to the equation of motion and where $[x^\mu, p^\nu] = -i\eta^{\mu\nu}$. Furthermore, we introduce a vacuum $|0\rangle$ such that

$$p^\mu |0\rangle = 0 \tag{6.2}$$

Then

$$p^\mu e^{ik \cdot x(\tau)}|0\rangle = k^\mu e^{ik \cdot x(\tau)}|0\rangle \tag{6.3}$$

The natural way to write a scattering amplitude for a diagram of the first type in Figure 6.1 is then as a correlation function

$$A^N \sim \lambda^{N-2} \int \Pi \, d\tau_i \langle 0| \exp[ik_1 \cdot x(\tau_1)]$$

$$\times \exp[ik_2 \cdot x(\tau_2)] \cdots \exp[ik_N \cdot x(\tau_N)]|0\rangle \tag{6.4}$$

From the Baker-Hausdorff formula we derive (assuming $k^2 = 0$)

$$e^{ik \cdot (x + p\tau)} = \exp\left(\frac{i}{2}\tau p^2\right) e^{ik \cdot x} \exp\left(-\frac{i}{2}\tau p^2\right) \tag{6.5}$$

Since we interpret τ as a real time which can vary from $-\infty$ to ∞, it is seen that the exponents in relation (6.5) are badly defined. This is typical of a relativistic quantum theory. To get the proper pole structure one is forced to continue analytically to imaginary time (perform a Wick rotation), perform the computations, and in the result continue back. This means also that p^0 must be imaginary, and finally we have to Wick-rotate it back. We hence let $\tau \to i\tau$.

We can now write

$$\exp(-\tau_i) = z_i, \qquad d\tau_i = -dz_i/z_i \qquad (6.6)$$

where z_i run along the real axis. Since we are interested in S-matrix elements we start with particle 1 at time $-\infty$ and end with particle N at time $+\infty$. Hence we choose

$$z_1 = \infty, \qquad z_N = 0$$

In expression (6.4) z_1 and z_N then disappear. The remaining z_i are ordered and we scale them such that $z_2 = 1$. It is now straightforward to change variables such that $x_i = z_i/z_{i-1}$ and integrate over x_3 to x_{N-1} between 0 and 1. The result is then

$$A_N \sim \lambda^{N-2} \frac{1}{(k_1 + k_2)^2} \cdot \frac{1}{(k_1 + k_2 + k_3)^2} \cdots \frac{1}{(k_1 + \cdots + k_{N-2})^2} \qquad (6.7)$$

which we recognize as a typical S-matrix element. When we Wick-rotate p^0 back we obtain poles with the proper Feynman prescription. We could have performed the calculations with real τ_i, if we had inserted damping factors in expression (6.5) to make the amplitude (6.4) properly defined. This procedure would also lead to the amplitude (6.7) with Feynman-prescripted poles. In practice, when we use an operator formalism we will use the z variables and keep p^0 real, and remember that poles must have the Feynman prescription.

Relation (6.7) allows one to easily read off the Feynman rules. We can, in principle, construct operator vertices for internal vertices too, such as the one occurring in the second diagram in Figure 6.1. However, in a point-particle theory it is easier to invoke unitarity and use the Feynman rules following from relation (6.7).

We note that the operator formalism has not really been derived. In fact, we can put it in a one-to-one correspondence with the path integral formalism and in this sense we can derive it, if we accept the latter formalism [51].

For point-particle theories, quantum field theory has turned out to be the superior instrument to use. It not only defines the proper perturbation expansion, but allows also for nonperturbative studies. When we turn to string theories, which are more complex, it turns out to be advantageous not to follow the most ambitious line initially. The path-integral method has been thoroughly developed and it does admit a well-defined way to study perturbation expansions. It will not be discussed here but the reader is referred to Mandelstam [52]. Instead, the remainder of this chapter will be devoted to the operator formalism.

One key ingredient in the operator formalism for point particles was the assumption that only three-particle interactions occur. For strings, it is even more natural to assume that strings split into two strings. (Try to cut a piece of cord into three pieces at one time.) Consider first bosonic strings. Since we are going to construct an S matrix, we are interested in amplitudes, where the strings emit specific states. For simplicity we start with an amplitude with N tachyons (see Figure 6.2). The tachyon is just a point particle, a specific state of an open string, and to have locality we must assume it is emitted from an end point of the string. A natural generalization of the point-particle vertex operator is then [53]

$$V(k, \tau) = \, : e^{ik \cdot x(\sigma=0, \tau)} : \qquad (6.8)$$

with x^μ as in equation (2.25), where normal ordering has been introduced to make sense of vacuum expectation values. To conform with notation from the Veneziano model we define

$$Q^\mu(z) \equiv x^\mu(\sigma = 0, \tau = -i \ln z) \qquad (6.9)$$

Then

$$V(k, z) = \exp\left(k \cdot \sum_{n=1}^{\infty} \frac{\alpha_{-n} z^n}{n}\right) e^{ik \cdot x + k \cdot p \ln z} \cdot \exp\left(-k \cdot \sum_{n=1}^{\infty} \frac{\alpha_n z^{-n}}{n}\right) \qquad (6.10)$$

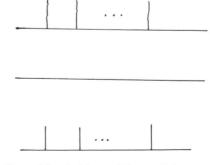

Figure 6.2. A string emitting particles.

6 • The Operator Formalism

In analogy with the discussion of interacting point particles we attempt an amplitude

$$A_N \sim g^{n-2} \int \frac{\Pi \, dz_i}{z_i} \langle 0| V(k_1, z_1) \cdots V(k_N, z_N) |0\rangle \qquad (6.11)$$

with the z_i ordered along the real axis such that $z_i > z_{i+1}$. We assume that this is the amplitude where particle 1 and particle 2 interact and then chop off particles i in order. This will be a correct amplitude if the amplitude has poles at intermediate channels, and where the particles coupling in the residues are physical states, i.e., satisfying

$$L_n |\text{intermediate state}\rangle = 0 \qquad (6.12)$$

and

$$(L_0 - 1)|\text{intermediate state}\rangle = 0 \qquad (6.13)$$

We note that the amplitude (6.10) has *not* been derived. We have only made the amplitude plausible.

To investigate the amplitude we first compute the vacuum expectation value in equation (6.10). This can be done either by commuting all annihilation operators to the right of all creation operators using the Baker–Hausdorff formula or by operator-product methods. The result is

$$A_N = g^{N-2} \int \prod_{i=1}^{N} dz_i \prod_{i<j} (z_i - z_j)^{k_i \cdot k_j} \qquad (6.14)$$

In principle, we have already seen on physical grounds that we should be able to put $z_1 = \infty$, $z_2 = 1$, and $z_N = 0$. However, in string theory it is easily seen that we must fix three of the z_i because the amplitude, as it stands, is invariant under the Möbius transformations

$$z_i \to z_i' = \frac{az + b}{cz + d} \qquad (6.15)$$

where a, b, c, and d are real and $ad - bc = 1$. (This is only true if $k^2 = 2$, i.e., the emitted particles are on-shell.) In the operator expressions such transformations are performed by L_1, L_0, and L_{-1} (which constitute a subalgebra of the Virasoro algebra).

The Möbius invariance allows us to transform three z_i to fixed values and the integration over these three variables is redundant. In order for the amplitude not to select three particles we have to divide with an invariant

(Haar) measure. This is done by writing the amplitude as

$$A_N = g^{n-2} \int \prod_{i=1}^{N} \frac{dz_i}{z_i \, dV} \langle 0 | V(k_1, z_1) \cdots V(k_N, z_N) | 0 \rangle \qquad (6.16)$$

where

$$dV = \frac{dz_a \, dz_b \, dz_c}{(z_a - z_b)(z_b - z_c)(z_c - z_a)}$$

for any choice of three z_i.

For the amplitude (6.16) we may in fact relax the reality condition on a, b, c, and d in transformation (6.15). Then we can transform the real axis to a line in the complex plane. By choosing the transformation

$$z \to \frac{z - i}{-iz + 1} \qquad (6.17)$$

we map the real axis onto a unit circle. The amplitude (6.14) with the Haar measure introduced with z along the unit circle is the original Koba-Nielsen amplitude for N-particle scattering [54]. From equation (6.14) it is easily seen that the amplitude has the correct pole structure. It remains to be proven that only physical states couple in the residues. To prove this fact the operator formalism was once introduced. Hence we return to expression (6.16). An explicit calculation gives

$$[L_n, V(k, z)] = z^n \left(z \frac{d}{dz} + \frac{n}{2} k^2 \right) V(k, z) \qquad (6.18)$$

An operator transforming in this way is said to have conformal spin $k^2/2$. In particular we find

$$[L_0, V(k, z)] = z \frac{d}{dz} V(k, z) \qquad (6.19)$$

which amounts to

$$V(k, z) = z^{L_0} V(k, 1) z^{-L_0} \qquad (6.20)$$

We now consider the amplitude (6.16) in the Möbius frame, $z_1 = \infty$, $z_2 = 1$,

$z_N = 0$:

$$A_N = g^{N-2} \int \prod_{i=3}^{N-1} \frac{dz_i}{z_i} e^{ik_1 \cdot x} \langle 0| V(k_2, 1) \cdots V(k_{N-1}, z_{N-1})|0\rangle e^{ik_N \cdot x}$$

$$= e^{ik_1 \cdot x} \langle 0| V(k_2, 1) \frac{1}{L_0 - 1} V(k_3, 1) \cdots \frac{1}{L_0 - 1} V(k_{N-1}, 1)|0\rangle e^{ik_N \cdot x}$$
(6.21)

where we used equation (6.20) and changed variables to "Chan variables" [55]

$$x_i = z_i/z_{i-1} \qquad (6.22)$$

and performed the integration over these variables. Again the pole structure is clear. It remains to prove that only states satisfying the Virasoro constraint (6.13) couple in the residues. For this purpose we use

$$[L_n - L_0, V(k)] = \frac{n}{2} k^2 V(k) \qquad (6.23)$$

and

$$(L_n - L_0 + 1) \frac{1}{L_0 - 1} = \frac{1}{L_0 + n - 1} (L_n - L_0 + 1 - n) \qquad (6.24)$$

where the first condition follows from equation (6.18) and the latter from the Virasoro algebra. Their combination yields

$$(L_n - L_0 + 1) \frac{1}{L_0 - 1} V(k) = \frac{1}{L_0 + n - 1} V(k)(L_n - L_0 + 1) \qquad (6.25)$$

Hence the operator $(L_n - L_0 + 1)$ effectively commutes through the chain of propagators and vertices in an amplitude to act on the "vacuum" $|0\rangle \exp(ik_N \cdot x)$, on which it is zero. This fact shows that the states coupling in the residues are physical states and we have proven that the amplitude has the correct pole structure we set out to prove. We note that it has been tacitly assumed that the emitted particles are on-shell. The amplitudes only work as S-matrix elements. A great problem in dual models was to find consistent off-shell amplitudes. One solution to this problem will be found in the field-theoretic formulation.

A remarkable property of the amplitude (6.16) is the fact that it is cyclically symmetric in the external legs. If we take, say, the last vertex operator we can essentially commute it past the other vertices. An explicit calculation gives

$$V(k_1, z_1) V(k_2, z_2) = V(k_2, z_2) V(k_1, z_1) \exp[i\pi k_1 \cdot k_2 \varepsilon(z_1 - z_2)] \quad (6.26)$$

Commuting $V(k_N, z_N)$ past the other vertices amounts to a phase factor, which is one, when momentum conservation (and mass-shell conditions) are used. Hence we find that

$$A_N = g^{N-2} \int \prod_{i=1}^{N} \frac{dz_i}{z_i \, dV} \langle 0 | V(k_N, z_N) V(k_1, z_1) \cdots V(k_{N-1}, z_{N-1}) | 0 \rangle \quad (6.27)$$

The commutation rule (6.26) is mathematically ill-defined, but it is easy to check that equation (6.27) is correct by using equation (6.14) with the z_i on the unit circle. The mathematically correct way to treat the commutator is to have the momenta span a lattice, a fact used for the heterotic string in Chapter 5.

This cyclicity property is the *duality* property on which the dual models were originally based. This means that the amplitude has poles in many more channels than Figure 6.2 shows, namely, in all channels with momenta $(k_i + k_{i+1} + \cdots + k_{i+n})^2 = -2M$, where $n \leq N$ and M an integer ≥ -1. This considerably reduces the number of terms needed in a full amplitude. To get a complete amplitude one simply sums over all noncyclic permutations. This is what makes the operator formalism so useful for strings. The amplitudes (6.16) carry all information about vertices for the other physical states. Owing to the duality properties, an amplitude can be factorized as in Figure 6.3. By letting $(k_i + k_{i+1})^2 = -2M$ the residue gives an expression for an amplitude with one external particle with mass^2 = $2M$. This can be expressed with an "excited" vertex. Further information is available elsewhere [56].

If we go back to the starting point, Figure 6.2, we notice that only particles emitted at $\sigma = 0$ were considered. For tree-graph amplitudes this is in fact enough, because of the duality property. However, when we deal with loops we must also allow emissions at the other end of the string, at

Figure 6.3. A factorized string amplitude.

6 • The Operator Formalism

$\sigma = \pi$. Such diagrams can be topologically different and must be taken into account. From our formalism it is easy to construct a vertex for emission at $\sigma = \pi$. Following relation (6.8) we define it as

$$\bar{V}(k,\tau) = : e^{ik\cdot x(\sigma=\pi,\tau)} : = (-1)^N V(k,\tau)(-1)^N \qquad (6.28)$$

with N the number operator as in equation (2.57). The operator $\Omega = (-1)^N$ is called the "twist operator."

With such vertices we can draw diagrams such as those in Figure 6.4.

The ordering among the integration variables is now such that the variables (times) for particles emitted at $\sigma = 0$ and for particles emitted at $\sigma = \pi$ are ordered separately. One end of the string does not know what the other one is doing.

All one-loop diagrams have been constructed in the operator formalism. This is essentially done by taking traces over tree amplitudes. It is easily seen that the naive prescription is not enough. Unphysical states couple in the residues. To remedy this fact a physical state projector operator was introduced. Reference 57 gives the explicit calculations. A more modern way is to introduce Feynman-Fadde'ev-Popov ghosts [58]. It is straightforward to see that this leads to the correct result [59].

If we proceed to higher loops the situation is more complicated, since such diagrams necessarily involve vertices with three strings off-shell. Even so, beautiful results were obtained based on the topology of such diagrams, although all details were not proven. Again it would digress too far to discuss it here, so the reader is referred to the old literature [60].

An operator formalism for the Ramond-Neveu-Schwarz model is constructed by following the same procedure as in the Veneziano model. The emission vertex for a tachyon (in the open-string case) is obtained by considering a string emitting particles from an end point. It is natural to use the superspace formulation and consider a vertex operator [39]

$$V(k,z,\theta) = :\exp[ik\cdot\phi(\sigma=0, z=e^{i\tau}, \theta = \theta_1 = -\theta_2]: \qquad (6.29)$$

where the appropriate boundary conditions have to be chosen for the Ramond sector and for the Neveu-Schwarz sector. We can now write an

Figure 6.4. A string diagram with particles emitted from both ends.

N-point amplitude as a correlation function over such vertices,

$$A_N = g^{N-2} \int \frac{\prod_{i=1}^{N} dz_i}{dV_{abc}} \prod_{i=1}^{N} d\theta_i \langle 0| V(k_1, z_1, \theta_1) \cdots V(k_N, z_N, \theta_N)|0\rangle \quad (6.30)$$

In the Ramond case the vacuum carries a spinor (with no mass or momentum). If we perform the θ integrations, which are in fact trivial (apart from giving an overall factor $(-1)^{N/2}$), and use the equations of motion for x^μ and λ^μ we obtain the following vertices:

$$V_{NS}(k, z) = k \cdot H(z) : e^{ik \cdot Q(z)} : \quad (6.31)$$

$$V_R(k, z) = k \cdot \Gamma(z) : e^{ik \cdot Q(z)} : \quad (6.32)$$

with

$$H^\mu(z) = \sum_{r=-\infty}^{\infty} b_r^\mu z^{-r} \quad (6.33)$$

$$\Gamma^\mu(z) = \frac{\gamma^\mu}{\sqrt{2}} + i\gamma_{11} \sum_{n \neq 0} d_r^\mu z^{-n} \quad (6.34)$$

and $\gamma_{11} = \gamma_0 \gamma_1 \cdots \gamma_9$. We note that for the Ramond vertex a scalar particle is emitted. Also a pseudoscalar particle can be emitted.

To check these amplitudes, again we have to show that only physical states couple in the residues. This is done in complete analogy to the Veneziano case. The z integrals are integrated out and one can check that L_n, G_r, or F_n are essentially zero in a residue. The cyclicity can also be proven as in the previous case by commuting the vertices around. The easiest way again is to go to the Koba–Nielsen form. In the Neveu-Schwarz case the expectation value in amplitude (6.30) can be computed and one finds

$$A_N = g^{N-2} \int \frac{\prod_{i=1}^{N} dz_i}{dV_{abc}} \prod_{i=1}^{N} d\theta_i \prod_{i<j} (z_i - z_j + \theta_i \theta_j)^{k_i \cdot k_j} \quad (6.35)$$

a form first obtained by Fairlie and Martin [61]. Also, the duality properties here are easily read off.

As in the Veneziano case one can introduce a twist operator by considering emissions from the other end of the string and one can construct loops with external tachyons. To obtain the correct unitarity structure, projection

6 • The Operator Formalism

operators were used [57]. In the more modern formalism ghosts are introduced and the computation of the one-loop graphs is straightforward. For more details, see the literature.

The vertices so far considered are vertices, where one string chops off particles while remaining the same type of string. However, if a propagating fermionic string emits a fermionic particle, the string becomes bosonic. Such a vertex is much more complicated, since the two strings are described by different oscillators.

The generic form of a fermion emission vertex must be

$$V_\psi(k, z) = {}_2\langle 0|_b \langle 0|O(k, z)|0\rangle_1|0\rangle_d \qquad (6.36)$$

The explicit form of the operator $O(k, z)$ is quite complex. It was first constructed by Corrigan and Olive [62], to whose paper the reader is referred for the details. After a huge effort the four-fermion tree-graph amplitude was computed [63]. To proceed further seemed quite difficult then, but modern techniques with ghost states included may improve the situation.

This formalism is also quite useful for superstring theories. Since these can be obtained from the Ramond-Neveu-Schwarz model by appropriate projections, they can easily be implemented in the operator formalism.

No covariant supersymmetric operator formalism exists yet for superstrings and heterotic strings, although one can use an operator formalism in the light-cone gauge [64]. If we introduce the light-cone expression (2.42) for $x^\mu(\sigma, \tau)$ into the vertex operator (6.8) and the propagator, we obtain a formalism with only physical states. The drawback is the complicated form of x^-, which is quadratic in x^i. However, if the number of external particles is $\leq d$, we can choose a Lorentz frame with $k_i^+ = 0$ to write the vertex operator in terms of transverse operators and momenta. This means that the results, even for the S-matrix elements, will be noninvariant and the invariance has to be restored by hand.

A formalism like this was used to construct the one-loop graphs for superstrings [65] and, more recently, a similar formalism has been constructed for the heterotic string [30].

Chapter 7

Field Theory for Free Superstrings

In order to describe the interactions among pointlike particles, it is mostly advantageous to go over to a second-quantized formalism, a quantum field theory. We consider a scalar particle described by its phase space, x^μ and p^μ. A field ϕ will associate a value to each operator in a commuting set of the phase space, say x^μ. The Lorentz generators act on a field $\phi(x^\mu)$ such that x^μ multiplies and $p^\mu \to -i\partial/\partial x^\mu$. The variable x^μ can now be interpreted as a c number. (It can be taken as the eigenvalue of the operator x^μ.) A second-quantized formalism can be constructed in this manner by also introducing a momentum $\pi(x)$ canonically conjugate to $\phi(x)$, and then the generators of the Lorentz algebra can be expressed as

$$G \sim \int d^3x \, \pi(x) g \phi(x) \tag{7.1}$$

It is straightforward to write an action for $\phi(x)$ which describes any number of free, propagating particles. Interactions are introduced by adding polynomials in $\phi(x)$ which are Lorentz-invariant.

We will now follow the same lines to construct a field theory for superstrings. We start by considering the super-Poincaré algebra for type IIb strings, (4.10)–(4.12). We need to write it in terms of a maximal set of commuting coordinates. For the bosonic coordinates this is an easy task;

we can simply choose to use $x^i(\sigma)$, x^-, x^+ and then represent $p^i(\sigma)$ as

$$p^i(\sigma) = -i\frac{\delta}{\delta x^i(\sigma)} \equiv -i\delta^i, \qquad p^+ = i\frac{\partial}{\partial x^-} \equiv i\partial_-$$

The original representation of the anticommuting coordinates is, however, not suitable since the S_i do not anticommute with themselves. Instead we choose

$$\theta^a = \frac{1}{\sqrt{2p^+\pi}}(S_1^a + iS_2^a) \tag{7.2}$$

and

$$d^a = \sqrt{\frac{p^+}{2\pi}}(S_1^a - iS_2^a) \tag{7.3}$$

They satisfy

$$\{\theta^a(\sigma), \theta^b(\sigma')\} = \{d^a(\sigma), d^b(\sigma')\} = 0 \tag{7.4}$$

and

$$\{\theta^a(\sigma), d^b(\sigma')\} = \delta^{ab}\delta(\sigma - \sigma') \tag{7.5}$$

We can now use $\theta^a(\sigma)$ as the coordinate and represent $d^a(\sigma)$ as $\delta/\delta\theta^a(\sigma)$. It is not completely trivial to rewrite the algebra in terms of θ^a and d^a since they involve p^+, which has nontrivial commutations with generators j^{i-} and j^{+-}. The correct new representation is [66, 67]

$$p^- = h = \frac{i}{2\partial_-}\int d\sigma\left(\pi\delta^i\delta^i - \frac{1}{\pi}\dot x^i\dot x^i + d\dot d\frac{1}{\partial_-} - \theta\dot\theta\partial_-\right) \equiv \int d\sigma\, h(\sigma)$$

$$p^+ = i\partial_-, \qquad p^i = -i\int d\sigma\,\delta^i$$

$$j^{ij} = -i\int d\sigma(x^i\delta^j - x^j\delta^i + \tfrac{1}{2}\theta\gamma^{ij}d)$$

$$j^{+i} = -i\int d\sigma(x^+\delta^i + x^i\partial_-)$$

$$j^{+-} = x^+h - ix^-\partial_- - \frac{i}{4}\int d\sigma[\theta^a, d^a] + 2i \tag{7.6}$$

7 • Field Theory for Free Superstrings

$$j^{-i} = -\frac{i}{2}\int d\sigma \left(\{x^-(\sigma), \delta^i\} - i\{x^i, h(\sigma)\} \right.$$

$$\left. + \frac{1}{2\sqrt{\pi}} \left([(\gamma^i\theta)^{\dot a},(\gamma^j d)^{\dot a}] \frac{\pi\delta^j}{\partial_-} - \left(d\gamma^{ij}d\frac{1}{\partial_-} - \theta\gamma^{ij}\theta\partial_-\right)\frac{\acute x^j}{\partial_-} \right) - 4\frac{\delta^i}{\partial_-}\right)$$

$$q_1^{+a} = i\sqrt{2}\int d\sigma\, \theta^a\partial_-, \qquad q_1^{-\dot a} = -\frac{i}{\sqrt{\pi}}\int d\sigma\left[(\gamma^i\theta)^{\dot a}\pi\delta^i - (\gamma^i d)^{\dot a}\frac{\acute x^i}{\partial_-}\right]$$

$$q_2^{+a} = \sqrt{2}\int d\sigma\, d^a, \qquad q_2^{-\dot a} = -\frac{1}{\sqrt{\pi}}\int d\sigma\left[(\gamma^i d)^{\dot a}\frac{\pi\delta^i}{\partial_-} + (\gamma^i\theta)^{\dot a}\acute x^i\right]$$

The $x^-(\sigma)$ occurring in j^{-i} is determined by

$$\acute x^-(\sigma) = -\frac{\pi}{\partial_-}(\acute x^i\delta^i + \acute\theta d) \tag{7.7}$$

The appropriate field is $\Psi[x^+, x^-, x^i(\sigma), \theta^a(\sigma)]$, where the coordinates are c numbers. This turns out to be a reducible representation. We define the Fourier-transformed field

$$\Psi[x,\lambda] = \int D^8\theta(\sigma)\exp\left(\int d\sigma\, \lambda^a\theta^a\right)\Psi[x,\theta] \tag{7.8}$$

and impose the constraint

$$\Psi[x,\lambda] = \Psi[x,\theta]^* \tag{7.9}$$

where we identify $\theta^* = (1/p^+)\lambda$. This constraint hence relates Ψ^* to Ψ, which means that in a variation of fields in an action, Ψ^* should not be varied independently of Ψ. Furthermore, we impose the constraint that any point on the string can be selected as the origin, i.e.,

$$\Psi[x(\sigma+\sigma_0), \theta(\sigma+\sigma_0)] = \Psi[x(\sigma),\theta(\sigma)] \tag{7.10}$$

Checking it in a mode basis shows that $N = \tilde N$ on a field.

The next step is to second-quantize the theory by imposing a commutator between the field and its conjugate momentum. In the light-cone gauge one can in fact essentially use the field itself or, more precisely, $\partial_-\Psi$ as the momentum. (We note that ∂_- is a space derivative.) We call a configuration x^-, x^i, $\theta^a \equiv \Sigma$ and then choose

$$[\partial_-\Psi[\Sigma_1], \Psi[\Sigma_2]]_{x_1^+=x_2^+} = -\frac{i}{2}\Delta^{17}[\Sigma_1, \Sigma_2] \qquad (7.11)$$

where

$$\Delta^{17}[\Sigma_1, \Sigma_2] \equiv \delta(x_1^- - x_2^-) \int d\sigma_0 \, \Delta^8[x_1^i(\sigma) - x_2^i(\sigma + \sigma_0)]$$

$$\times \Delta^8[\theta_1^a(\sigma) - \theta_2^a(\sigma + \sigma_0)] \qquad (7.12)$$

The ∂_- in the commutator is essential to obtain the correct symmetry. It is only in the light-cone gauge that we can use a space derivative of the field as a momentum. In a covariant description, there is no such covariant expression.

Given the canonical commutation rule (7.11), one can represent the super-Poincaré generators as

$$G = i \int D\Sigma \partial_- \Psi[\Sigma] g \Psi[\Sigma] \qquad (7.13)$$

with g as in representation (7.6). From the Hamiltonian p^- we can write the equation of motion

$$\partial_+ \Psi = -ih\Psi \qquad (7.14)$$

and construct the action

$$S = \int \partial_-\Psi(\partial_+\Psi + ih\Psi) D\Sigma \, dx^+$$

$$= \int \partial_+\Psi \partial_-\Psi D\Sigma \, dx^+ + i \int H \, dx^+ \qquad (7.15)$$

If a similar analysis is performed for the underlying point-particle field theory, one can show that there is an infinity of possible representations of the super-Poincaré algebra [67], apart from the one obtained by truncating the string to a point. These representations can be interpreted as corresponding to terms with $\Box^2$, $\Box^3$, and so on in the action and are possible counterterms to the kinetic term. In an interacting quantum theory these terms will be generated unless there is a symmetry forbidding them. Such a symmetry does not seem to exist and this is a strong indication that the quantum theory does not make sense. In the superstring theory one can prove [67]

7 • Field Theory for Free Superstrings

that *the representation (7.13) is unique*, leaving no possible counterterm to the kinetic term.

The expressions (7.13) and (7.15) are quite formal and must be given a precise meaning. This can be done by relating them to infinite mode expansions. The coordinates and functional derivatives can be expanded as

$$x^i(\sigma) = x^i + \frac{i}{2} \sum_{n \neq 0} \frac{1}{n} (\alpha_n^i e^{2in\sigma} + \tilde{\alpha}_n^i e^{-2in\sigma})$$

$$\equiv x^i + \sum_{n=1}^{\infty} \frac{1}{\sqrt{n}} (x_n^i \cos 2n\sigma + \tilde{x}_n^i \sin 2n\sigma) \tag{7.16}$$

$$\delta^i(\sigma) = \frac{1}{\pi} \left\{ \frac{\partial}{\partial x^i} + i \sum_{n \neq 0} (\alpha_n^i e^{2in\sigma} + \tilde{\alpha}_n^i e^{-2in\sigma}) \right\}$$

$$\equiv \frac{1}{\pi} \left\{ \frac{\partial}{\partial x^i} + \sum_{n=1}^{\infty} 2\sqrt{n} \left(\frac{\partial}{\partial x_n^i} \cos 2n\sigma + \frac{\partial}{\partial \tilde{x}_n^i} \sin 2n\sigma \right) \right\} \tag{7.17}$$

where

$$x_n^i = \frac{i}{2\sqrt{n}} (\alpha_n^i - \alpha_{-n}^i + \tilde{\alpha}_n^i - \tilde{\alpha}_{-n}^i) \tag{7.18}$$

$$\tilde{x}_n^i = -\frac{1}{2\sqrt{n}} (\alpha_n^i + \alpha_{-n}^i - \tilde{\alpha}_n^i - \tilde{\alpha}_{-n}^i) \tag{7.19}$$

and

$$\theta^a(\sigma) = \frac{1}{\sqrt{2p^+}\pi} \sum_{n=-\infty}^{\infty} (S_n^{1a} + iS_{-n}^{2a}) e^{2in\sigma}$$

$$\equiv \frac{1}{\sqrt{\pi}} \sum_{n=-\infty}^{\infty} (\theta_n^{1a} + i\theta_{-n}^{2a}) e^{2in\sigma}$$

$$\equiv \theta_0^a + \sum_{n=1}^{\infty} (\theta_n^a \cos 2n\sigma + \tilde{\theta}_n^a \sin 2n\sigma) \tag{7.20}$$

$$d^a(\sigma) = \sqrt{\frac{p^+}{2\pi}} \sum_{n=-\infty}^{\infty} (S_n^{1a} - iS_{-n}^{2a}) e^{2in\sigma}$$

$$\equiv \frac{1}{\pi} \left\{ \frac{\partial}{\partial \theta_0^a} + \sum_{n=1}^{\infty} 2 \left(\frac{\partial}{\partial \theta_n^a} \cos 2n\sigma + \frac{\partial}{\partial \tilde{\theta}_n^a} \sin 2n\sigma \right) \right\} \tag{7.21}$$

where

$$\theta_n^a = \frac{1}{\sqrt{2p^+ n}} (S_n^{1a} + S_{-n}^{1a} + iS_n^{2a} + iS_{-n}^{2a}) \tag{7.22}$$

and

$$\tilde{\theta}_n^a = \frac{i}{\sqrt{2p^+ \pi}} (S_n^{1a} - S_{-n}^{1a} - iS_n^{2a} + iS_{-n}^{2a}) \tag{7.23}$$

The coordinates x_n^i, $\tilde{x}_n^i$, θ_n^a, and $\tilde{\theta}_n^a$ are harmonic-oscillator coordinates. A field $\Psi[x(\sigma), \theta(\sigma)]$ can then be expanded in component fields using a complete set of harmonic-oscillator wave functions $\psi_n(x)$ and corresponding wave functions $\chi_{n,n'}(\theta, \tilde{\theta})$ for the anticommuting variables [66]:

$$\Psi[x(\sigma), \theta(\sigma)] = \sum_{n_k, n'_l, m_s, m'_t} \psi_{(n_k, n'_l, m_s, m'_t)}(x, \theta_0)$$

$$\times \prod_k \psi_{n_k}(x_k) \prod_l \psi_{n'_l}(\tilde{x}_l) \prod_{s,t} \chi_{m_s, m'_t}(\theta_s, \tilde{\theta}_t) \tag{7.24}$$

The functional measure in action (7.15) is then defined such that when the integrations over the harmonic-oscillator coordinates are performed the action is

$$S = \int d^{10}x \, d^8\theta \left[-\tfrac{1}{2} \sum_{\{n\}} \phi_{\{n\}}(x, \theta_0)(\Box - N_\phi) \phi_{\{n\}}(x, \theta_0) \right.$$

$$\left. + \frac{i}{2} \sum_{\{m\}} \psi_{\{m\}}(x, \theta_0) \frac{\Box - M_\psi}{\partial_-} \psi_{\{m\}}(x, \theta_0) \right] \tag{7.25}$$

In this expression $\{n\}$ incorporates number labels for every bosonic and fermionic oscillator made with the understanding that the bosonic fields are included in the first sum and fermionic fields in the second sum. The action is hence a sum of ordinary kinetic actions for all component fields of the string.

When we turn to open and type IIa strings, the above construction cannot be used. As in the construction of the massless supermultiplets in Chapter 4, we can find an SO(8) covariant formalism with vector or tensor superfields. These fields will, however, be highly reducible and have to satisfy quite complex constraints, which will make the formalism inaccessible. In a superfield formalism one should try to use scalar superfields to avoid a high redundancy in field components. The way to achieve it here is to break the explicit SO(8) invariance down to SU(4) × U(1) [68]. We consider two SO(8) spinors A^a and B^a. Then in SU(4) notation $A^a \to A_A$,

7 • Field Theory for Free Superstrings

$\bar{A}^A$, $A = 1, \ldots, 4$, and similarly for B^a:

$$A^a B^a = \tfrac{1}{2}(A_A \bar{B}^A + \bar{A}^A B_A) \tag{7.26}$$

The chiral theories containing spinors S^{1a} and S^{2a} decompose as

$$S^{1a} \to S_A, \bar{S}^A \tag{7.27}$$

$$S^{2a} \to \tilde{S}_A, \bar{\tilde{S}}^A \tag{7.28}$$

and the type IIa case as

$$S^{1a} \to S_A, \bar{S}^A \tag{7.29}$$

$$S^{2\dot{a}} \to \tilde{S}^A, \bar{\tilde{S}}_A \tag{7.30}$$

Since the anticommutator is

$$\{S_A(\sigma), \bar{S}^B(\sigma')\} = \pi \delta_A^B \delta(\sigma - \sigma') \tag{7.31}$$

and similarly for the other S's, we use

$$\theta_A = \frac{1}{\sqrt{\pi p^+}} S_A \tag{7.32}$$

$$\frac{\delta}{\delta \theta_A} = \sqrt{\frac{p^+}{\pi}} \bar{S}^A \tag{7.33}$$

$$\tilde{\theta}_A = \frac{1}{\sqrt{\pi p^+}} \tilde{S}_A \tag{7.34}$$

$$\frac{\delta}{\delta \tilde{\theta}_A} = \sqrt{\frac{p^+}{\pi}} \bar{\tilde{S}}^A \tag{7.35}$$

as coordinates and derivatives and rewrite the algebra (4.10)–(4.12). Sometimes it is also convenient to rewrite the SO(8) vector x^i in SU(4) × U(1) notation:

$$x^i \to (x^I, x^R, x^L), \quad I = 1, \ldots, 6$$

$$x^R = \frac{1}{\sqrt{2}} (x^7 + ix^8)$$

$$x^L = \frac{1}{\sqrt{2}} (x^7 - ix^8)$$

For the open strings a scalar field $\Phi^{ab}[x(\sigma), \theta(\sigma), \tilde{\theta}(\sigma)]$ can be used, and for closed strings $\Psi[x(\sigma), \theta(\sigma), \tilde{\theta}(\sigma)]$.

The appropriate boundary conditions for the coordinates distinguish the various cases. Also, for this case one finds the fields to be reducible and a "reality constraint" as in relation (5.9) must be imposed. Type I strings must also satisfy a nonorientation constraint

$$\Phi^{ab}[x(\sigma), \theta(\sigma), \tilde{\theta}(\sigma)] = -\Phi^{ba}[x(\pi - \sigma), \tilde{\theta}(\pi - \sigma), \theta(\pi - \sigma)] \quad (7.36)$$

$$\Psi[x(\sigma), \theta(\sigma), \tilde{\theta}(\sigma)] = \Psi[x(-\sigma), \tilde{\theta}(-\sigma), \theta(-\sigma)] \quad (7.37)$$

Again one can introduce a canonical commutation relation and rewrite the super-Poincaré algebra in a second-quantized form and find the Hamiltonian and the Lagrangian. As in the previous case one can show that the representations are unique [67]. There are no possible counterterms (apart from the action itself) to the kinetic terms.

At this stage let us digest the notion of a quantum string. In expansion (7.24) we have given a local pointlike structure to each quantum state in which the string can appear. One way of understanding this fact is to consider each point along the string as a separate pointlike object, which can be excited to give a specific state. (This will not be a mass eigenstate and hence not a true state, but it is easier to comprehend than the harmonic modes, which are the true states.) However, when a string propagates, all the states (or the points along the string) propagate coherently and we must consider an extended object. In the next chapter we examine interactions and consider three-string vertices. If each string is projected to specific states, we obtain three-point functions.

A further remark is that, in a superfield $\Psi[x(\sigma), \theta(\sigma)]$, the Grassmann coordinates are more than just bookkeeping devices. We cannot Taylor-expand to get a finite number of fields $\Psi[x(\sigma)]$ as could be done for point-particle superfields.

Chapter 8

Interaction Field Theory for Type IIB Superstrings

The light-cone gauge formulation is quite advantageous for the construction of interactions. The generator p^- is the Hamiltonian and we can simply ask if we can add interaction terms to it and still keep an algebra that closes. It turns out that all generators that transform the system out of the quantization plane, $x^+ = $ const, will have to contain interaction terms. These generators are called Hamiltonians or dynamical generators (in contrast to the linearly realized generators, which are called kinematical generators) in the notation of Dirac [69].

The precise form of the interaction terms will be obtained by working in two steps. First, we represent the generators in a functional form. A general form for the dynamical generators will be assumed and then the closure of the algebra must be imposed. In this process we will be able to pin down the expressions to a unique form. Then, to give the precise meaning of this form, we must go over to a mode basis to check that each coupling is properly defined. After that we can go back to a functional form again.

We write a general three-string contribution to a generator (in type IIb) as

$$G_3 = i \int D\Sigma_1 D\Sigma_2 \partial_- \Psi[\Sigma_1 + \Sigma_2] g(\sigma_1, \sigma_2) \Psi[\Sigma_1] \Psi[\Sigma_2] \tag{8.1}$$

where the configurations Σ_1 and Σ_2 have one point in common. The configuration $\Sigma_1 + \Sigma_2$ is the union of the configurations Σ_1 and Σ_2. To make the interaction local, the operator $g(\sigma_1, \sigma_2)$ acts at points σ_1 and σ_2 infinitesimally close to the point common to Σ_1 and Σ_2, the interaction point. When we transcribe to the mode basis we will find that convergence factors must be inserted to damp singular behavior near this point, but they are of no importance for the closure of the super-Poincaré algebra.

When trying to close the algebra, we will deal with commutators between two- and three-string operators. If we let

$$A = i \int D\Sigma \partial_- \Psi[\Sigma] \int d\sigma \, a(\sigma) \Psi[\Sigma] \tag{8.2}$$

$$B = i \int D\Sigma_1 D\Sigma_2 \partial_- \Psi[\Sigma_1 + \Sigma_2] b_1(\sigma_1) \Psi[\Sigma_1] b_2(\sigma_2) \Psi[\Sigma_2] \tag{8.3}$$

and use the fact that functional operators can be integrated partially between $\Psi[\Sigma_1]$ or $\Psi[\Sigma_2]$ and $\Psi[\Sigma_1 + \Sigma_2]$ (with appropriate changes of sign), we get the result

$$[A, B] = i \int D\Sigma_1 D\Sigma_2 \partial_- \Psi[\Sigma_1 + \Sigma_2]$$

$$\times \int d\sigma \{ [b_1(\sigma_1), a(\sigma)] \Psi[\Sigma_1] b_2(\sigma_2) \Psi[\Sigma_2]$$

$$+ b_1(\sigma_1) \Psi[\Sigma_1] [b_2(\sigma_2), a(\sigma)] \Psi[\Sigma_2] \} \tag{8.4}$$

under the condition that $a(\sigma)$ does not contain ∂_-. The ∂_- structure must be handled carefully; consider a simple example:

$$\left[\int D\Sigma \partial_- \Psi[\Sigma] \int d\sigma \frac{a(\sigma)}{\partial_-} \Psi[\Sigma], i \int D\Sigma_1 D\Sigma_2 \partial_- \Psi[\Sigma_1 + \Sigma_2] \Psi[\Sigma_1] \Psi[\Sigma_2] \right]$$

$$= \int D\Sigma_1 D\Sigma_2 \partial_- \Psi[\Sigma_1 + \Sigma_2]$$

$$\times \int d\sigma \frac{1}{\partial_-} \left\{ \frac{a(\sigma)}{\partial_-} \Psi[\Sigma_1] \partial_- \Psi[\Sigma_2] \partial_- \Psi[\Sigma_1] \frac{a(\sigma)}{\partial_-} \Psi[\Sigma_2] \right\} \tag{8.5}$$

One would expect the three-string vertex to contain terms with at most two transverse functional derivatives, since the three-point couplings involving massless states contain the three-graviton vertex, but here we will allow a general expression in the Hamiltonian.

8 • Interaction Fields for Type IIB

We will follow the procedure used in constructing ordinary light-cone superfield theory. We start with the three-string part of the dynamical supersymmetry generators $Q_A^{-\dot{a}}$ and the Hamiltonian p^- and consider their mutual commutators as well as those with the kinematical generators. We introduce the notation

$$Q_{A_3}^{-\dot{a}} = \mathrm{i} \int D\Sigma_1 D\Sigma_2 \partial_- \Psi[\Sigma_1 + \Sigma_2] q_A^{\dot{a}}(\sigma_1, \sigma_2) \Psi[\Sigma_1] \Psi[\Sigma_2] \tag{8.6}$$

and

$$H_3 = \mathrm{i} \int D\Sigma_1 D\Sigma_2 \partial_- \Psi[\Sigma_1 + \Sigma_2] h(\sigma_1, \sigma_2) \Psi[\Sigma_1] \Psi[\Sigma_2] \tag{8.7}$$

The procedure now is to go through the various (anti-)commutators in order of simplicity. The commutators $\{Q_A^{+a}, Q_B^{-\dot{a}}\}$, $[Q_A^{+a}, H]$, $[J^{+i}, Q_A^{-\dot{a}}]$, and $[J^{+i}, H]$ force the functional derivatives to enter only via $\mathbf{d}^a = d_1^a/\partial_{-1} - d_2^a/\partial_{-2}$ and $\boldsymbol{\delta}^i = \delta_1^i/\partial_{-1} - \delta_2^i/\partial_{-2}$. From $\{Q_1^{-\dot{a}}, Q_2^{-\dot{b}}\}$ it follows that x_i appears only through $\mathbf{x}^i = x_1^i/\partial_{-1} - x_2^i/\partial_{-2}$. From this anticommutator it also follows that there is no explicit θ-dependence in the functional operators. Furthermore, the commutator $[J^{+-}, Q_A^{-\dot{a}}]$ yields $\#\partial_- + \frac{1}{2}\#d = \frac{3}{2}$.

With this knowledge one can attempt a more restricted form for the operator $q_A^{\dot{a}}(\sigma_1, \sigma_2)$. In the commutators $\{Q_A^{-\dot{a}}, Q_B^{-\dot{b}}\}$ the fact that they are proportional to $\delta^{\dot{a}\dot{b}}$ is a very strong restriction. By starting with the part in $q_A^{\dot{a}}$ with no spinor derivatives we can then work upward in the number of spinor derivatives. After a long and painstaking calculation, with abundant use of SO(8) properties, one finds, in fact, a unique result for $q_A^{\dot{a}}$ and h:

$$h = \kappa \partial_-^3 \sum_{n=0,2,4,6,8} c_{a_1 \cdots a_n}^{ij} \mathbf{d}^{a_1} \cdots \mathbf{d}^{a_n} \bar{\mathbf{p}}^i \mathbf{p}^j \frac{\partial_{-1}^{n/2} \partial_{-2}^{n/2}}{\partial_-^{n/2}} \tag{8.8}$$

where

$$\mathbf{p}^i = -\mathrm{i}\boldsymbol{\delta}^i + \dot{\mathbf{x}}^i, \qquad \bar{\mathbf{p}}^i = \mathrm{i}\boldsymbol{\delta}^i + \dot{\mathbf{x}}^i \tag{8.9}$$

and

$$C^{ij} = \delta^{ij}$$

$$C_{a_1 a_2}^{ij} = -\frac{\mathrm{i}}{2} \gamma_{ab}^{ij}$$

$$C^{ij}_{a_1 a_2 a_3 a_4} = -\frac{1}{4!} t^{ij}_{a_1 a_2 a_3 a_4}$$

$$C^{ij}_{a_1 \cdots a_6} = \frac{i}{2 \cdot 6!} \varepsilon_{a_1 \cdots a_8} \gamma^{ij}_{a_7 a_8}$$

$$C^{ij}_{a_1 \cdots a_8} = \frac{1}{8!} \varepsilon_{a_1 \cdots a_8} \delta^{ij} \qquad (8.10)$$

The matrices used are defined by

$$u^{i\dot{a}}_{abc} = -\gamma^{ij}_{[ab} \gamma^{j}_{c]\dot{a}}, \qquad t^{ij}_{abcd} = \gamma^{ik}_{[ab} \gamma^{jk}_{cd]} \qquad (8.11)$$

What remains is to find the exact ∂_- structure. This can be obtained from the commutators with J^{i-}. This is a very long and intricate calculation, but one which should be done to check the whole algebra. To find just the ∂_- structure one can try a simpler approach, such as checking a four-point amplitude. This has been done in the open-string case [68].

For completeness, we now investigate the Hamiltonian in the mode basis. Consider the Hamiltonian (8.7) with h given by expression (8.8). The integration

$$\int D\Sigma_3 \Delta^{17}[\Sigma_3 - \Sigma_1 - \Sigma_2] = 1 \qquad (8.12)$$

is introduced to make the expression symmetric in the strings 1, 2, and 3. It is here advantageous to rescale the σ-variables to be proportional to p^+. In these new variables the length of string 3 is the lengths of string 1 + string 2, and the condition (8.12) just says that strings 1 + 2 lie on top of string 3 at the interaction or that string 3 just splits into strings 1 and 2 such that their lengths in σ match. This is clearly a very geometric interaction! Unfortunately, in this formulation there is the operator $h(\sigma_1, \sigma_2)$, acting at the point of splitting, that somewhat shadows the pure geometric interpretation of splitting. To perform the explicit integrations over all higher modes is an arduous task. Instead we consider the expression

$$E = \kappa \int D\Sigma_1 D\Sigma_2 D\Sigma_3 \Delta^{17}[\Sigma_3 - \Sigma_1 - \Sigma_2] \Psi[\Sigma_1] \Psi[\Sigma_2] \Psi[\Sigma_3] \qquad (8.13)$$

8 • Interaction Fields for Type IIB

If one expands all coordinates in their normal modes and integrates over all nonzero modes, the result is formally

$$E = \int \prod_{r=1}^{3} d^9 x_r \, d^8\theta_r \delta^{17}(z_1 - z_2) \delta^{17}(z_2 - z_3)$$

$$\times \sum_{\{n^{(1)}, n^{(2)}, n^{(3)}\}} C(\{n^{(1)}, n^{(2)}, n^{(3)}\}) \prod_{r=1}^{3} \Psi_{\{n^{(r)}\}}(z_r) \tag{8.14}$$

where $\{n^{(r)}\}$ is short for the infinite set of mode numbers $\{n_k^{(r)}, n_l'^{(r)}, m_s^{(r)}, m_t'^{(r)}\}$ in equation (7.24) and z_r is short for (x_r, θ_r). This is a horrendous expression and in order to avoid this morass of indices we define the number-basis vector by

$$|V\rangle = \sum_{\{n^{(1)}, n^{(2)}, n^{(3)}\}} C(\{n^{(1)}, n^{(2)}, n^{(3)}\}) |n^{(1)}, n^{(2)}, n^{(3)}\rangle \tag{8.15}$$

A coupling of three specific fields is then given by

$$C_{N_1, N_2, N_3} = \langle N_1, N_2, N_3 | V \rangle \tag{8.16}$$

where $|N_1, N_2, N_3\rangle$ is a Fock-space vector with excitation number given by the three sets of integers N_1, N_2, and N_3. From $|V\rangle$ one can in principle obtain the starting expression E.

To obtain $|V\rangle$ by direct computation is quite tedious. A faster way is to note that if we insert $x_3^i(\sigma) - x_1^i(\sigma) - x_2^i(\sigma)$ in the integrand of expression (6.14) we get zero, because of the Δ functional. On following through the steps from (6.14) to (6.16) we deduce that $|V\rangle$ must satisfy

$$(x_3^i(\sigma) - x_1^i(\sigma) - x_2^i(\sigma))|V\rangle = 0 \tag{8.17}$$

and

$$(\theta_3^a(\sigma) - \theta_1^a(\sigma) - \theta_2^a(\sigma))|V\rangle = 0 \tag{8.18}$$

as well as the corresponding momentum conditions. All these conditions determine $|V\rangle$ up to certain overall factors that do not involve the oscillators. Such factors will be determined eventually by the continued analysis. By setting up a general expression for $|V\rangle$ in terms of creation operators, the conditions (8.17) and (8.18) lead to

$$|V\rangle = \exp(E_\alpha + E_\theta)|0\rangle \delta^{17}(z_1^0 - z_2^0) \delta^{17}(z_2^0 - z_3^0) \tag{8.19}$$

with $z^0 = (x, \theta_0)$, the zero modes, and

$$E_\alpha = \frac{1}{2}\left\{\sum_{r,s=1}^{3}\sum_{m=1}^{\infty}(\alpha_{-m}^{(r)}\bar{N}_{mn}^{rs}\alpha_{-n}^{(s)} + \tilde{\alpha}_{-m}^{(r)}\bar{N}_{mn}^{rs}\tilde{\alpha}_{-n}^{(s)}\right.$$

$$\left. + \mathbb{P}\sum_{r=1}^{3}\sum_{m=1}^{\infty}\bar{N}_m^r(\alpha_{-m}^{(r)} + \tilde{\alpha}_{-m}^{(r)}) - \frac{\tau_0}{\alpha}\mathbb{P}^2\right\} \quad (8.20)$$

$$E_\theta = \frac{1}{2}\sum_{r,s=1}^{3}\sum_{m,n=1}^{\infty}\frac{1}{\alpha_r}[\theta_{-m}^{1(r)}(C\bar{N}^{rs})_{mn}\theta_{-n}^{1(s)} + \theta_{-m}^{2(r)}(C\bar{N}^{rs})_{mn}\theta_{-n}^{2(s)}]$$

$$+ \frac{i}{2}\alpha\left(\sum_{r=1}^{3}\sum_{m=1}^{\infty}\theta_{-m}^{1(r)}\frac{C}{\alpha_r}\bar{N}_m^r\right)\left(\sum_{s=1}^{3}\sum_{n=1}^{\infty}\theta_{-n}^{2(s)}\frac{C}{\alpha_s}\bar{N}_n^s\right)$$

$$- \Lambda\sum_{r=1}^{3}\sum_{m=1}^{\infty}\frac{1}{\alpha_r}(\bar{N}^rC)_m(e^{-i\pi/4}\theta_{-m}^{1(r)} + e^{i\pi/4}\theta_{-m}^{2(r)}) \quad (8.21)$$

with $\theta_{-m}^{1,2}$ defined in relation (7.20) and where

$$\alpha_r = 2p_r^+, \qquad \alpha = \alpha_1\alpha_2\alpha_3, \qquad \tau_0 = \sum_{r=1}^{3}\alpha_r \ln \alpha_r$$

$$\mathbb{P}^i = \alpha_1 p_2^i - \alpha_2 p_1^i, \qquad \Lambda^a = \alpha_1\frac{\partial}{\partial \theta_{0_2}^a} - \alpha_2\frac{\partial}{\partial \theta_{0_1}^a}, \qquad C_{mn} = m\delta_{m,n}$$

$$\bar{N}_{mn}^{rs} = -\frac{mn\alpha}{n\alpha_r + m\alpha_s}\bar{N}_m^r\bar{N}_n^s, \qquad \bar{N}_m^r = \frac{1}{\alpha_r}\frac{(-1)^{m+1}}{m!}\frac{\Gamma\left(m\left(1 + \frac{\alpha_{r+1}}{\alpha_r}\right)\right)}{\Gamma\left(1 - m\frac{\alpha_{r+1}}{\alpha_r}\right)}$$

$$(8.22)$$

We can now consider the Hamiltonian (8.7) with $h(\sigma_1, \sigma_2)$ as in expression (8.8). By use of the δ functional (8.12) h can be symmetrized in an expression $h(\sigma_1, \sigma_2, \sigma_3)$. Following the steps (8.13) to (8.15) we find that it corresponds to a mode-basis vertex vector

$$|H\rangle = h(\sigma_1, \sigma_2, \sigma_3)|V\rangle \quad (8.23)$$

with h expressed in terms of oscillators. If the operator in h is commuted through the exponential in $|V\rangle$, one finds that

$$p_1^i(\sigma)|V\rangle \xrightarrow[\sigma \to \sigma_1]{} \frac{1}{\pi}(\sigma_1 - \sigma)^{-1/2} Z^i |V\rangle \qquad (8.24)$$

and

$$d_1^a(\sigma)|V\rangle \xrightarrow[\sigma \to \sigma_1]{} \frac{1}{\pi}(\sigma_1 - \sigma)^{-1/2} Y^a |V\rangle \qquad (8.25)$$

where $Z^i|V\rangle$ and $Y^a|V\rangle$ are vectors of finite norm.

Finally we can now transform back to the functional expression in order to write the correctly normalized three-string Hamiltonian in the form (8.7) and (8.8), where we let

$$\mathbf{p}^i(\sigma) \to \sqrt{\sigma_1 - \sigma}\,\mathbf{p}^i(\sigma)$$
$$\bar{\mathbf{p}}^i(\sigma) \to \sqrt{\sigma_1 - \sigma}\,\bar{\mathbf{p}}^i(\sigma) \qquad (8.26)$$
$$\mathbf{d}^a(\sigma) \to \sqrt{\sigma_1 - \sigma}\,\mathbf{d}^a(\sigma)$$

and consider the full expression in the limit $\sigma \to \sigma_1$.

A further check on the Hamiltonian may be obtained by considering the couplings of three massless particles. They are most easily obtained from the vertex vector (8.23) for $|H\rangle$ by taking its matrix elements with three ground states. Alternatively one lets $\sigma \to 0$ properly to recover pointlike particles in equations (8.7) and (8.8). It can be shown that these expressions correspond to the cubic couplings of $N = 2$ supergravity.

Chapter 9

Other String Interactions and the Possible Occurrence of Anomalies

The construction of interactions involves finding nonlinear representations of the super-Poincaré algebra using second-quantized functional fields. For open strings it is also rather straightforward to construct the three-string interaction. The natural interaction to try is when two end points on two strings join to make one string. We start with the field representation of Chapter 7 and attempt a Hamiltonian of the form

$$H_3 = i \int D\Sigma_1 D\Sigma_2 h(\sigma_1, \sigma_2) \operatorname{Tr}[\partial_- \Phi[\Sigma_1 + \Sigma_2]\Phi[\Sigma_1]\Phi[\Sigma_2]] \quad (9.1)$$

where the trace is taken over the SO(N) or Sp($2N$) indices of the fields Φ. This time the configurations Σ_1 and Σ_2 match at two end points and we use quantities σ such that the length of a string is $\pi\alpha \equiv 2p^+\pi$.

Again one can grind through the algebra to find a unique answer for the Hamiltonian with h given by

$$h = g\partial_- \sum_{n=0,2,4} C^i_{A_1,\ldots,A_n} \frac{\delta}{\delta\theta_{A_1}} \cdots \frac{\delta}{\delta\theta_{A_n}} \mathbf{p}^i \frac{\partial^{n/2}_{-1}\partial^{n/2}_{-2}}{\partial^{n/2}_-} \quad (9.2)$$

with notation as in expression (8.8). (For further information on notation, see the Appendix at the end of Part I.) Also

$$c^i = 1/\sqrt{2} \qquad \text{for } i = L, \text{ otherwise zero}$$

$$c^i_{AB} = \rho^I_{AB} \qquad \text{for } i = I, \text{ otherwise zero} \qquad (9.3)$$

$$c^i_{ABCD} = (\sqrt{2}/3)\varepsilon_{ABCD} \qquad \text{for } i = R, \text{ otherwise zero}$$

As in the case of the closed string, a detailed investigation in the mode basis shows that certain convergence factors must be inserted, namely, for each operator $\delta/\delta\boldsymbol{\theta}_A$ or $\mathbf{p}^i$ a factor $\lim_{\sigma \to \pi\alpha_1}(\pi\alpha_1 - \sigma)^{1/2}$ must multiply the operator.

The construction of the open-string vertex can also be used to construct closed-string vertices, which are obtained by a "stuttering process" by direct products of open-string vertices. Since we have seen that left-going and right-going modes are separated in all generators, we can construct the closed-string vertex by a product of two open-string vertices, one containing left-going and the other right-going modes. This can also be translated into the functional form as a product of left-going and right-going operators.

In this way we know the three-string interactions. Are there higher-string interactions as there are higher-point interactions for point particles? The answer is most probably no! One can check that the four-string interaction term expected from the commutator $\{Q_{A_3}^{-\dot a}, Q_{B_3}^{-b}\}$ is indeed zero [68] (apart from a possible nonzero term in the forward direction) and that an explicit computation of a four-particle amplitude using only three-string vertices yields a Lorentz-invariant expression. Hence the algebra closes with only the three-string terms. Whether it is still possible to include higher-string interactions has not yet been excluded, but it is highly unlikely.

The two types of interaction introduced are both local. To have a consistent theory we must demand that the interactions can occur as soon as two end points touch or two intermediate points meet. See Figures 9.1 and 9.2.

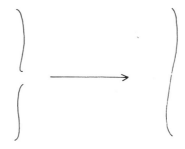

Figure 9.1. Two end points on open strings meet to join.

9 • Other String Interactions

Figure 9.2. Two internal points on the strings meet to exchange segments.

In the case of an open-string theory with a natural three-string "Yang–Mills" coupling as in expression (9.2), one string can curl up to a closed string leading to a coupling $g\Psi\Phi^{aa}$. Similarly for the gravity coupling of expression (8.8) two open strings can touch, exchange segments, and scatter into two new strings giving a coupling $\kappa\Phi^4$. We note that this coupling differs from an ordinary four-string coupling where $1 \to 3$ and $3 \to 1$ can occur.

With the gravity coupling an open string can also decay into a closed string and an open string giving a coupling $\kappa\Psi\Phi^2$.

Finally, in type I theories both an open and a closed string can double on themselves to open up to a new open or closed string, leading to couplings $\kappa\Phi^2$ and $\kappa\Psi^2$.

All the functional actions for the respective strings are unique [67]. This can also be proven for the heterotic string [70]. Hence there are no possible counterterms other than possibly the action itself. If one can prove that all quantum corrections respect the super-Poincaré invariance, then the theories are at least renormalizable. Mandelstam [71] has shown that for superstrings the S matrix is covariant, so it should at least be renormalizable. To finally settle the question of finiteness, more insight into higher loops is needed.

The one-loop graphs have been computed using a light-cone operator formalism [65]. It was then found that for the type II models these loops are indeed finite, while for the type I case the loops are seemingly infinite but renormalizable. However, a detailed check shows that in the case of gauge group SO(32) the loops are indeed finite [72]. Also, for the heterotic string the finiteness of the one-loop graphs have been shown.

The above super-Poincaré invariance at the interaction level has only been checked for the classical theory. Checking the full quantum theory can only further constrain the theory. At the one-loop level a condition $\kappa \sim g^2$ emerges. Furthermore, anomalies can emerge in the quantum theory for the chiral theories. Remarkably, for type I strings with gauge group SO(32) [28], for the type IIb strings [27], and for the heterotic string there are no anomalies at the one-loop level although they contain chiral fermions!

The absence of anomalies for this case came as a big surprise. In a ten-dimensional gauge theory with chiral fermions, such as a type I model, we expect an anomalous divergence of the gauge current arising from hexagon diagrams, since such diagrams can contribute a term

$$\partial \cdot J^a \sim \varepsilon^{\mu_1 \cdots \mu_{10}} \mathrm{Tr}(\Lambda^a F_{\mu_1 \mu_2} \cdots F_{\mu_9 \mu_{10}}) \tag{9.4}$$

where the matrices are in a representation dictated by the chiral fermions that circulate in the loop. Green and Schwarz [28] computed the hexagon diagram in the projected operator formalism of the Ramond-Neveu-Schwarz diagram. They then found that the anomaly is cancelled if the gauge group is SO(32).

Since it is essentially the massless fermions that cause anomalies, one should be able to understand this result for the underlying point-particle theory. We start with the complete string field theory for type I strings and expand the generic fields Φ and Ψ as infinite series of point-particle fields and integrate out all massive superfields. Then (in principle) we are left with an effective action in terms of the superfields ϕ_0 and ψ_0 for the massless modes:

$$S_{\mathrm{eff}} = S_1(\phi_0, \psi_0) + S_2(\phi_0, \psi_0) \tag{9.5}$$

where S_1 consists of local terms and S_2 of an infinite series of nonlocal terms. Again, in principle, we undo all gauge fixings and obtain a covariant action in terms of the supergravity multiplet, coupled to the Yang-Mills multiplet. Although this is a hypothetical calculation one can find individual terms by considering string amplitudes with massless external states and perform an expansion in α'.

One would then expect the action S_1 to lowest order in α' to coincide with that of minimal supergravity coupled to Yang-Mills theory. However, it does not! The new terms include a term involving the antisymmetric field $B_{\mu\nu}$ which, in the case of a gauge group SO(32), looks like

$$S_c \sim \int B_\Lambda [\tfrac{1}{24} \mathrm{Tr}(F_\Lambda F_\Lambda F_\Lambda F) - \tfrac{1}{7200} \mathrm{Tr}(F_\Lambda F)_\Lambda \mathrm{Tr}(F_\Lambda F)$$

$$- \tfrac{1}{240} \mathrm{Tr}(F_\Lambda F)_\Lambda \mathrm{tr}(R_\Lambda R) + \tfrac{1}{8} \mathrm{tr}(R_\Lambda R_\Lambda R_\Lambda R)$$

$$- \tfrac{1}{32} \mathrm{tr}(R_\Lambda R)_\Lambda \mathrm{tr}(R_\Lambda R)] \tag{9.6}$$

This term is such that it leads to tree-graph anomalies cancelling all the ones from one-loop graphs in the minimal theory. This is a most remarkable result and shows how string theories have an inherent degree of consistency.

The above field-theoretic result also works for the gauge group $E_8 \times E_8$ and this fact led to the discovery of the heterotic string.

For type I superstrings and the heterotic string all the one-loop graphs are finite. Hence, there is no regularization involved and no anomalies could occur.

We have seen a deep connection between finiteness and absence of anomalies. This most certainly continue at higher-loop orders and it strengthens our belief that a fundamental string theory must be finite.

Chapter 10

Outlook

The prospect for string theory as a fundamental unique theory of Nature is quite bright. As we have seen, it seemingly solves the quantum gravity problem, the gauge group is selected, and the number of possible, theoretically consistent models is limited. At this stage we have five models which seem consistent. In the best of all worlds only one of these models should be fully consistent. It is of the utmost importance to understand three issues:

1. We need to understand the finiteness of the quantum corrections. As discussed above there is a good chance that the finiteness of the perturbation expansion will be shown in detail in the near future. The approach by Mandelstam [71] using the first-quantized path-integral formalism might be the most fruitful, but the field-theoretic approach discussed above should also be feasible. With some more insight, the counterterm argument should give a decisive result.

    However, it is insufficient to show that each term in the perturbation expansion is finite. We must also be able to sum the perturbation expansion or at least show that it is an asymptotic series. We might find a problem here, since the expansion parameter is κ, which has a dimension. In principle, the higher loops will dominate at higher energies, just where string theory is supposed to take over from ordinary point-particle theories! In the old S-matrix language, the loops give Regge cuts, which dominate over the Regge pole if the intercept of the pole is greater than 1, which is the case here. This fact casts some doubt on string theory as a truly fundamental theory to infinite energies and the resolution of this dilemma will certainly be instructive.

2. The second issue is to find a geometric interpretation of string theory. The string theories have been discussed essentially only in the light-cone gauge. A similar description of ordinary gravity in the light-cone gauge will not reveal easily the equivalence principle or the general covariance underlying Einstein's theory. We must find the generally covariant theories which, when gauge-fixed to the light-cone gauge, lead to the results above. Some progress has been made here [73]. This is a really challenging problem. Finding the general principle on which to build the correct string theory might mean finding the principle underlying the theory of Nature! This would also have enormous consequences for mathematics. Einstein's gravity has influenced modern mathematics in differential geometry, topology, and many other areas; the string theory will most probably use more sophisticated mathematics, mathematics that might not yet be fully developed.

3. Third, we must understand how (or rather if) the string theory compactifies to four dimensions. Great progress has been made here. Candelas *et al.* [74] have shown that by demanding an $N = 1$ supersymmetry for $d = 4$, for the heterotic string with gauge group $E_8 \times E_8$, one can find classical solutions that compactify to $M_4 \times K$, where M_4 is the four-dimensional Minkowski space and K is a six-dimensional space. Both from string arguments and from an investigation of the underlying field theory they found that this space should be a Ricci-flat space with holonomy group SU(3). Such "Calabi-Yau spaces" [75] have been constructed. There might be some 10,000 such spaces possible. One hope here would be that a further investigation of the quantum theory will reduce the number of possible solutions. The ultimate hope is that only one such space is fully consistent. The real dream would be that the quantum theory not only selects one string theory, but that it also demands this theory to be compactified to four dimensions in a unique way. Doubts have been cast on the consistency of the Calabi-Yau solutions, but I think we need to await the complete solution for the quantum theory before we can fully understand compactification.

Phenomenology based on the Calabi-Yau solutions has been considered by Witten [76] and it looks quite promising. The four-dimensional theory starts out as an $E_6 \times E_8$ model. We can consider the E_8 as part of a shadow world, since it only interacts with the E_6

10 • Outlook

part gravitationally. The GUT based on the group E_6 has all the good properties of the old, quite successful E_6 GUT, but differs from it for the Yukawa couplings, which had problems in the old model. It is quite nontrivial that such a promising phenomenology comes out! Much remains to be done here and the final conclusions again have to await an understanding of the quantum theory.

If string theory really is as successful as I hope, it should also explain some of the fundamental problems in physics as well as influence other fields in physics:

1. It should explain why the cosmological constant is zero. There does not seem to be any possibility to explain this fact in conventional physics. The hope in string theory is that the general covariance is so big that even when the theory is considered for the energies at which we investigate Nature today, there is a remnant of the invariance demanding that the cosmological constant be zero.
2. It should have important consequences in cosmology. The possibility of a shadow world has been mentioned. The very early universe should be governed by string physics. It should also affect the universe at later stages and here it is important that it not destroy the beauty of the standard model of cosmology.
3. It should affect small-scale gravity. Since string theory is much smoother at high energies, black-hole physics will be changed. The singularity structure will most probably be altered and the Schwarzschild singularity might disappear.

The whole program described here would have sounded too far-fetched only some years ago. Now I think this is a realistic program and I expect developments in fundamental physics to follow the above lines. Several times before the physics community has believed that physics is essentially solved. Again we have reached such a stage. The future will certainly carry surprises but I feel strongly that string physics will lead us to a unique fundamental description of Nature.

Appendix

Some Notations and Conventions

The algebra of SO(8) has three inequivalent real eight-dimensional representations, one vector, and two spinors. We use 8-valued indices $i, j, \ldots$ corresponding to the vector, $a, b, \ldots$ corresponding to one spinor, and $\dot{a}, \dot{b}, \ldots$ corresponding to the other spinor. Dirac matrices $\gamma^i_{a\dot{a}}$ may be regarded as Clebsch–Gordan coefficients for combining the three eights into a singlet. A second set of matrices $\tilde{\gamma}^i_{\dot{a}a}$ is also introduced. We choose

$$\tilde{\gamma} = \gamma^T \tag{A.1}$$

$$\{\gamma^i, \tilde{\gamma}^j\} = 2\delta^{ij} \tag{A.2}$$

The 16 × 16 matrices

$$\begin{bmatrix} 0 & \gamma^i_{a\dot{a}} \\ \tilde{\gamma}^i_{\dot{b}b} & 0 \end{bmatrix}$$

form a Clifford algebra. We also define

$$\gamma^{ij}_{ab} = \tfrac{1}{2}[\gamma^i_{a\dot{a}} \tilde{\gamma}^j_{\dot{a}b} - \gamma^j_{a\dot{a}} \tilde{\gamma}^i_{\dot{a}b}] \tag{A.3}$$

These matrices are seen to be antisymmetric in a and b using relation (A.1).

We can also define

$$\gamma^{ij}_{\dot a \dot b} = \tfrac{1}{2}[\tilde\gamma^i_{\dot a a}\gamma^j_{a\dot b} - \tilde\gamma^j_{\dot a a}\gamma^i_{a\dot b}] \tag{A.4}$$

which, in a similar fashion, is antisymmetric in $\dot a$ and $\dot b$.

To span the whole 8×8 dimensional matrix spaces we also define

$$\gamma^{ijkl}_{ab} \equiv (\gamma^{[i}\tilde\gamma^j\gamma^k\tilde\gamma^{l]})_{ab} \tag{A.5}$$

and

$$\gamma^{ijkl}_{\dot a \dot b} \equiv (\tilde\gamma^{[i}\gamma^j\tilde\gamma^k\gamma^{l]})_{\dot a \dot b} \tag{A.6}$$

These matrices are symmetric.

The general Fierz formula is

$$M_{ab} = \tfrac{1}{8}\delta_{ab}\,\text{tr}\,M - \tfrac{1}{16}\gamma^{ij}_{ab}\,\text{tr}(\gamma^{ij}M) + \tfrac{1}{384}\gamma^{ijkl}_{ab}\,\text{tr}(\gamma^{ijkl}M) \tag{A.7}$$

In the case of SU(4), the six-vector can be obtained as the antisymmetric tensor product of two 4s or two $\bar 4$s. The corresponding Clebsch–Gordan coefficients (or Dirac matrices) are denoted by $\rho^I{}_{AB}$ and ρ^{IAB}. They are normalized as usual so that

$$\rho^{IAB}\rho^J{}_{BC} + \rho^{JAB}\rho^I{}_{BC} = 2\delta^A{}_C\delta^{IJ} \tag{A.8}$$

We also define

$$\rho^{IJ}{}_A{}^B = \tfrac{1}{2}(\rho^I{}_{AC}\rho^{JCB} - \rho^J{}_{AC}\rho^{ICB}) \tag{A.9}$$

References

1. T. Regge, *Nuovo Cimento* **14** (1959) 951; **18** (1960) 947.
2. R. Dolen, D. Horn, and C. Schmid, *Phys. Rev.* **166** (1968) 1768.
3. G. Veneziano, *Nuovo Cimento* **57A** (1968) 190.
4. M. A. Virasoro, *Phys. Rev.* **D1** (1970) 2933.
5. R. C. Brower, *Phys. Rev.* **D6** (1972) 1655; P. Goddard and C. B. Thorn, *Phys. Lett.* **40B** (1972) 235.
6. Y. Nambu, in: *Proc. Int. Conf. on Symm. and Quark Model*, Wayne State Univ. 1969, Gordon and Breach, London (1970); H. B. Nielsen, several talks (1969); 15th International Conference on High Energy Physics, Kiev (1970); L. Susskind, *Phys. Rev.* **D1** (1970) 1182.
7. P. M. Ramond, *Phys. Rev.* **D3** (1971) 2415.
8. A. Neveu and J. H. Schwarz, *Nucl. Phys.* **B31** (1971) 86; *Phys. Rev.* **D4** (1971) 1109.
9. L. Brink, D. I. Olive, C. Rebbi, and J. Scherk, *Phys. Lett.* **45B** (1973) 379.
10. P. Goddard, J. Goldstone, C. Rebbi, and C. B. Thorn, *Nucl. Phys.* **56** (1973) 109.
11. Y. Nambu, Lectures at Copenhagen Symposium (1970).
12. O. Hara, *Prog. Theor. Phys.* **46** (1971) 1549; T. Gotō, *Prog. Theor. Phys.* **46** (1971) 1560.
13. L. Brink and H. B. Nielsen, *Phys. Lett.* **45B** (1973) 332.
14. A. M. Polyakov, *Phys. Lett.* **103B** (1981) 207, 211.
15. L. Brink, P. Di Vecchia, and P. S. Howe, *Phys. Lett.* **65B** (1976) 471; S. Deser and B. Zumino, *Phys. Lett.* **65B** (1976) 369.
16. J.-L. Gervais and B. Sakita, *Nucl. Phys.* **B34** (1971) 477.
17. M. Ademollo, L. Brink, A. D'Adda, R. D'Auria, E. Napolitano, S. Sciuto, E. Del Giudice, P. Di Vecchia, S. Ferrara, F. Gliozzi, R. Musto, and R. Pettorino, *Phys. Lett.* **62B** (1976) 105.
18. M. Ademollo, L. Brink, A. D'Adda, R. D'Auria, E. Napolitano, S. Sciuto, E. Del Giudice, P. Di Vecchia, S. Ferrara, F. Gliozzi, R. Musto, R. Pettorino, and J. H. Schwarz, *Nucl. Phys.* **B111** (1976) 77.
19. C. Lovelace, *Phys. Lett.* **34B** (1971) 500.

20. H. B. Nielsen and P. Olesen, *Nucl. Phys.* **B57** (1973) 367.
21. A. Neveu and J. Scherk, *Nucl. Phys.* **B36** (1972) 155.
22. T. Yoneya, *Prog. Theor. Phys.* **51** (1974) 1907.
23. J. Scherk and J. H. Schwarz, *Nucl. Phys.* **B81** (1974) 118.
24. F. Gliozzi, J. Scherk, and D. I. Olive, *Phys. Lett.* **65B** (1976) 282; *Nucl. Phys.* **B122** (1977) 253.
25. M. B. Green and J. H. Schwarz, *Nucl. Phys.* **B181** (1981) 502.
26. See J. H. Schwarz, *Phys. Rep.* **69** (1982) 223; M. B. Green, *Surveys in High Energy Physics* **3** (1983) 127; L. Brink, in: *Supersymmetry* (K. Dietz, R. Flume, G. v. Gehlen, and V. Rittenberg, eds.), p. 89, Plenum Press, New York (1984).
27. L. Alvarez-Gaumé and E. Witten, *Nucl. Phys.* **B234** (1984) 269.
28. M. B. Green and J. H. Schwarz, *Phys. Lett.* **149B** (1984) 117; *Nucl. Phys.* **B255** (1985) 93.
29. N. Marcus and A. Sagnotti, *Phys. Lett.* **119B** (1982) 97.
30. D. J. Gross, J. A. Harvey, E. Martinec, and R. Rohm, *Phys. Rev. Lett.* **54** (1985) 502; *Nucl. Phys.* **B256** (1985) 502; Princeton preprint (1985).
31. M. Goroff and A. Sagnotti, *Phys. Lett.* **160B** (1985) 81.
32. P. S. Howe and U. Lindström, *Nucl. Phys.* **B181** (1981) 487; R. E. Kallosh, *Phys. Lett.* **99B** (1981) 122.
33. B. Zumino (unpublished).
34. See *Proceedings from the Workshop on Unified String Theories, Santa Barbara*, 1985 (M. B. Green and D. J. Gross, eds.), World Scientific Publishing Company, Singapore (1985).
35. P. A. M. Dirac, *Can. J. Math.* **2** (1950) 129.
36. J. H. Weis (unpublished).
37. F. Gliozzi (unpublished).
38. B. Zumino, in: *Renormalization and Invariance in Quantum Field Theory* (E. Caianiello, ed.), pp. 367 and 383, Plenum Press, New York (1974).
39. L. Brink and J.-O. Winnberg, *Nucl. Phys.* **B103** (1976) 445.
40. M. B. Green and J. H. Schwarz, *Phys. Lett.* **109B** (1982) 444.
41. M. B. Green and J. H. Schwarz, *Nucl. Phys.* **B181** (1981) 502.
42. L. Brink, O. Lindgren, and B. E. W. Nilsson, *Nucl. Phys.* **B212** (1983) 401.
43. J. Paton and C. Hong-Mo, *Nucl. Phys.* **B10** (1969) 519.
44. J. H. Schwarz, *Phys. Rep.* **69** (1982) 233.
45. M. B. Green and J. H. Schwarz, *Phys. Lett.* **136B** (1984) 367.
46. I. Bengtsson and M. Cederwall, ITP-Göteborg 84-21 (1984).
47. W. Siegel, UCB-PTH-83/22 (1983).
48. E. Cremmer and J. Scherk, *Nucl. Phys.* **B103** (1976) 399; M. B. Green, J. H. Schwarz, and L. Brink, *Nucl. Phys.* **B198** (1982) 474.
49. P. Goddard and D. Olive, in: *Vertex Operators in Mathematics and Physics* (J. Lepowski et al., eds.), MSRI Publication No. 3, p. 419, Springer-Verlag, Berlin (1984).
50. I. B. Frenkel and V. G. Kac, *Invent. Math.* **62** (1980) 23; G. Segal, *Commun. Math. Phys.* **80** (1982) 301; P. Goddard and D. Olive, see Ref. 49.
51. R. Casalbuoni, J. Gomis, and G. Longhi, *Nuovo Cimento* **24A** (1974) 249.
52. See S. Mandelstam, in: *Recent Development in Quantum Field Theory* (J. Ambjörn, B. J. Duurhuus, and J. L. Petersen, eds.), p. 251, North-Holland, Amsterdam (1985).
53. S. Fubini and G. Veneziano, *Nuovo Cimento* **67A** (1970) 29.
54. Z. Koba and H. B. Nielsen, *Nucl. Phys.* **B10** (1969) 633.
55. C. Hong-Mo, *Phys. Lett.* **28B** (1968) 425.
56. D. Amati, V. Alessandrini, M. La Bellac, and D. Olive, *Phys. Rep.* IC. No. 6 (1971).
57. L. Brink and D. Olive, *Nucl. Phys.* **B56** (1973) 256; *Nucl. Phys.* **B58** (1973) 237.
58. M. Kato and K. Ogawa, *Nucl. Phys.* **B212** (1983) 443; S. Hwang, *Phys. Rev.* **D28** (1983) 2614.

References

59. I. Bengtsson, *Classical and Quantum Gravity*, **3** (1986) L31.
60. C. Lovelace, *Phys. Lett.* **32B** (1970) 703; V. Alessandrini, *Nuovo Cimento* **2A** (1971) 321; V. Alessandrini and D. Amati, *Nuovo Cimento* **4A** (1971) 793.
61. D. B. Fairlie and D. Martin, *Nuovo Cimento* **18A** (1973) 373.
62. E. Corrigan and D. Olive, *Nuovo Cimento* **11A** (1972) 749.
63. E. F. Corrigan, P. Goddard, R. A. Smith, and D. Olive, *Nucl. Phys.* **B67** (1973) 477.
64. M. B. Green and J. H. Schwarz, *Nucl. Phys.* **B198** (1982) 252.
65. M. B. Green and J. H. Schwarz, *Nucl. Phys.* **B198** (1982) 441.
66. M. Green, J. H. Schwarz, and L. Brink, *Nucl. Phys.* **B219** (1983) 437.
67. A. K. H. Bengtsson, L. Brink, M. Cederwall, and M. Ögren, *Nucl. Phys.* **B254** (1985) 625.
68. M. B. Green and J. H. Schwarz, *Nucl. Phys.* **B243** (1984) 475.
69. P. A. M. Dirac, *Rev. Mod. Phys.* **26** (1949) 392.
70. L. Brink, M. Cederwall, and M. B. Green (to appear).
71. S. Mandelstam, in: *Proceedings from the Workshop on Unified String Theories, Santa Barbara*, 1985 (M. B. Green and D. J. Gross, eds.), p. 46, World Scientific Publishing Company, Singapore (1985).
72. M. B. Green and J. H. Schwarz, *Phys. Lett.* **151B** (1985) 21.
73. W. Siegel, *Phys. Lett.* **151B** (1985) 391, 396; W. Siegel and B. Zweibach, Berkeley preprint (1985); T. Banks and M. Peskin, SLAC preprint (1985); K. Itoh, T. Kugo, H. Kunimoto, and H. Oogori, Kyoto preprint (1985); A. Neveu and P. West, CERN preprints (1985); D. Friedan, University of Chicago preprints (1985).
74. P. Candelas, G. T. Horowitz, A. Strominger, and E. Witten, *Nucl. Phys.* **B258** (1985) 46.
75. E. Calabi, in: *Algebraic Geometry and Topology: A Symposium in Honor of S. Lefschetz*, p. 78, Princeton University Press (1957); S.-T. Yau, *Proc. Natl. Acad. Sci. USA* **74** (1977) 1798.
76. E. Witten, *Nucl. Phys.* **B258** (1985) 75.

Part II

Lectures on String Theory
With Emphasis on Hamiltonian and BRST Methods

Marc Henneaux

Chapter 11

Introduction

With the discovery by Green and Schwarz of remarkable anomaly cancellations in superstring models with gauge group SO(32) or $E_8 \times E_8$, interest in string theory has increased tremendously. This interest is further motivated by the strong belief that superstring theory, which unifies gravity with the other fundamental interactions, may be finite and hence free from the infinities which have plagued all attempts at quantizing Einstein theory.

Even if string models of nature do not turn out to yield the "ultimate theory," the fact that they illustrate almost all difficulties and concepts of modern field theory makes them worthy of study and has largely contributed to the attraction which they have for researchers. This is even true for the first-quantized models, which make use of such concepts as anomalies (quantum breaking of classical symmetries or gauge invariances), constrained Hamiltonian systems, nonpositive-definite "Hilbert" spaces, Becchi-Rouet-Stora-Tyutin (BRST) symmetry, local and global supersymmetry, Kaluza-Klein ideas, Kac-Moody algebras, σ-model techniques, topological effects, and Teichmüller spaces.

This review is based on lectures given in Brussels and Santiago to a nonexpert audience, for which this was the first contact with strings. The main emphasis was on a pedagogical exposition of the foundations of the first-quantized free theory.

The idea was to discuss in depth the bosonic model, which was used as the prototype illustrating all string techniques (the only missing important ingredient being supersymmetry). The other models were then simply surveyed briefly, since it was felt that the tools necessary to understand them had been provided.

The Hamiltonian approach was adopted throughout the lectures. One of its advantages is that it clearly relates the Virasoro conditions to two-dimensional reparametrization invariance and exhibits their intrinsic character.

One of the aims of the lectures was to show that the bosonic string theory entirely follows from the Nambu–Goto action (or its quadratic reformulation), without further input—in both the open and closed cases. Classical material which is not easily accessible, such as the validity of the light-cone gauge, was included as well.

The different methods for computing the critical dimension and the intercept parameter were explained and the spectrum was described. Particular attention was paid to the BRST approach, which seems to play an important role in the second-quantized string. How gauge invariance is enforced in the quantum theory was also analyzed in detail.

The inclusion of local supersymmetry was then discussed, again with special emphasis on BRST methods. The superstring and the heterotic model were finally briefly described.

This review possesses the same motivations as the lectures on which it is based, and hence the same shortcomings. The main one is the omission of many important topics of current research, which include discussions of the interactions, the zero-slope limit, and the second-quantized models. These subjects are well covered in the general references listed below.

The author is grateful to Claudio Teitelboim for giving him the opportunity to lecture on strings at the Centro de Estudios Científicos de Santiago. Helpful discussions with him and Luca Mezincescu are also acknowledged.

GENERAL REFERENCES

L. Brink, Superstrings, in: *Supersymmetry*, Plenum Press, New York, p. 89 (1984).
M. Green, *Surveys in High Energy Physics* **3** (1983) 127.
M. Green and J. H. Schwarz, *Nucl. Phys.* **B243** (1984) 45 (light-cone gauge field theory of superstrings).
M. Jacob, *Dual Theory*, Physics Reports Reprint Book Series, North-Holland, Amsterdam (1974).
M. Kaku and K. Kikkawa, *Phys. Rev.* **D10** (1974) 1110, 1823 (light-cone gauge field theory of bosonic strings).
J. Scherk, *Rev. Mod. Phys.* **47** (1975) 123.
J. H. Schwarz, *Phys. Rep.* **89** (1982) 223.
J. H. Schwarz, *Superstrings* (The First 15 Years of Superstring Theory), 2 vols., World Scientific Publishing Company, Singapore (1985).

Chapter 12

The Nambu–Goto String: Classical Analysis

Many of the concepts used in string theory are already present in the simpler bosonic model. Therefore this model will be discussed first.

12.1. ACTION PRINCIPLE

12.1.1. Nambu–Goto Action

A free relativistic particle moves in space time so that its world line is timelike and also has maximum proper length. The appropriate action is

$$S = -m \int ds \qquad (12.1.1.1)$$

where m is the mass and ds is the line element along the world line. One of the key properties of the action (12.1.1.1) is its reparametrization invariance.

A string is a one-dimensional extended object. The world history of the string is a two-dimensional surface in space time. The natural generalization of the particle case is to postulate that a free string (with free boundaries if it is open) is described by a surface with the following properties:

1. The surface is timelike, i.e., it possesses everywhere timelike and spacelike directions (except possibly at the boundaries).
2. It has extremum area, i.e., it is an "extremum surface."

It is convenient to use a parametric description of the string. This has, among other things, the virtue of yielding a manifestly covariant formalism. The parametric equations of the string world surface are

$$X^A = X^A(x^\alpha), \qquad A = 0, \ldots, d - 1, \qquad \alpha = 0, 1 \qquad (12.1.1.2)$$

We assume that $x^\alpha \equiv (\tau, \sigma)$ provides a good parametrization, in the sense that the tangent vectors $\partial X^A/\partial \tau$ and $\partial X^A/\partial \sigma$ are everywhere nonzero and linearly independent. Furthermore, it is assumed that $\partial X^A/\partial \tau$ is timelike (or null) and that $\partial X^A/\partial \sigma$ is spacelike,

$$\eta_{AB} \frac{\partial X^A}{\partial \tau} \frac{\partial X^B}{\partial \tau} \leq 0 \qquad (12.1.1.3\text{a})$$

$$\eta_{AB} \frac{\partial X^A}{\partial \sigma} \frac{\partial X^B}{\partial \sigma} > 0 \qquad (12.1.1.3\text{b})$$

where signature $\eta_{AB} = (-, +, +, \ldots, +)$.

The embedding of the string history induces a metric on the surface $X^A(\tau, \sigma)$, given explicitly by

$$g_{\alpha\beta} = \frac{\partial X^A}{\partial x^\alpha} \frac{\partial X^B}{\partial x^\beta} \eta_{AB} \qquad (12.1.1.4)$$

The metric η_{AB} of the background d-dimensional space time in which the string is embedded is taken to be flat. It could be curved without altering the classical theory in an essential way. However, curvature leads to serious complications in the quantum theory and these have not been solved completely in the general case of an arbitrary background. This motivates the above choice. Some of the effects of curvature on the critical dimension will be briefly discussed later.

The area of the surface swept out by the string is equal to

$$A[X^A(x^\alpha)] = \int d^2x \sqrt{-{}^{(2)}g} \qquad (12.1.1.5)$$

where $^{(2)}g$ is the determinant of $g_{\alpha\beta}$ and is negative, since $g_{\alpha\beta}$ has signature $(-,+)$.

The action of the free string is proportional to the area (12.1.1.5) and hence given by†

$$S[X^A(x^\alpha)] = -\frac{1}{2\pi\alpha'} \int_{\tau_1}^{\tau_2} d\tau \int_0^{\pi(\text{open}) \text{ or } 2\pi(\text{closed})} d\sigma \sqrt{-^{(2)}g} \qquad (12.1.1.6)$$

For closed strings, equation (12.1.1.6) is supplemented by the periodicity conditions

$$X^A(\tau, 0) = X^A(\tau, 2\pi) \qquad \text{(closed strings)} \qquad (12.1.1.7)$$

(In that case, the integral over σ can actually be taken over any interval of length 2π.)

Equation (12.1.1.6) is just the Nambu-Goto action [1, 2], where α' is a constant with dimension of length-squared in units of $\hbar$. Hence, there is a mass scale $\alpha'^{-1/2}$ in the theory. This mass used to be taken of the order of 1 GeV when dual models were applied to the hadronic world. With the radical change in perspective introduced by Scherk and Schwarz [3], α' is now taken of the order of the Planck mass ($\sim 10^{19}$ GeV) since it is believed that string theory unifies all interactions.

Remark. One can further generalize the particle action to higher dimensions, i.e., one can consider extended objects of higher dimensionality ("membrane"), with an action proportional to their space-time volume. Very little is known about the quantum theory of relativistic membranes. It is not even clear whether it is consistent. For this reason, membranes have not attracted much interest (but see elsewhere [4] for interesting developments).

Exercises

1. Derive explicitly the equations of motion following from the Nambu-Goto action principle. Show that they can be written as

$$\Box X^A = 0$$

   where $\Box$ is the covariant Laplacian formed with the induced metric $g_{\alpha\beta}$ given by expression (12.1.1.4).

2. a. Show that the projections of the equations of motion along the tangent vectors $X^A_{,\alpha}$ are actually identities, $X^A_{,\alpha} \Box X_A \equiv 0$, so that only $d-2$ equations are independent. This redundancy is a consequence of the reparametrization invariance of the action.

---
† The choice of the upper and lower bounds of the σ interval is a matter of convention. We retain here the original usage.

b. Let $\xi^A_{(\Delta)}$, $\Delta = 1, 2, \ldots, d-2$, be $d-2$ orthonormal vectors, normal to the string space-time history,

$$\xi^A_{(\Delta)} X_{A,\alpha} = 0$$

$$\xi^A_{(\Delta)} \xi_{(\Lambda)A} = \delta_{(\Delta)(\Lambda)}$$

One defines the "second fundamental forms $\Omega_{(\Delta)\alpha}{}^\beta$" of the surface $X^A = X^A(x^\alpha)$ by the equations

$$X^B_{,\alpha} \xi^A_{(\Delta),B} = -\Omega_{(\Delta)\alpha}{}^\beta X^A_{,\beta} + \mu_{(\Delta)}{}^{(\Sigma)} \xi^A_{(\Sigma)}$$

(see Ref. 5, Section 47). As one parallel transports the normal vector $\xi^\alpha_{(\Delta)}$ with respect to the d-dimensional background geometry, this normal vector rotates. The amount of rotation is parametrized by $\Omega_{(\Delta)\alpha}{}^\beta$ and $\mu_{(\Delta)}{}^{(\Sigma)}$ and measures how the 2-surface $X^A = X^A(x^\alpha)$ is curved in the background.

If $d - 2 = 1$, the last term is absent from the above equality. In general, one finds that $\mu_{(\Delta)}{}^{(\Sigma)}$ defines an infinitesimal $SO(d-2)$ rotation. Under a change of the normals $\xi^A_{(\Delta)}$ [which are determined up to an $SO(d-2)$ transformation], $\Omega_{(\Delta)\alpha}{}^\beta$ transforms homogeneously as an $SO(d-2)$ vector and $\mu_{(\Delta)}{}^{(\Sigma)}$ transforms inhomogeneously ([5]).

Show that

$$\Omega_{(\Delta)\alpha\beta} = \xi^A_{(\Delta)} X_{A;\alpha\beta}$$

where the semicolon stands for covariant differentiation in the induced metric $g_{\alpha\beta}$. Hence, for each value of Δ, $\Omega_{(\Delta)\alpha\beta}$ is a two-dimensional symmetric tensor.

c. The mean curvatures $\Omega_{(\Delta)}$ are defined as $\Omega_{(\Delta)} = \Omega_{(\Delta)\alpha\beta} g^{\alpha\beta}$. Infer from the above that the string equations of motion are completely equivalent to

$$\Omega_{(\Delta)} = 0$$

i.e., to vanishing mean curvature in every normal direction.

3. Repeat the discussion of Exercises 1 and 2 when the background metric is curved, $\eta_{AB}(X^C)$. [Note: simply insert Christoffel symbols $\Gamma^A{}_{BC}$ in the equations of motion and in the other relations so as to guarantee covariance with respect to the background! $X^A{}_{,\alpha}$ transforms as a d-vector, 2-covector (see Ref. 5, Section 52).]

12.1.2. Quadratic Form of the Action

The action $S[X^A(x^\alpha)]$ has one drawback: it is not quadratic in the fields. This is a serious problem if one wants to quantize the string by means of path-integral methods, for the usual Lagrangian form of the path integral is only valid for quadratic actions.

One can improve this state of affairs by introducing auxiliary fields, i.e., fields without degrees of freedom, obeying algebraic (as opposed to differential) equations, the purpose of which here is just to render the action quadratic.

The appropriate auxiliary fields turn out to be proportional to the metric tensor $g_{\alpha\beta}$ of the two-dimensional surface swept out by the string.

12 • Nambu–Goto String: Classical Analysis

The quadratic form of the action is given by

$$S[X^A(x^\alpha), \gamma_{\alpha\beta}(x^\gamma)] = -\frac{1}{4\pi\alpha'} \int d^2x \sqrt{-\gamma}\gamma^{\alpha\beta}\partial_\alpha X^A \partial_\beta X_A \quad (12.1.2.1)$$

That the new action principle (in which X^A and $\gamma_{\alpha\beta}$ are varied independently) is equivalent to the old will be seen shortly. Before proceeding with the proof, we digress briefly and examine a new, more "field-theory-like" interpretation of the action.

One can think of equation (12.1.2.1) as the action describing d massless scalar fields X^A in two dimensions (one, X^0, having the wrong sign for the kinetic term), moving on a curved background $\gamma_{\alpha\beta}$. Furthermore, because the metric components $\gamma_{\alpha\beta}$ are varied in equation (12.1.2.1), the two-dimensional "gravitational field" $\gamma_{\alpha\beta}$ is treated not as a given background field, but rather as an adjustable quantity coupled to the scalar fields.

This suggests that maybe one should add to equation (12.1.2.1) a kinetic term for $\gamma_{\alpha\beta}$, i.e., the Einstein–Hilbert action, and also a cosmological term. However, the Hilbert action in two dimensions is trivial ($R\sqrt{-\gamma}$ is a total divergence), while the cosmological term leads to inconsistencies because it conflicts with Weyl invariance. Accordingly, the action (12.1.2.1) is all right as it stands, even in the light of the new interpretation.

That the new action principle (12.1.2.1) is equivalent to the previous one can be seen as follows. The equations of motion obtained by varying action (12.1.2.1) with respect to $\gamma_{\alpha\beta}$ are

$$\delta S/\delta\gamma_{\alpha\beta} = \tfrac{1}{2}\sqrt{-\gamma}T^{\alpha\beta}(x) + \frac{\lambda}{2}\sqrt{-\gamma}\gamma^{\alpha\beta} = 0 \quad (12.1.2.2)$$

where we have added a cosmological term to show in a moment that it must vanish, and where $T^{\alpha\beta}(X)$ is the energy-momentum tensor of the d two-dimensional scalar fields,

$$T_{\alpha\beta}(X) = -\frac{1}{2\pi\alpha'}[\tfrac{1}{2}\gamma_{\alpha\beta}\gamma^{\rho\sigma}X^A_{,\rho}X_{A,\sigma} - X^A_{,\alpha}X_{A,\beta}] \quad (12.1.2.3)$$

In two dimensions, a massless scalar field is conformally invariant, from which it follows that the trace of $T_{\alpha\beta}$ vanishes identically, as can be checked explicitly from equation (12.1.2.3). By tracing equation (12.1.2.2), one gets

$$0 + \lambda\sqrt{-\gamma} = 0 \quad \Rightarrow \quad \lambda = 0$$

so that we set $\lambda = 0$ from now on.

With $\lambda = 0$, the $\gamma_{\alpha\beta}$ equations express that the total energy-momentum of the d scalar fields X^A vanishes,

$$T_{\alpha\beta}(X) = 0 \qquad (12.1.2.4)$$

(two equations instead of three owing to the tracelessness of $T_{\alpha\beta}$). The general solution to equation (12.1.2.4) is

$$\gamma_{\alpha\beta} = \beta \partial_\alpha X^A \partial_\beta X_A \qquad (12.1.2.5)$$

where β is an arbitrary function. The auxiliary field $\gamma_{\alpha\beta}$ is thus conformally related to the induced metric $\partial_\alpha X^A \partial_\beta X_A$, with an undetermined conformal factor.

The precise value of the conformal factor is actually irrelevant because of Weyl invariance, i.e., invariance of action (12.1.2.1) under "Weyl rescalings,"

$$\gamma_{\alpha\beta} \to \phi^2 \gamma_{\alpha\beta} \qquad (12.1.2.6a)$$

and

$$X^A \to X^A \qquad (12.1.2.6b)$$

Hence, one can assume $\gamma_{\alpha\beta} = g_{\alpha\beta}$. If the auxiliary field $\gamma_{\alpha\beta}$ is eliminated from the action (12.1.2.1) by using its own equations of motion, namely equations (12.1.2.5), one gets the correct Nambu-Goto action. In that sense, the theories based on actions (12.1.1.6) and (12.1.2.1) are equivalent, since they give the same weight to any $X^A(x^\alpha)$ history, even "off-shell" [but the auxiliary field must obey solution (12.1.2.5) to make the comparison possible].

Remarks. a. In order to prove the equivalence of the theories based on actions (12.1.1.6) and (12.1.2.1), one must do more than just check that the corresponding classical equations of motion have the same solutions. One must also show, as done here, that the actions are the same, in the sense that if one eliminates the auxiliary field from action (12.1.2.1) using its own equation of motion, one gets equation (12.1.1.6). If this were not the case, the Poisson brackets of the Xs—and hence also their quantum commutators and the quantum theory—would be different, even though the classical equations of motion are the same in both cases. This question is closely related to the inverse problem of the calculus of variations (see, e.g., Ref. [6] and work cited therein).

b. The elimination of the auxiliary field $\gamma_{\alpha\beta}$ is not as straightforward in the quantum theory. Only for $d = 26$ does one recover equivalence of action (12.1.2.1) with action (12.1.1.6) [7a].

Exercises

1. **a.** Derive the quadratic form of the action for the free particle, starting from $-m \int ds$. Show that a "cosmological term" is needed and relate it to the mass m.

    b. Derive the Hamiltonian form S_H of the action $-m \int ds$.

    c. Show that if one integrates out (from S_H) the canonical momentum p_A conjugate to X^A, one just obtains the quadratic form of the action of (a). Relate the Lagrange multiplier N associated with the "on-the-mass-shell" Hamiltonian constraint $p^2 + m^2 \approx 0$ to the metric g_{00} on the particle world line.

    d. Note that both the quadratic action and S_H describe correctly the massless case in the limit $m = 0$.

2. Derive the quadratic form of the action for the relativistic membrane. Show again that a cosmological term is needed (no Weyl invariance).

12.1.3. σ-Model Interpretation of the Action

The action (12.1.2.1) refers simultaneously to two different spaces: the two-dimensional space of the coordinates $x^\alpha \equiv (\tau, \sigma)$, and the d-dimensional Minkowski space M^d of the scalar fields. That latter space is the quotient space of the Poincaré group by the Lorentz group, $M^d =$ Poincaré/Lorentz.

Any field history $X^A(\tau, \sigma)$ provides a mapping from the first to the second space. The mappings which extremize action (12.1.2.1) are called "harmonic maps" in the mathematical literature, expression (12.1.2.1) being referred to as the "energy functional." In the physics literature, such a model is called a "σ model." (The role of harmonic maps in physics was discussed by Misner [7b] and references cited therein.)

The σ-model interpretation of the action underplays the geometrical significance of the Nambu–Goto action (area) and focuses more on its field-theory aspects. It will prove particularly useful when we discuss strings in curved backgrounds and superstrings.

12.1.4. Gauge Invariances

The Nambu–Goto action is obviously invariant under changes of coordinates, since the area is a geometric invariant.

Infinitesimal changes of coordinates lead to the following changes in the fields,

$$\delta X^A = \mathfrak{L}_{\delta\xi} X^A = \delta\xi^\alpha X^A_{,\alpha} \qquad (12.1.4.1)$$

where $\mathcal{L}_{\delta\xi}$ is the Lie derivative operator along the two-dimensional vector field $\delta\xi^\alpha$. Relation (12.1.4.1) expresses the fact that X^A are two-dimensional scalars.

When the fields X^A are modified according to relation (12.1.4.1), the variation of the Lagrangian is given by

$$\delta\mathcal{L} = \mathcal{L}_{\delta\xi}\mathcal{L} = (\delta\xi^\alpha \mathcal{L})_{,\alpha} \qquad (12.1.4.2)$$

($\mathcal{L}$ is a two-dimensional scalar density). Hence, the variation of the action reduces to the surface (line) integral

$$\delta S = \oint \delta\xi^\alpha \mathcal{L}\, dS_\alpha \qquad (12.1.4.3)$$

and is indeed zero for coordinate transformations that vanish at the boundary.

The quadratic action (12.1.2.1) possesses the same invariance, the change in the metric $\gamma_{\alpha\beta}$ being given by

$$\delta\gamma_{\alpha\beta} = \mathcal{L}_{\delta\xi}\gamma_{\alpha\beta} = \delta\xi^\gamma \gamma_{\alpha\beta,\gamma} + \delta\xi^\gamma_{,\alpha}\gamma_{\gamma\beta} + \delta\xi^\gamma_{,\beta}\gamma_{\alpha\gamma} \qquad (12.1.4.4)$$

Furthermore, we have seen that it is also Weyl invariant [transformations (12.1.2.6)].

12.1.5. Global Symmetries

The d-dimensional space time in which the string is embedded is just flat, so one has Poincaré invariance.

Poincaré invariance looks here like an internal symmetry (a σ-model type of symmetry), which only relates field variables evaluated at the same point x^α:

$$\delta X^A = a^A + \Lambda^A{}_B X^B \qquad (12.1.5.1a)$$

$$\delta\gamma_{\alpha\beta} = 0 = \delta g_{\alpha\beta} \qquad (12.1.5.1b)$$

[no derivatives of the fields in equations (12.1.5.1)].

By Noether methods, one gets $d(d+1)/2$ conserved currents obeying the continuity equation $\partial_\alpha j^\alpha = 0$. These are, explicitly,

$$j^\alpha_A = \frac{\partial \mathcal{L}}{\partial X^A_{,\alpha}} = -\frac{1}{2\pi\alpha'}\sqrt{-g}\, g^{\alpha\beta} \partial_\beta X_A \qquad (12.1.5.2)$$

(currents associated with translations) and

$$j_{AB}^\alpha = -j_{BA}^\alpha = \frac{\partial \mathscr{L}}{\partial X_\alpha^{[A}} X^{B]} = -\frac{1}{2\pi\alpha'}\sqrt{-g}g^{\alpha\beta}\partial_\beta X_{[A} X_{B]} \qquad (12.1.5.3)$$

(currents associated with rotations and boosts).
The charges are obtained by integration,

$$Q_A = \int_0^{\pi \text{ or } 2\pi} j_A^0 \, d\sigma \qquad (12.1.5.4)$$

(energy momentum) and

$$Q_{AB} = \int_0^{\pi \text{ or } 2\pi} j_{AB}^0 \, d\sigma \qquad (12.1.5.5)$$

(angular momentum).

For the closed string, the charges are obviously conserved as a result of the continuity equation, since there cannot be an ingoing or outgoing flux through nonexisting boundaries. In the open case, the charges will only be conserved if one imposes appropriate boundary conditions at $\sigma = 0$ and π, ensuring that there is no flux through the sides there (see Section 12.1.7 below).

12.1.6. Conformal Symmetry

Among the coordinate transformations $x^\alpha \to x'^\alpha = f^\alpha(x^\beta)$, those which are such that the new form of the metric $g_{\alpha\beta}$ is the same as the old one up to a local rescaling are called pseudoconformal transformations.†

$x'^\alpha = f^\alpha(x^\beta)$ is a conformal transformation

⇕

$$g_{\alpha\beta}(x)\frac{\partial x^\alpha}{\partial x'^\lambda}\frac{\partial x^\beta}{\partial x'^\mu} = g_{\lambda\mu}(x')\Lambda^2(x') \qquad (12.1.6.1)$$

Infinitesimal conformal transformations $x'^\alpha = x^\alpha + \delta\xi^\alpha$ obey

$$\mathscr{L}_{\delta\xi}g_{\alpha\beta} = \lambda g_{\alpha\beta} \qquad (12.1.6.2)$$

as can be seen by straightforward expansion of relation (12.1.6.1).

† One might obviously adopt a coordinate-independent language to define the conformal group, but this is unnecessary here for we consider (in this section at least) manifolds with trivial topology ($\sim R^2$) and use global coordinates.

In order to analyze the properties of the conformal group, it is convenient to rewrite equation (12.1.6.1) in "the conformal gauge," i.e., in the coordinate systems in which the metric components $g_{\alpha\beta}(x)$ are just proportional to the two-dimensional Minkowski tensor,

$$g_{\alpha\beta}(x) = \phi^2(x)\eta_{\alpha\beta} \quad \text{("conformal gauge")} \quad (12.1.6.3)$$

$$\Leftrightarrow g_{\alpha\beta} = \sqrt{-g}\,\eta_{\alpha\beta}$$

The existence of such coordinate systems is a well-known result of differential geometry (see, e.g., Eisenhart's book [5] and Section 12.5 on the light-cone gauge).

Relation (12.1.6.3) enables equation (12.1.6.1) to be expressed in the form

$$\eta_{\alpha\beta}\frac{\partial x^\alpha}{\partial x'^\lambda}\frac{\partial x^\beta}{\partial x'^\mu} = \eta_{\lambda\mu}\bar{\Lambda}^2(x) \quad (12.1.6.4)$$

while its infinitesimal version reads

$$\delta\xi^0_{,0} = \delta\xi^1_{,1}, \quad \delta\xi^0_{,1} = \delta\xi^1_{,0} \quad (12.1.6.5)$$

("pseudo"-Cauchy–Riemann conditions).

These equations are easier to handle in lightlike coordinates (u, v), with $ds^2 = -2\phi^2\,du\,dv$. A conformal transformation $U(u, v)$, $V(u, v)$ then appears as a coordinate transformation subject to

$$\frac{\partial U}{\partial u}\frac{\partial V}{\partial u} = 0 \quad (12.1.6.6)$$

and

$$\frac{\partial U}{\partial v}\frac{\partial V}{\partial v} = 0 \quad (12.1.6.7)$$

Hence, up to the exchange transformation $U = v$, $V = u$, a general conformal transformation is the direct product of two coordinate transformations in one dimension,

$$U = U(u), \quad V = V(v) \quad (12.1.6.8)$$

The conformal group is accordingly the direct product of twice the diffeomorphism group in one dimension.† That direct-product structure is manifest in lightlike coordinates.

In terms of Minkowskian coordinates x^0, x^1 one finds

$$x'^0 = f(x^0 + x^1) + g(x^0 - x^1)$$
$$x'^1 = f(x^0 + x^1) - g(x^0 - x^1)$$
(12.1.6.9)

The new coordinates x'^0 and x'^1 obey the free wave equation.

The conformal group plays an important role in string theory for various reasons. One reason is that in the much-used conformal gauge (12.1.6.3), it is just the residual gauge group.

More importantly, the algebra of the energy-momentum tensor components of the scalar fields X^A is isomorphic to the conformal algebra (its central extension is called the Virasoro algebra). It is this property which we now explain.

Consider a two-dimensional massless scalar field $X(x^\alpha)$, described by the usual action

$$S[X] = -\tfrac{1}{2} \int \sqrt{-g}\, g^{\alpha\beta} \partial_\alpha X \partial_\beta X\, d^2 x$$
(12.1.6.10)

Because equation (12.1.6.10) only contains the Weyl-invariant combinations $\sqrt{-g}\, g^{\alpha\beta}$, it is invariant under changes of coordinates which belong to the conformal group.

By applying the Noether theorem, one gets from this invariance an infinite number of conserved currents. These are

$$j^\alpha(\xi) = T^{\alpha\beta} \xi_\beta$$
(12.1.6.11)

where ξ^β is a conformal Killing vector, i.e., a solution of equation (12.1.6.2). Here, $T^{\alpha\beta}$ are the components of the symmetric, traceless energy momentum of the scalar field [$\sim \delta \mathscr{L}^M / \delta g_{\alpha\beta}$; see expression (12.1.2.3)]:

$$T^{\alpha\beta} = T^{\beta\alpha}, \qquad T^{\alpha\beta}{}_{;\beta} = 0, \qquad T^\alpha{}_\alpha = 0$$
(12.1.6.12)

($T^{\alpha\beta}$ is traceless as a result of Weyl invariance, as discussed in Section 12.1.2; the semicolon stands for covariant differentiation.)

By using relations (12.1.6.12) and the conformal Killing equation (12.1.6.2), one indeed checks that the currents $j^\alpha(\xi)$ are conserved,

$$j^\alpha(\xi)_{;\alpha} = 0$$
(12.1.6.13)

† This group is denoted by Diff(S^1) when the one-dimensional manifold is a circle.

The corresponding charges are given by

$$Q(\xi) = \int \xi^\alpha T_\alpha{}^\beta \, d\sigma_\beta \qquad (12.1.6.14)$$

where the integral is along a spacelike line. From general arguments, one knows that the charges $Q(\xi)$, reexpressed as phase-space functions of X and its conjugate momentum, close in the Poisson bracket according to the conformal algebra

$$[Q(\xi), Q(\eta)] = Q([\xi, \eta]) \qquad (12.1.6.15)$$

Here, $[\xi, \eta]$ is the conformal Killing vector obtained by taking the Lie bracket of ξ and η,

$$[\xi, \eta]^\alpha = \eta^\beta \xi^\alpha_{,\beta} - \xi^\beta \eta^\alpha_{,\beta} \qquad (12.1.6.16)$$

The relation (12.1.6.15) will be checked explicitly below, and simply states that the conformal symmetry is realized in phase space through the Poisson brackets.

In order to go from relation (12.1.6.15) to the Poisson brackets of the energy-momentum components, one considers particular vectors ξ^α. For example, taking ξ^α and η^α at time $x^0 = 0$ in the form

$$\xi^0(0, x^1) = \delta(x^1 - \sigma), \qquad \xi^1(0, x^1) = 0 \qquad (12.1.6.17a)$$

$$\eta^0(0, x^1) = \delta(x^1 - \sigma'), \qquad \eta^1(0, x^1) = 0 \qquad (12.1.6.17b)$$

(in Minkowski coordinates), one finds that the charges reduce to

$$Q(\xi) = T_0{}^0(\sigma), \qquad Q(\eta) = T_0{}^0(\sigma') \qquad (12.1.6.18)$$

and algebra (12.1.6.15) yields the Poisson bracket $[T_0{}^0(\sigma), T_0{}^0(\sigma')]$. To compute the components of the Lie brackets (12.1.6.16), one must integrate the conformal Killing equations for ξ^α and η^α with the above initial condition (12.1.6.17). This yields

$$\xi^0(x^0, x^1) = \tfrac{1}{2}[\delta(x^0 + x^1 - \sigma) + \delta(x^0 - x^1 + \sigma)]$$
$$\xi^1(x_0, x^1) = \tfrac{1}{2}[\delta(x^0 + x^1 - \sigma) - \delta(x^0 - x^1 + \sigma)] \qquad (12.1.6.19)$$

A similar expression holds for η^α. [It is, of course, a key property that ξ^α and η^α are completely determined by the initial conditions (12.1.6.17) and the conformal Killing equations. Without this property, algebra (12.1.6.15) would be meaningless for it would relate a quantity well-defined on $x^0 = 0$ (in its left-hand side) to a quantity which contains the vector fields ξ^α, η^α off the surface $x^0 = 0$ (in its right-hand side). This property of the conformal group is not shared by the full diffeomorphism group.] The Lie bracket of ξ^α and η^α is given by

$$[\xi, \eta]^0(0, x^1) = 0$$

$$[\xi, \eta]^1(0, x^1) = -\delta(x^1 - \sigma)\delta'(x^1 - \sigma') + \delta(x^1 - \sigma')\delta'(x^1 - \sigma) \quad (12.1.6.20)$$

(where ' denotes the ordinary spatial derivative). Hence, the "equal time" Poisson bracket $[T_{00}(\sigma), T_{00}(\sigma')]$ reads

$$[T_{00}(\sigma), T_{00}(\sigma')] = (T_{01}(\sigma) + T_{01}(\sigma'))\delta'(\sigma - \sigma') \quad (12.1.6.21a)$$

Similarly, one finds

$$[T_{00}(\sigma), T_{01}(\sigma')] = (T_{00}(\sigma) + T_{00}(\sigma'))\delta'(\sigma - \sigma') \quad (12.1.6.21b)$$

$$[T_{01}(\sigma), T_{01}(\sigma')] = (T_{01}(\sigma) + T_{01}(\sigma'))\delta'(\sigma - \sigma') \quad (12.1.6.21c)$$

This completes the computation of the energy-momentum tensor-component algebra, which is just the conformal algebra written in the basis of "δ-function vectors." (For a related discussion, see Fubini *et al.* [8] and references cited therein.)

What is true for a massless scalar field remains true for the string, described by d such fields. However, because the conformal group is now a subgroup of the gauge group, its generators are constrained to vanish. But the important property remains that these generators, i.e., the energy-momentum components, still close according to the conformal algebra.

Exercises

1. a. Show that equation (12.1.6.4) implies

$$\dot{Z}^0 = \varepsilon Z^{1\prime}, \quad Z^{0\prime} = \varepsilon \dot{Z}^1, \quad \varepsilon^2 = 1 \quad (i)$$

for finite conformal transformations $x'^\alpha = Z^\alpha(x^\gamma)$. Infer that any finite conformal transformation is a product of the spatial reflection $Z^0 = x^0$, $Z^1 = -x^1$ times a transformation which obeys the pseudo-Cauchy-Riemann conditions.

b. Point out the similarities between the light-cone variables u, v used in the study of the pseudoconformal group and the complex variables $z, \bar{z} = x \pm iy$ used in the study of the Euclidean conformal group.

2. a. Show that the traceless condition $T^\alpha{}_\alpha = 0$ reads $T^{uv} = 0$ in lightlike coordinates.

b. Show that the conservation of energy momentum implies that T^{uu} is a function of v only while T^{vv} is a function of u alone, $T^{uu} = T^{uu}(v)$, $T^{vv} = T^{vv}(u)$.

12.1.7. Boundary Conditions

In the closed case, the equations of motion

$$\Box X^A = 0 \qquad (12.1.7.1a)$$

and

$$T_{\alpha\beta}(X) = 0 \qquad (12.1.7.1b)$$

are enough to guarantee that the action (12.1.2.1) is an extremum on the classical trajectory. This is not true in the open case, where the action is an extremum only if the equations of motion (12.1.7.1) are supplemented by appropriate boundary conditions at $\sigma = 0$ and π. These boundary conditions are necessary for the vanishing of unwanted surface terms in the variation of S.

Let $X_1^A(\sigma)$ and $X_2^A(\sigma)$ be two configurations of the open string at times τ_1 and τ_2, respectively.

We want to find an extremum of the action in the class of all string histories $X^A(\tau, \sigma)$ which match $X_1^A(\sigma)$ at time τ_1 and $X_2^A(\sigma)$ at time τ_2. We do not impose that $X^A(\tau, \sigma)$ take prescribed values on the vertical sides $\sigma = 0$ and $\sigma = \pi$, and hence we allow arbitrary values of the variations $\delta X^A(\tau, 0)$ and $\delta X^A(\tau, \pi)$ there. Two "competing" field histories are depicted in Figure 12.1.

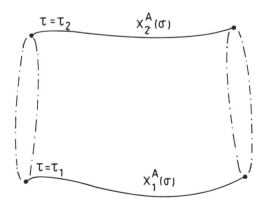

Figure 12.1. In the open string variational principle, the motion of the endpoints $X^A(\tau, 0)$ and $X^A(\tau, \pi)$ is also varied.

12 • Nambu–Goto String: Classical Analysis

The reason why we treat the time boundaries differently from the spatial ones is because we consider here strings with free ends, whose motion should not be specified in advance by "external agents" but rather should follow from the action principle itself. The absence of interactions with the outside through the boundaries is necessary for conservation of the string momentum and angular momentum.

We shall compute the variation of the action for arbitrary variations of the fields compatible with the above requirements. For simplicity, we work with the quadratic form of the action. One easily finds, keeping all terms, that

$$\delta S = -\frac{1}{2\pi\alpha'}\left[\int_0^\pi d\sigma \sqrt{-g}\, g^{\alpha 0}\partial_\alpha X^A \delta X_A\right]_{\tau_1}^{\tau_2}$$

$$-\frac{1}{2\pi\alpha'}\left[\int_{\tau_1}^{\tau_2} d\tau \sqrt{-g}\, g^{1\alpha}\partial_\alpha X^A \delta X_A\right]_0^\pi$$

$$+\frac{1}{2\pi\alpha'}\int_{\tau_1}^{\tau_2} d\tau \int_0^\pi d\sigma(\sqrt{-g}\,\Box X^A \delta X_A + \sqrt{-g}\, T^{\alpha\beta}\delta g_{\alpha\beta}) \qquad (12.1.7.2)$$

The first term in the right-hand side of equation (12.1.7.2) vanishes, since it is required that the string be described by given $X_1^A(\sigma)$ and $X_2^A(\sigma)$ at τ_1 and τ_2. Demanding then that the remaining terms in δS be zero for all allowed histories leads not only to the equations of motion (12.1.7.1), but also to the conditions

$$\sqrt{-g}\, g^{1\alpha}\partial_\alpha X^A = 0 \qquad \text{at } \sigma = 0, \pi \qquad (12.1.7.3)$$

These are the announced boundary conditions for the open string [for the closed string there is, of course, no spatial boundary and no need to impose conditions (12.1.7.3) to get $\delta S = 0$].

Before discussing the geometrical meaning of relations (12.1.7.3), we note that the boundary conditions are more fundamental than just "on-shell" equations, as the above derivation might suggest, but must actually be imposed even off-shell if one wants a consistent variational principle. Indeed, at field histories which do not obey conditions (12.1.7.3) the action is in fact "undifferentiable," in the sense that its variation is not given by a two-dimensional integral containing only undifferentiated field variations,

$$\delta S \neq \int \left(\frac{\delta S}{\delta X^A}\delta X^A + \frac{\delta S}{\delta g_{\alpha\beta}}\delta g_{\alpha\beta}\right) d\tau\, d\sigma \qquad (12.1.7.4)$$

(the boundary terms at $\sigma = 0, \pi$ remain).

Since we only know how to deal with theories with a differentiable action (which leads to a well-defined canonical structure), we will from now on restrict the space of open-string histories to string configurations which fulfill conditions (12.1.7.3). As we have shown, these boundary conditions are the natural ones for the variational principle at hand, since they allow the motion of the string endpoints $X^A(\tau, 0)$ and $X^A(\tau, \pi)$ to be varied in the variational principle.

The geometrical meaning of relations (12.1.7.3) is easily understood once it is realized that, by hypothesis, the two d-dimensional vectors $\partial_\alpha X^A = \{\partial X^A/\partial\tau, \partial X^A/\partial\sigma\}$ are linearly independent. Hence, if some linear condition governing them vanishes, the coefficients of the linear combination are zero too. So conditions (12.1.7.3) are equivalent to

$$\sqrt{-g}\, g^{1\alpha} = 0 \qquad (12.1.7.5)$$

or, by expressing $g^{1\alpha}$ in terms of $g_{\alpha\beta}$, also to

$$\frac{g_{00}}{\sqrt{-g}} = 0 = \frac{g_{01}}{\sqrt{-g}} \qquad (12.1.7.6)$$

Accordingly, one sees that g_{00} and g_{01} vanish at the boundaries [$g_{\alpha\beta}$ and g remain bounded (as Minkowskian scalar products of regular vectors)], i.e., the induced metric is singular at the ends of the string!

$$\frac{\partial X^A}{\partial\tau}\frac{\partial X_A}{\partial\tau} = 0 = \frac{\partial X^A}{\partial\tau}\frac{\partial X_A}{\partial\sigma} \qquad \text{at } \sigma = 0, \pi \qquad (12.1.7.7)$$

Moreover, g_{11} is strictly positive and g_{01}^2 must go to zero faster than g_{00} does, so that the ratios (12.1.7.6) indeed vanish:

$$g_{11} > 0 \qquad (12.1.7.8a)$$

$$g_{01}^2/g_{00} \to 0 \qquad \text{at } \sigma = 0, \pi \qquad (12.1.7.8b)$$

(If $g_{01}^2 \sim g_{00}$ or $g_{00} \ll g_{01}^2$, $g_{01}/\sqrt{-g} \nrightarrow 0$ at the end points.)

The first equality in conditions (12.1.7.7) expresses that the vector $\partial X^A/\partial\tau$ is lightlike. The second means that $\partial X^A/\partial\tau$ is not only null, but also orthogonal to all directions tangent to the string world surface. Hence one sees that the ends of the string move with the speed of light, at right angles to the string; the string history is tangent to a null plane at the boundaries.

It should be pointed out that the degeneracy of the metric at the end points seems to conflict at first sight with the choice of a conformal gauge ($g_{00} = 0$, $g_{11} \neq 0$ versus $g_{\alpha\beta} = \phi^2 \eta_{\alpha\beta}$). This is not a serious conflict, however, and will be discussed later (it is solved by taking, in an appropriate manner, coordinates such that $\partial X^A/\partial\sigma = 0$ at $\sigma = 0, \pi$).

The physical interpretation of this peculiar motion of the string ends is as follows. The string is free and needs angular momentum (motion of the end points at right angles to the string) to balance its inner tension, which tends to make it collapse. With the above boundary conditions, the effective tension vanishes at $\sigma = 0, \pi$ [9]. At the same time, the boundary conditions imply that the energy-momentum and angular-momentum fluxes, j^1_A and j^1_{AB}, given by expressions (12.1.5.2) and (12.1.5.3), are zero at the end points, so that the Poincaré charges are all conserved, as they should if the string is indeed completely free.

Exercises

1. Problems with the variational principle in the absence of boundary conditions.

    a. Consider a two-dimensional scalar field X in flat space, with the usual Klein–Gordon action. Write the reduced action for configurations of the form

    $$X(\tau, \sigma) = a(\tau)\cos 2\sigma + b(\tau)\sin \sigma, \quad 0 \leq \sigma \leq \pi$$

    Show that extremization of the reduced action with respect to a and b does not lead to the correct differential equations.

    b. Trace the difficulty to nonvanishing surface terms.

2. Rigidly rotating string [10].

    a. Show that the history

    $$X^0 = \tau$$

    $$X^1 = A(\sigma - \pi/2)\cos \omega\tau$$

    $$X^2 = A(\sigma - \pi/2)\sin \omega\tau$$

    $$X^3 = 0$$

    where A and ω are constants related by $\frac{1}{2}\pi\omega A = 1$, is a classical solution to the string equations in four dimensions.

    b. Show that the boundary conditions are also satisfied.

    c. Study lightlike curves on the string world surface. Compute the amount of Minkowskian time required for one light signal to reach one end of the string starting from the other. Note that it is finite (even though the metric degenerates at the boundaries) so that the ends of the string are not causally disconnected from the inside.

    d. Compute the conserved currents. Show that the mass and angular momentum are respectively given by

    $$M = A\pi/4\alpha' \quad \text{and} \quad J = A^2\pi^2/16\alpha'$$

and hence

$$J = \alpha' M^2$$

(linear Regge trajectory).

    e. Are the currents j_A^α, j_{AB}^α everywhere timelike? Interpret.

    f. Consider the intersections of the string history with the null planes $X^0 + X^3 = $ const. Are these intersections timelike, lightlike, or spacelike?

3. a. Show that there is no solution to the equations of motion and the *boundary conditions* in two-dimensional Minkowski space: the metric induced at the ends of the string cannot be degenerate (there is a need for one further spatial direction into which the end points move, at right angles to the string).

    b. Discuss the geometrical significance of the result. Show that there is no way (in two dimensions) to make the area extremum for variations $\delta X^A(\tau, \sigma)$ which do not vanish at the boundary.

    c. What happens if one demands that $X^A + \delta X^A$ be nonspacelike at $\sigma = 0$, $\sigma = \pi$? Show how one might construct a (rather trivial) string theory in two dimensions.

    d. Conclusion: the string in two (background) dimensions is a poor representative of what happens for $d > 2$.

12.2. HAMILTONIAN FORMALISM

12.2.1. Constraints

Owing to the gauge invariance of the action, there are constraints in the canonical formalism. Not all momenta are independent functions of the velocities. How to handle this problem without fixing the gauge has been shown by Dirac in classic papers [11].

The situation is not unfamiliar from electromagnetism, where Gauss's law appears in the canonical formalism as a constraint related to gauge invariance. The associated Lagrange multiplier turns out to be A_0, which is free in the absence of any gauge fixing. As is known, the role played by Gauss's law is essential. It guarantees that the quantum theory is gauge invariant. All approaches to the quantum theory must, one way or the other, incorporate it. In contrast, the gauge conditions (Coulomb gauge, temporal gauge, Lorentz gauge, and so on) are less fundamental since there is a freedom in their choice. Different gauge conditions can be adopted, but there is only one Gauss's law.

The main purpose of this section is to emphasize that the "Virasoro conditions" appearing in the canonical formulation of the string are to be compared with Gauss's law, and not with gauge conditions. To be more precise, these constraints follow directly from the coordinate invariance of the string action. They thus play a fundamental role in the theory. The discovery of an action principle (12.1.1.6), which reproduces all the Virasoro conditions, was actually a major development [1, 2, 12].

12 • Nambu–Goto String: Classical Analysis

In order to fully display the meaning of the "Virasoro conditions," one needs to retain all gauge invariances of the action and adopt the Dirac formalism [11, 13]. Since the detailed Hamiltonian treatment of the Nambu–Goto form of the action can be found in the literature [10, 13], we rather adopt here the quadratic form (12.1.2.1) as starting point.

Straightforward evaluation of the canonical momenta yields

$$\mathcal{P}_A = \frac{\partial \mathcal{L}}{\partial \dot{X}^A} = -\frac{1}{2\pi\alpha'} \sqrt{-\gamma}\, \gamma^{0\beta} \partial_\beta X_A \qquad (12.2.1.1)$$

and

$$p^{\alpha\beta} = \frac{\partial \mathcal{L}}{\partial \dot{\gamma}_{\alpha\beta}} = 0 \qquad (12.2.1.2)$$

The equations $\phi^{\alpha\beta} = p^{\alpha\beta} \approx 0$ are called primary constraints. They follow from the absence of a kinetic term for the metric.

The next step in the Dirac method consists in rewriting the Hamiltonian

$$H = \int (\mathcal{P}_A \dot{X}^A + p^{\alpha\beta} \dot{\gamma}_{\alpha\beta} - \mathcal{L})\, d\sigma$$
$$\approx \int (\mathcal{P}_A \dot{X}^A - \mathcal{L})\, d\sigma \qquad (12.2.1.3)$$

in terms of the canonical variables only. This is not difficult, since relation (12.2.1.3) is just the familiar Hamiltonian for d scalar fields in a curved background.

In order to simplify the equations and also in order to take advantage of the Weyl invariance, it is convenient to reparametrize the metric components in terms of the "lapse" and the shift (see Figure 12.2). To that end one decomposes the vector $\partial/\partial\tau$ tangent to the lines of constant σ (i.e., to the τ-coordinate lines) in the frame $(\mathbf{n}, \partial/\partial\sigma)$. Here, $\mathbf{n}$ is the normed normal to the lines $\tau = $ const,

$$\mathbf{n} \cdot \frac{\partial}{\partial \sigma} = 0, \qquad \mathbf{n} \cdot \mathbf{n} = -1 \qquad (12.2.1.4)$$

while $\partial/\partial\sigma$ is the vector tangent to the lines of constant τ.†

The (rescaled) lapse N and shift N^1 are defined by

$$\partial/\partial\tau = N\sqrt{\gamma}\,\mathbf{n} + N^1 \partial/\partial\sigma \qquad (12.2.1.5)$$

† The d space-time components of the vector $\partial/\partial\sigma$ are just $\partial X^A/\partial\sigma$; similarly, $(\partial/\partial\tau)^A = \partial X^A/\partial\tau$, and so on; we note that $\mathbf{n}$ is of course taken to be tangent to the string history.

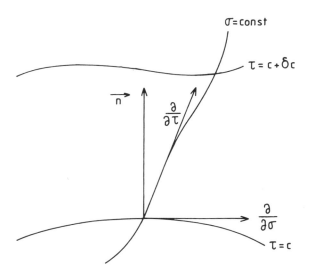

Figure 12.2. The vector $\partial/\partial\tau$ can be decomposed in the frame $\{\mathbf{n}, \partial/\partial\sigma\}$.

where γ is the determinant of the spatial (one-dimensional!) metric γ_{11} induced on the lines $\tau = \text{const.}$†

From now on, quantities without superscript $^{(2)}$ will systematically be one-dimensional. For example, γ^{11} stands for the inverse of the one-dimensional metric, $\gamma^{11} = 1/\gamma_{11}$, and does not coincide, when $N^1 \neq 0$, with $^{(2)}\gamma^{11}$ appearing in $^{(2)}\gamma^{\alpha\beta}\gamma_{\beta\sigma} = \delta^\alpha_\sigma$ (although $^{(2)}\gamma_{11} = \gamma_{11}$ so that the superscript $^{(2)}$ is superfluous for γ_{11}). Everything will have one-dimensional covariance. We note the rules that in one dimension, density of weight one = covector, and so on (for infinitesimal coordinate transformations).

One easily derives the following useful relations, which enable one to go back and forth between $\gamma_{\alpha\beta}$ on the one hand and (N, N^1, γ_{11}) on the other:

$$\gamma_{00} = -N^2\gamma + (N^1)^2\gamma_{11} \qquad (12.2.1.6a)$$

$$\gamma_{01} = N^1\gamma_{11}, \qquad N^1 = \gamma^{11}\gamma_{01} \qquad (12.2.1.6b)$$

$$N = (-\gamma^{00}\gamma_{11})^{-1/2} \qquad (12.2.1.6c)$$

† Relation (12.2.1.5) is equivalent to

$$\frac{\partial X^A}{\partial \tau} = N\sqrt{\gamma}\, n^A + N^1 \frac{\partial X^A}{\partial \sigma}$$

in terms of space-time components.

$$(-^{(2)}\gamma)^{1/2} = N\gamma_{11} \qquad (12.2.1.6d)$$

$$\gamma^{00} = -\frac{1}{N^2\gamma}, \qquad \gamma^{01} = \frac{N^1}{N^2\gamma}, \qquad {}^{(2)}\gamma^{11} = \gamma^{11} - \frac{(N^1)^2}{N^2\gamma} \qquad (12.2.1.6e)$$

Equations (12.2.1.6a–c) show that the changes of variables $\gamma_{\alpha\beta} \to N, N^1, \gamma_{11}$ is invertible, so that it is permissible to adopt the second set as new independent variables.

The advantage of working with the rescaled lapse function N (which has weight minus one) instead of the more usual function $\alpha = N\gamma$ is that N is Weyl invariant, as is N^1. Hence, the only non-Weyl-invariant variable is γ_{11}, which one expects to drop out from the formalism by Weyl invariance.

Indeed, one finds by standard manipulation that the Hamiltonian (12.2.1.3) reads

$$H = \int (N\mathcal{H} + N^1\mathcal{H}_1)\,d\sigma \qquad (12.2.1.7)$$

with

$$\mathcal{H} = \frac{1}{2}\left(2\pi\alpha'\mathcal{P}_A\mathcal{P}^A + \frac{1}{2\pi\alpha'}X'^A X'_A\right) \qquad (12.2.1.8a)$$

$$\mathcal{H}_1 = \mathcal{P}_A X'^A \qquad (12.2.1.8b)$$

$(X'^A \equiv \partial X^A/\partial\sigma)$.

The coefficient $\mathcal{H}$ of the lapse function in the Hamiltonian is equal to the energy density of the scalar fields in the normal frame (more precisely, $\mathcal{H} = \gamma T_{\alpha\beta}n^\alpha n^\beta$). It is a density of weight two and is sometimes called the super-Hamiltonian by analogy with gravity (see elsewhere [13] for a discussion of the Hamiltonian form of the Einstein equations).

The function $\mathcal{H}_1$ is equal to $\sqrt{\gamma}\,T_{1\alpha}n^\alpha$ and is sometimes called the supermomentum. It is a vector density of weight one.

According to Dirac [11], the equations of motion are generated by the total Hamiltonian, obtained by adding to H the primary constraints multiplied by Lagrange multipliers,

$$H_T = H + \int (\lambda p_N + \lambda^1 p_{N^1} + \mu p_{\gamma_{11}})\,d\sigma \qquad (12.2.1.9)$$

To determine whether the Lagrange multipliers are arbitrary functions of σ and τ, or rather are determined by the theory, one must now proceed to the "consistency algorithm."

The demand that the primary constraints $p_N \approx 0$, $p_{N^1} \approx 0$, $p_{\gamma_{11}} \approx 0$ [equivalent to equation (12.2.1.2)] be preserved in time leads to the conditions

$$\dot{p}_N \approx 0 \quad \Rightarrow \quad \mathcal{H} \approx 0 \qquad (12.2.1.10a)$$

$$\dot{p}_{N_1} \approx 0 \quad \Rightarrow \quad \mathcal{H}_1 \approx 0 \qquad (12.2.1.10b)$$

$$\dot{p}_{\gamma_{11}} \approx 0 \quad \Rightarrow \quad \text{no further condition} \quad (H \text{ does not contain } \gamma_{11}) \qquad (12.2.1.10c)$$

The new "secondary" constraints $\mathcal{H} \approx 0$ and $\mathcal{H}_1 \approx 0$ are just the conditions (12.1.2.4) that the energy-momentum components $T_{\alpha\beta}(X)$ vanish. They imply that the Hamiltonian (12.2.1.9) is weakly zero, as in general relativity.

One easily checks that $\mathcal{H}$ and $\mathcal{H}_1$ commute with the primary constraints. Furthermore, as already pointed out and as explicitly verified in Section 12.3, they obey the conformal algebra:

$$[\mathcal{H}(\sigma), \mathcal{H}(\sigma')] = (\mathcal{H}_1(\sigma) + \mathcal{H}_1(\sigma'))\delta'(\sigma, \sigma') \qquad (12.2.1.11a)$$

$$[\mathcal{H}(\sigma), \mathcal{H}_1(\sigma')] = (\mathcal{H}(\sigma) + \mathcal{H}(\sigma'))\delta'(\sigma, \sigma') \qquad (12.2.1.11b)$$

$$[\mathcal{H}_1(\sigma), \mathcal{H}_1(\sigma')] = (\mathcal{H}_1(\sigma) + \mathcal{H}_1(\sigma'))\delta'(\sigma, \sigma') \qquad (12.2.1.11c)$$

[with canonical Poisson brackets for $X^A(\sigma)$ and $\mathcal{P}_A(\sigma)$]. Hence, they are preserved in time ($\dot{\mathcal{H}} \approx 0$, $\dot{\mathcal{H}}_1 \approx 0$) and the consistency algorithm ends. We have found all the constraints, these are "first class" [11], and the Lagrange multipliers in expression (12.2.1.9) are undetermined.

Exercise. Derive relations (12.2.1.6).

12.2.2. Meaning of the Constraints—Simplification of the Formalism

First-class constraints are generically associated with gauge invariance. The association is made explicit in this subsection and the appearance of undetermined functions of τ in the Hamiltonian equations is related to the possibility of performing arbitrary gauge transformations in the course of the evolution of the system. In the absence of gauge-fixing conditions, such an arbitrariness must of course be present.

The constraint $p_{\gamma_{11}} = 0$ or, more exactly, $\gamma_{11} p_{\gamma_{11}} = 0$ generates Weyl rescaling of the canonical variables. Indeed, one finds

$$\delta \gamma_{11}(\sigma) = \left[\gamma_{11}(\sigma), \int \mu(\sigma') p^{11}(\sigma') \gamma_{11}(\sigma') \, d\sigma' \right]$$

$$= \mu(\sigma) \gamma_{11}(\sigma) \qquad (12.2.2.1)$$

$$\delta p^{11}(\sigma) = -\mu(\sigma) p^{11}(\sigma)$$

For the other canonical variables y^i,

$$\left[y^i, \int \mu(\sigma') p^1{}_1(\sigma') \, d\sigma' \right] = 0$$

(with $p_{y^{11}} \equiv p^{11}$). By a Weyl transformation one can give any (positive) value to γ_{11}, which is thus an arbitrary function of time. This is why the constraint $p_{11} = 0$ appears in the Hamiltonian multiplied by an undetermined function.

The pair (γ_{11}, p_{11}) does not correspond to any true degree of freedom (γ_{11} is arbitrary, p^{11} is constrained to vanish). Moreover, it does not occur anywhere in $\mathcal{H}$ or $\mathcal{H}_1$. Hence, one can just omit it from the theory, together with the Lagrange multiplier μ. This does not modify the equations of motion of the other, Weyl-invariant variables.

The remaining constraints $\mathcal{H} = 0$, $\mathcal{H}_1 = 0$, $p_N = 0$, and $p_{N^1} = 0$ are related to the other gauge invariance of the string, namely, reparametrization invariance. From the equations

$$\dot{N} = \lambda \quad \text{and} \quad \dot{N}^1 = \lambda^1 \qquad (12.2.2.2)$$

one learns that N and N^1 are arbitrary, since they can be changed at will by appropriate choice of λ and λ^1. This was expected, for the lapse N and shift N^1 describe the slicing of the two-dimensional string history by the $\tau = $ constant surfaces (see Figure 12.2). Since this slicing is gauge-dependent in a generally covariant theory, the functions which characterize it cannot be determined by the equations of motion.

Moreover, the canonically conjugate momenta p_N and p_{N^1} are constrained to vanish, so that N, p_N, N^1, and p_{N^1} are not true degrees of freedom. The situation is very similar to what we found with γ_{11} and p^{11}. But this time, one must keep N and N^1 in the action because they appear as Lagrange multipliers enforcing the super-Hamiltonian and supermomentum constraints $\mathcal{H} = 0$ and $\mathcal{H}_1 = 0$ on the other canonical variables. Therefore, one can only forget about p_N, p_{N_1}, λ, and λ^1 and the simplified action becomes

$$S_H[X^A, \mathcal{P}_A, N, N^1] = \int d\tau \left(\int d\sigma\, \mathcal{P}_A \dot{X}^A - H \right) \quad (12.2.2.3)$$

$$H = \int d\sigma\, (N\mathcal{H} + N^1 \mathcal{H}_1) \quad (12.2.2.4)$$

One easily checks that the equations obtained from (12.2.2.3) are completely equivalent to those which follow from the Nambu–Goto action.

The similarities between the canonical formulations of the string model [as in equation (12.2.2.3)] and of Einstein theory (see Hanson et al. [13] and references cited therein) cannot be stressed enough. In both cases, one finds that the Hamiltonian is weakly vanishing and has the structure (12.2.2.4), with arbitrary lapse and shift functions. These multiply the constraints $\mathcal{H} = 0$ and $\mathcal{H}_1 = 0$, which generate the changes in the canonical variables under an arbitrary deformation of the line (hypersurface) $\tau =$ const.

The common features of the string model and of gravity are not accidental. They have the same origin, namely, reparametrization invariance in two or four dimensions.

Finally, it should be pointed out that reference to Weyl invariance has completely disappeared from equation (12.2.2.3). This is not because we have fixed the Weyl gauge in equation (12.2.2.3)—no gauge fixation condition has been imposed—but rather it follows from the use of Weyl-invariant variables.

Remark. The precise sense in which $\mathcal{H}(\sigma)$ and $\mathcal{H}_1(\sigma)$ generate two-dimensional reparametrizations of the string history $X^A(\tau, \sigma)$ is as follows.

Let $X^A(\tau, \sigma)$ and $\mathcal{P}_A(\tau, \sigma)$ be a phase-space trajectory. Consider the τ-dependent generator

$$H_\tau[\xi] = \int \{\xi^\perp(\tau, \sigma) \mathcal{H}(\sigma) + \xi^1(\tau, \sigma) \mathcal{H}_1(\sigma)\}\, d\sigma \quad (12.2.2.5)$$

which acts on phase-space functions through the (equal-time) Poisson bracket. It generates the following canonical transformation

$$\delta X^A(\tau, \sigma) = [2\pi\alpha' \xi^\perp \mathcal{P}^A + \xi^1 X^{A\prime}](\tau, \sigma) \quad (12.2.2.6a)$$

$$\delta \mathcal{P}_A(\tau, \sigma) = \left[\frac{(\xi^\perp X'_A)'}{2\pi\alpha'} + (\mathcal{P}_A \xi^1)' \right](\tau, \sigma) \quad (12.2.2.6b)$$

The transformation (12.2.2.6a) coincides with an ordinary two-dimensional diffeomorphism provided (1) $\mathcal{P}^A$ is related to $\dot{X}^A$ by the "first" Hamiltonian equation; and (2) $\xi^\perp$ and ξ^1 are related to ξ^μ by the "lapse-shift decomposition," i.e., $\xi^\perp = \xi^0 N g^{1/2}$ and $\xi^1 = \xi^0 N^1 + {}^{(2)}\xi^1$. [Here, N and N^1 are functions of the canonical variables, given by $N = (-g^{00}g)^{-1/2}$ and $N_1 = g_{01}$, as can be seen from the constraints, also assumed to hold.] These conditions reduce relation (12.2.2.6a) to the form

$$\delta X^A = \xi^\mu X^A_{,\mu} \tag{12.2.2.7}$$

In order for relation (12.2.2.6b) to reproduce the appropriate change in $\mathcal{P}_A$ viewed as a function of $\dot{X}^A$, one must assume that the "second" Hamiltonian equation holds as well.

Because it is necessary to redefine the infinitesimal parameters and to use the equations of motion for identifying the canonical transformations (12.2.2.6) with the two-dimensional diffeomorphisms, there is no guarantee that the algebra of the generators $\mathcal{H}, \mathcal{H}_1$ is isomorphic to the diffeomorphism algebra—actually, it is not—nor even that the transformations generated by relation (12.2.2.5) close off-shell. (In dimensions greater than 2, they do not.)

This, however, will not prevent us from calling the functions $H_\tau[\xi]$ the generators of changes of coordinates in two dimensions, since transformation (12.2.2.6) coincides with relation (2.7) on-shell. This appears to be the best way one can "represent" the diffeomorphisms in the canonical formalism and hence also in quantum mechanics. A similar problem arises in general relativity.

A curious feature of two dimensions is that the subclass of functions (12.2.2.5) with constant $\xi^\perp(\sigma)$ and $\xi^1(\sigma)$, namely

$$H[\xi] = \int (\xi^\perp(\sigma)\mathcal{H}(\sigma) + \xi^1(\sigma)\mathcal{H}_1(\sigma)) \, d\sigma \tag{12.2.2.8}$$

form a closed algebra under the (equal-time) Poisson bracket.† This algebra is isomorphic to the two-dimensional conformal algebra, i.e., to twice the diffeomorphism group algebra in one dimension.

In quantum mechanics, one imposes the constraints on the physical states,

$$\mathcal{H}(\sigma)|\psi\rangle = \mathcal{H}_1(\sigma)|\psi\rangle = 0 \tag{12.2.2.9}$$

If equations (12.2.2.9) make sense, namely, if they remain first-class quantum mechanically (no anomaly), then the physical states are not only invariant

---

† This is not true in higher dimensions, where the structure "constants" turn out to involve the canonical variables.

under the conformal group (12.2.2.8),

$$\left[\exp i \int (\xi^\perp(\sigma)\mathcal{H}(\sigma) + \xi^1(\sigma)\mathcal{H}_1(\sigma))\, d\sigma \right]|\psi\rangle = |\psi\rangle \quad (12.2.2.10)$$

but they are also invariant under arbitrary changes of coordinates $\xi^\mu(\tau, \sigma)$,

$$\left[\exp i \int \xi^\perp(\tau, \sigma)\mathcal{H}(\sigma) + \xi^1(\tau, \sigma)\mathcal{H}_1(\sigma)\, d\sigma \right]|\psi\rangle = |\psi\rangle \quad (12.2.2.11)$$

This is how gauge invariance is enforced in the quantum theory. We note that invariance under arbitrary conformal transformations $\xi^\perp(\sigma)$ and $\xi^1(\sigma)$ implies equations (12.2.2.9) and hence also relation (12.2.2.11).

If the constraint algebra develops an anomaly at the quantum level, one cannot impose equation (12.2.2.8). This means that one loses not only conformal invariance, but also invariance under changes of coordinates, since the physical states would no longer obey relation (12.2.2.11). This is what happens in string theory, although it turns out that in the critical dimensions, invariance under transformations generated by the constraints is restored when the ghosts are taken into account.

Exercises

1. Derive equation (12.2.2.3) by applying the Dirac method to the Nambu-Goto form of the action.

2. Derive the Hamiltonian formalism for the relativistic membrane, starting either from the square root or the quadratic form of the action. In the latter case, second-class constraints appear. Get rid of them by defining the appropriate Dirac brackets.

12.2.3. Hamiltonian Form of the Boundary Conditions (Open Case)

We have seen that in the open case, the equations of motion must be supplemented by the boundary conditions (12.1.7.6). The purpose of this section is to translate (12.1.7.6) in terms of the canonical variables X^A, $\mathcal{P}_A$ and the Lagrange multipliers N, N^1.

From the definition of $\mathcal{P}_A$, N, and N^1 one easily obtains

$$\mathcal{P}_A \to \infty, \quad N \to 0, \quad N^1 \to 0 \quad \text{at } \sigma = 0, \pi \quad (12.2.3.1)$$

Moreover, $\mathcal{P}_A$ cannot blow up faster than N^{-1} does, since

$$\partial_0 X^A = [X^A, H]$$

$$= N\mathcal{P}^A + N^1 X^{A'} \quad (12.2.3.2)$$

must remain finite.

The form (12.2.3.1) of the boundary conditions is inconvenient. One can take advantage of the fact that $\mathcal{P}_A$ is a density to rewrite them in a more practical manner. The idea is that if one rescales the σ coordinate,

$$\sigma' = f(\tau, \sigma)$$

in such a way that $\partial\sigma/\partial\sigma' \to 0$ at the boundaries, one can make $\mathcal{P}_A$ finite. Then, N becomes also finite. This can be done without changing the range of σ.

In terms of the appropriately rescaled coordinates, the new form of the boundary conditions becomes

$$\left.\begin{array}{l} X^{A'} = 0 \\ \mathcal{P}_A \text{ finite} \\ N \text{ finite} \\ N^1 = 0 \end{array}\right\} \text{ at } \sigma = 0, \pi \qquad \begin{array}{l}(12.2.3.3a)\\(12.2.3.3b)\\(12.2.3.3c)\\(12.2.3.3d)\end{array}$$

Clearly, given functions which obey condition (12.2.3.3a) (and $X^{A'} \neq 0$ inside), one can go back to "regular" coordinates in which $X^{A'} \neq 0$ everywhere, simply by making a suitable coordinate transformation. In the process, one gets new $\mathcal{P}_A$, N, and N^1 which obey the boundary conditions in their original form. Accordingly, one can adopt conditions (12.2.3.3), which are more convenient since with them no variable blows up at the end points.

Exercise. Check explicitly the above statements. Study in more detail the behavior of the variables in the vicinity of $\sigma = 0, \pi$.

The equations of motion will preserve conditions (12.2.3.3) only if appropriate conditions are imposed on the higher-order spatial derivatives of the canonical variables.

The problem is most simply analyzed by assuming that N and N^1, which are gauge functions at our disposal, obey

$$N^{(2k+1)} = 0 \quad \text{and} \quad N^{1(2k)} = 0 \quad \text{at } \sigma = 0, \pi \qquad (12.2.3.4)$$

Here, $f^{(k)}$ stands for the kth derivative of f.† Then the equations of motion

$$\dot{X}^A = [X^A, H], \qquad \dot{\mathcal{P}}_A = [\mathcal{P}_A, H] \qquad (12.2.3.5)$$

imply, together with conditions (12.2.3.3), that all odd derivatives of X^A

---
† The assumption of C^∞ can of course be relaxed, but this will not concern us here.

and $\mathcal{P}_A$ vanish at the end points:

$$X^{A\,(2k+1)} = \mathcal{P}_A^{(2k+1)} = 0 \qquad \text{at } \sigma = 0, \pi \qquad (12.2.3.6)$$

The boundary conditions (12.2.3.4) at $\sigma = 0$ can be summarized in the concise statement that the variables may be extended smoothly on the interval $[-\pi, 0]$ by symmetry as follows:

$$X^A(-\sigma) = X^A(\sigma), \qquad \mathcal{P}_A(-\sigma) = \mathcal{P}_A(\sigma) \qquad (12.2.3.7a)$$

$$N(-\sigma) = N(\sigma), \qquad N^1(-\sigma) = -N^1(\sigma) \qquad (12.2.3.7b)$$

The boundary conditions at $\sigma = \pi$ then imply that the variables can be extended on the whole real line as smooth periodic functions of period 2π.

Exercise. Show that, with the above boundary conditions, the Hamiltonian has well-defined functional derivatives as it stands and does need to be "improved" by end-point terms at $\sigma = 0, \pi$ (see Refs. 13 and 14 in this context).

12.2.4. Hamiltonian Expression for the Poincaré Charges

Elimination of the time derivatives $\dot{X}^A$ from the charges (12.1.5.4) and (12.1.5.5) yields

$$Q_A = \int_0^{\pi \text{ or } 2\pi} \mathcal{P}_A(\sigma)\,d\sigma \qquad (12.2.4.1)$$

and

$$Q_{AB} = \tfrac{1}{2} \int_0^{\pi \text{ or } 2\pi} (\mathcal{P}_A(\sigma) X_B(\sigma) - \mathcal{P}_B(\sigma) X_A(\sigma))\,d\sigma \qquad (12.2.4.2)$$

which close in the Poisson bracket according to the Poincaré algebra.

Exercise. Verify that the charges are gauge invariant and conserved, i.e., that $[Q, H] = 0$ for any permissible choice of the lapse and shift.

12.3. A CLOSER LOOK AT THE CONSTRAINT ALGEBRA

12.3.1. Explicit Computation

The main results of the canonical analysis can be summarized as follows:

1. The Hamiltonian variables are the string coordinates $X^A(\sigma)$, their momenta $\mathcal{P}_A(\sigma)$, and the Lagrange multipliers $N(\sigma)$, $N^1(\sigma)$. One

has

$$[X^A(\sigma), \mathcal{P}_B(\sigma')] = \delta^A_B \delta(\sigma, \sigma') \qquad (12.3.1.1)$$

Furthermore, the variables are submitted to the boundary conditions (12.2.3.3) in the open case, and to periodicity conditions in the closed case.

2. All the dynamics is contained in the constraints,

$$\mathcal{H} = \frac{1}{2}\left(2\pi\alpha' \mathcal{P}^A \mathcal{P}_A + \frac{1}{2\pi\alpha'} X'^A X'_A\right) \approx 0 \qquad (12.3.1.2a)$$

$$\mathcal{H}_1 = \mathcal{P}_A X'^A \approx 0 \qquad (12.3.1.2b)$$

which restrict the allowed initial data. The Hamiltonian is

$$H = \int d\sigma (N\mathcal{H} + N^1 \mathcal{H}_1) \qquad (12.3.1.3)$$

and vanishes weakly; $\mathcal{H}(\sigma)$ generates deformations of the string $X^A(\sigma)$ normal to itself, since it is multiplied in expression (12.3.1.3) by the lapse; $\mathcal{H}_1(\sigma)$ generates tangential displacements. No gauge condition was necessary to arrive at equations (12.3.1.2).

3. The constraints close in the Poisson brackets according to the conformal or "Virasoro" algebra,

$$[\mathcal{H}(\sigma), \mathcal{H}(\sigma')] = (\mathcal{H}_1(\sigma) + \mathcal{H}_1(\sigma'))\delta'(\sigma, \sigma') \qquad (12.3.1.4a)$$

$$[\mathcal{H}(\sigma), \mathcal{H}_1(\sigma')] = (\mathcal{H}(\sigma) + \mathcal{H}(\sigma'))\delta'(\sigma, \sigma') \qquad (12.3.1.4b)$$

$$[\mathcal{H}_1(\sigma), \mathcal{H}_1(\sigma')] = (\mathcal{H}_1(\sigma) + \mathcal{H}_1(\sigma'))\delta'(\sigma, \sigma') \qquad (12.3.1.4c)$$

Owing to these latter relations, the constraints are "first class" and are preserved by the time evolution.

The explicit verification of relations (12.3.1.4b, c), postponed until now, is facilitated once it is realized that $\mathcal{H}_1(\sigma)$ generates one-dimensional coordinate transformations $\sigma \to \sigma' = f(\sigma)$, i.e.,

$$\left[F(\sigma), \int \mathcal{H}_1(\sigma')\xi^1(\sigma')\,d\sigma'\right] = \mathfrak{L}_\xi F \qquad (12.3.1.5)$$

where it is recalled that $\mathfrak{L}$ denotes the Lie derivative operator. Indeed, one finds that

$$\left[X^A(\sigma), \int \mathcal{H}_1(\sigma')\xi^1(\sigma')\,d\sigma'\right] = \xi^1(\sigma)X'^A(\sigma) = \mathfrak{L}_\xi X^A$$

(where X^A are d scalars) and

$$\left[\mathcal{P}_A(\sigma), \int \mathcal{H}_1(\sigma')\xi^1(\sigma')\,d\sigma'\right] = (\xi^1\mathcal{P}_A)' = \mathfrak{L}_\xi \mathcal{P}_A$$

(where $\mathcal{P}_A$ are d densities of unit weight).

Hence one knows that

$$\left[\mathcal{H}(\sigma), \int \mathcal{H}_1(\sigma')\xi^1(\sigma')\,d\sigma'\right] = \mathfrak{L}_\xi \mathcal{H}$$

$$= (\xi^1\mathcal{H})' + \xi^{1'}\mathcal{H} \quad (12.3.1.6\text{a})$$

(where $\mathcal{H}$ is a density of weight 2) and

$$\left[\mathcal{H}_1(\sigma), \int \mathcal{H}_1(\sigma')\xi^1(\sigma')\,d\sigma'\right] = \mathfrak{L}_\xi \mathcal{H}_1$$

$$= (\xi^1\mathcal{H}_1)' + \xi^{1'}\mathcal{H}_1 \quad (12.3.1.6\text{b})$$

(where $\mathcal{H}_1$ is a covector density of weight 1). Since relations (12.3.1.6a) and (12.3.1.6b) hold for any one-dimensional vector field $\xi^1(\sigma)$, the Poisson brackets (12.3.1.4b) and (12.3.1.4c) easily follow.

As for relation (12.3.1.4a), one gets

$$[\mathcal{H}(\sigma), \mathcal{H}(\sigma')] = -\mathcal{P}_A(\sigma)X'^A(\sigma')\frac{\partial}{\partial\sigma'}\delta(\sigma,\sigma') + X'^A(\sigma)\mathcal{P}_A(\sigma')\frac{\partial}{\partial\sigma}\delta(\sigma,\sigma')$$

$$= (\mathcal{H}_1(\sigma) + \mathcal{H}_1(\sigma'))\delta'(\sigma,\sigma') \quad (12.3.1.7)$$

where the rules

$$\frac{\partial}{\partial\sigma'}\delta(\sigma,\sigma') = -\frac{\partial}{\partial\sigma}\delta(\sigma,\sigma') \quad (12.3.1.8\text{a})$$

and

$$F(\sigma')\delta'(\sigma,\sigma') = F'(\sigma)\delta(\sigma,\sigma') + F(\sigma)\delta'(\sigma,\sigma') \quad (12.3.1.8\text{b})$$

have been used.

12.3.2. Virasoro Conditions

We have already pointed out that the algebra (12.3.1.4) of the energy-momentum tensor components is the conformal algebra in two dimensions. It is convenient to display its direct sum structure by going to lightlike coordinates. Hence one defines†

$$Q^{\pm}(\sigma) = 2\pi(\mathcal{H}(\sigma) \pm \mathcal{H}_1(\sigma))$$

$$= \left(\sqrt{2\alpha'}\,\pi\mathcal{P}_A \pm \frac{1}{\sqrt{2\alpha'}}X'_A\right)^2 \qquad (12.3.2.1)$$

The quantity $T^{\alpha\beta}$ is traceless ($T^{+-} = 0$ in light-cone coordinates), so $Q^{\pm}$ are just the T^{++} and T^{--} components of $T^{\alpha\beta}$. These new generators obey

$$[Q^+(\sigma), Q^+(\sigma')] = 4\pi(Q^+(\sigma) + Q^+(\sigma'))\delta'(\sigma, \sigma') \quad (12.3.2.2a)$$

$$[Q^+(\sigma), Q^-(\sigma')] = 0 \qquad (12.3.2.2b)$$

and

$$[Q^-(\sigma), Q^-(\sigma')] = -4\pi(Q^-(\sigma) + Q^-(\sigma'))\delta'(\sigma, \sigma') \quad (12.3.2.2c)$$

as expected.

In the open-string case, where the variables can be extended on the interval $[-\pi, 0]$ by symmetry, the constraints possess the following reflection properties:

$$\mathcal{H}(-\sigma) = \mathcal{H}(\sigma), \quad \mathcal{H}_1(-\sigma) = -\mathcal{H}_1(\sigma), \quad Q^+(-\sigma) = Q^-(\sigma)$$
$$(12.3.2.3)$$

Hence, the single condition

$$Q^+(\sigma) = 0, \quad -\pi \le \sigma \le \pi \qquad (12.3.2.4)$$

on the entire interval $[-\pi, \pi]$ summarizes all the constraints [10].

It is convenient to define the Virasoro generators $L[f]$ by

$$L[f] = \frac{1}{4\pi}\int_{-\pi}^{\pi} f(\sigma)Q^+(\sigma)\,d\sigma \qquad (12.3.2.5)$$

† The factor 2π has been inserted in relation (12.3.2.1) to conform with usual conventions.

One finds in particular that

$$H = L[N + N^1] \tag{12.3.2.6}$$

The Virasoro generators close according to the diffeomorphism algebra in one dimension

$$[L[f], L[g]] = L[fg' - f'g] \tag{12.3.2.7}$$

(with $fg' - f'g$ = Wronskian = one-dimensional Lie bracket).

It follows from algebra (12.3.2.7) that the Fourier components $L_n \equiv L[e^{in\sigma}]$ of the constraints $Q^+(\sigma)$ obey the Virasoro algebra in its original form†

$$[L_m, L_n] = i(n - m)L_{n+m} \tag{12.3.2.8}$$

The conditions

$$L_n = 0 \tag{12.3.2.9}$$

are completely equivalent to the canonical constraints, and are called the Virasoro conditions for the open string. One easily checks that they obey

$$L_{-n} = L_n^* \tag{12.3.2.10}$$

In the case of the closed string, the constraints $Q^+(\sigma) = 0$ and $Q^-(\sigma) = 0$ are independent. One defines accordingly two sets of Virasoro generators,

$$L[f] = \frac{1}{4\pi} \int_0^{2\pi} f(\sigma) Q^+(\sigma) \, d\sigma \tag{12.3.2.11a}$$

and

$$\bar{L}[f] = \frac{1}{4\pi} \int_0^{2\pi} f(\sigma) Q^-(\sigma) \, d\sigma \tag{12.3.2.11b}$$

Their Fourier components are taken as

$$L_n \equiv L[e^{in\sigma}] = L_{-n}^* \tag{12.3.2.12a}$$

and

$$\bar{L}_n \equiv \bar{L}[e^{-in\sigma}] = \bar{L}_{-n}^* \tag{12.3.2.12b}$$

† With an i because we use classical brackets, and without central charge.

(the minus sign in equations (12.3.2.12) has been inserted so as to maintain perfect symmetry between L_n and $\bar{L}_n$ below), and they close according to

$$[L_m, L_n] = i(n - m)L_{n+m} \qquad (12.3.2.13a)$$

$$[L_m, \bar{L}_n] = 0 \qquad (12.3.2.13b)$$

$$[\bar{L}_m, \bar{L}_n] = i(n - m)\bar{L}_{n+m} \qquad (12.3.2.13c)$$

The full set of Virasoro conditions is now

$$L_m = 0 \quad \text{and} \quad \bar{L}_m = 0 \qquad (12.3.2.14)$$

Equations (12.3.2.14) are completely equivalent to the Hamiltonian constraints.

Exercises

1. Open string:
   a. Owing to the relations $X^A(-\sigma) = X^A(\sigma)$, $\mathcal{P}_A(-\sigma) = \mathcal{P}_A(\sigma)$ the canonical variables $X^A(\sigma)$, $\mathcal{P}_A(\sigma)$ cannot fulfill

$$[X^A(\sigma), \mathcal{P}_B(\sigma')] = \delta^A_B \delta(\sigma, \sigma')$$

on the whole interval $[-\pi, +\pi]$. Write the correct bracket.
   b. Show that $[Q^+(\sigma), Q^+(\sigma')]$ is nevertheless given by relation (12.3.2.2a) on the whole interval.

2. Closed string: Show that $L_0 - \bar{L}_0$ generates constant translations $\sigma \to \sigma + a$ in the spatial parameter σ along the string.

12.4. FOURIER MODES

12.4.1. Open String

The constraints are quadratic in the fields. Fourier techniques are thus quite appropriate.

As a result of the boundary conditions at $\sigma = 0$, π, only "standing waves" of cosine type are allowed. Hence the Fourier analysis of the fields reads

$$X^A(\sigma) = X^A_0 + \sum_{n>0} X^A_n \cos n\sigma \qquad (12.4.1.1)$$

$$\mathcal{P}^A(\sigma) = \frac{p^A}{\pi} + \sum_{n>0} \mathcal{P}^A_n \cos n\sigma \qquad (12.4.1.2)$$

with

$$X_0^A = \frac{1}{\pi} \int_0^\pi X^A(\sigma) \, d\sigma, \qquad X_n^A = \frac{2}{\pi} \int_0^\pi X^A(\sigma) \cos n\sigma \, d\sigma \qquad (12.4.1.3)$$

$$p_A = \int_0^\pi \mathcal{P}_A(\sigma) \, d\sigma = \text{total energy momentum of the string}$$
$$(12.4.1.4)$$

$$\mathcal{P}_n^A = \frac{2}{\pi} \int_0^\pi \mathcal{P}^A(\sigma) \cos n\sigma \, d\sigma \qquad (12.4.1.5)$$

In the gauge $N = 1$, $N^1 = 0$, the center of mass variables (X_0^A, p_A) (= zero mode) appear as free particle coordinates $(X_0^A \sim \tau,\ p_A = \text{const})$ while the higher modes describe harmonic oscillators of frequency n. This suggests defining oscillator variables†

$$a_n^A = \frac{i\pi}{2} \sqrt{\frac{2\alpha'}{n}} \, \mathcal{P}_n^A + \frac{1}{2} \sqrt{\frac{n}{2\alpha'}} \, X_n^A \qquad (12.4.1.6\text{a})$$

and

$$a_n^{A*} = -\frac{i\pi}{2} \sqrt{\frac{2\alpha'}{n}} \, \mathcal{P}_n^A + \frac{1}{2} \sqrt{\frac{n}{2\alpha'}} \, X_n^A \qquad (12.4.1.6\text{b})$$

One finds that

$$[X_0^A, p_B] = \delta_B^A \qquad (12.4.1.7\text{a})$$

and

$$[a_n^A, a_{n'}^{B*}] = -i\eta^{AB} \delta_{n,n'} \qquad (n, n' > 0) \qquad (12.4.1.7\text{b})$$

(all other brackets vanish).

In terms of the new variables $(X_0^A, p_B, a_n^A, a_n^{B*})$, the Virasoro constraints take the form

$$L_m = -i\sqrt{2n\alpha'} p_A a_n^A + \sum_{m>0} \sqrt{(m+n)m} \, a_{Am}^* a_{m+n}^A$$

$$- \tfrac{1}{2} \sum_{m=1}^{n-1} \sqrt{(n-m)m} \, a_{Am} a_{n-m}^A, \qquad n > 0 \qquad (12.4.1.8\text{a})$$

† Although suggested by a particular gauge condition, the change of variables (12.4.1.6) can be made without any gauge fixing.

12 • Nambu–Goto String: Classical Analysis

$$L_n^* = L_{-n} \tag{12.4.1.8b}$$

$$L_0 = \alpha' p^2 + \sum_{n>0} n a_{An}^* a_n^A \tag{12.4.1.8c}$$

which is the starting point of the quantum theory. To arrive at the form (12.4.1.8) of the constraints presents no conceptual difficulty.

Finally, the Poincaré charges read

$$P_A = p_A \tag{12.4.1.9a}$$

$$M_{AB} = \tfrac{1}{2}(p_A X_{B0} - X_{A0} p_B) + \tfrac{1}{2} \sum_{n>0} i(a_{An}^* a_{Bn} - a_{Bn}^* a_{An}) \tag{12.4.1.9b}$$

Comparison of equations (12.4.1.8c) and (12.4.1.9a) shows that the constraint $L_0 = 0$ is a spectral equation relating the mass-squared $m^2 = -P^2 = -p^2$ of the string to its state of excitation.

Exercises

1. Verify constraints (12.4.1.8).

2. a. Define new α variables as

$$\alpha_0^A = -2\alpha' p^A, \qquad \alpha_n^A = i\sqrt{2\alpha'}\sqrt{n}\, a_n^A \qquad (n > 0)$$

$$\alpha_{-n}^A = (\alpha_n^A)^*$$

Compute their Poisson brackets.

  b. Show that the Virasoro functions L_n are given by

$$L_n = \frac{1}{4\alpha'} \sum_{-\infty}^{+\infty} \alpha_{n-m} \alpha_m = \frac{1}{4\alpha'} \sum_{-\infty}^{+\infty} \alpha_{-m} \alpha_{n+m}$$

12.4.2. Closed String

The closed string is treated in a manner very similar to the open one. This is not surprising in view of the similarity of the constraints.

There are only two technical differences:

1. Owing to the absence of boundary conditions, there are now twice as many oscillators. One can have standing waves of both types or, what is the same, independently traveling waves to the right and to the left. Hence the expansions of X^A and $\mathcal{P}_A$ are given by

$$X^A(\sigma) = X_0^A + \sqrt{\frac{\alpha'}{2}} \sum_{n=1}^{\infty} \frac{1}{\sqrt{n}} \{ c_n^A \exp -in\sigma + \bar{c}_n^A \exp in\sigma$$

$$+ \text{conjugate complex}\} \tag{12.4.2.1a}$$

and

$$\mathcal{P}^A(\sigma) = \frac{p^A}{2\pi} + \frac{1}{2\pi\sqrt{2\alpha'}} \sum_{n=1}^{\infty} \sqrt{n}$$

$$\times \{-ic_n^A \exp -in\sigma - i\bar{c}_n^A \exp in\sigma$$

$$+ \text{conjugate complex}\} \quad (12.4.2.1b)$$

The oscillator variables c_n^A and $\bar{c}_n^A$ are independent. The Poisson-bracket relations are

$$[X_0^A, p_B] = \delta_B^A, \qquad [c_n^A, c_{n'}^{*B}] = -i\eta^{AB}\delta_{n,n'} \quad (12.4.2.2a)$$

$$(n, n' > 0)$$

$$[\bar{c}_n^A, \bar{c}_{n'}^{B*}] = -i\eta^{AB}\delta_{n,n'} \quad (12.4.2.2b)$$

(all other Poisson brackets vanish).

2. There are twice as many constraints, because $Q^+(\sigma)$ and $Q^-(\sigma)$ are no longer dependent. The Virasoro generators are, in terms of the new oscillator variables,

$$L_n = -\frac{i}{2}\sqrt{2n\alpha'}p_A c_n^A + \sum_{m>0} \sqrt{m(m+n)} c_{Am}^* c_{m+n}^A$$

$$-\frac{1}{2}\sum_{m=1}^{n-1} \sqrt{m(n-m)} c_{Am} c_{n-m}^A, \quad n > 0 \quad (12.4.2.3a)$$

$$\bar{L}_n = \text{idem} \quad \text{with } c \text{ replaced by } \bar{c} \quad (12.4.2.3b)$$

$$L_{-n} = L_n^*, \qquad \bar{L}_{-n} = \bar{L}_n^* \quad (12.4.2.3c)$$

$$L_0 + \bar{L}_0 = \tfrac{1}{2}\alpha' p^2 + \sum_{n>0} n(c_{An}^* c_n^A + \bar{c}_{An}^* \bar{c}_n^A) \quad (12.4.2.3d)$$

$$L_0 - \bar{L}_0 = \sum_{n>0} n(c_{An}^* c_n^A - \bar{c}_{An}^* \bar{c}_n^A) \quad (12.4.2.3e)$$

In the constraints $L_n = 0$ and $\bar{L}_n = 0$, the right- and left-moving sectors, described respectively by the c and $\bar{c}$ oscillators, are almost decoupled. Only the constraint $L_0 - \bar{L}_0 = 0$, the spectral equation, and the fact that the zero mode p_A is simultaneously a right and left (non)mover relates the two sectors.

We note also that one passes from the open-string expressions (12.4.1.8) to the closed-string expressions (12.4.2.3) by replacing p by $\frac{1}{2}p$, and a by c (or $\bar{c}$).

Finally, we collect the expression for the Poincaré charges,

$$P_A = p_A \qquad (12.4.2.4a)$$

$$M_{AB} = \tfrac{1}{2}(p_A X_{B0} - p_B X_{A0}) + \tfrac{1}{2} \sum_{n>0} i(c^*_{An} c_{Bn} + \bar{c}^*_{An} \bar{c}_{Bn} - c^*_{Bn} c_{An} - \bar{c}^*_{Bn} \bar{c}_{An}) \qquad (12.4.2.4b)$$

Exercise. The range of σ for the closed string is sometimes taken to be $[0, \pi]$, the fields being periodic of period π. Devise a set of rules which enables one to translate the above formulas into these different conventions.

12.5. LIGHT-CONE GAUGE

The light-cone gauge plays an extremely important role in string theory. For instance, one knows how to quantize the superstring only in that gauge.

12.5.1. Conformal Gauges

The light-cone gauge belongs to the family of so-called conformal gauges.

We pointed out earlier that any two-dimensional metric is conformally flat. Hence, there exists a coordinate transformation which brings the metric to the form

$$g_{\alpha\beta} = \phi^2 \eta_{\alpha\beta} \qquad (12.5.1.1)$$

or, as it is also written,

$$g_{\alpha\beta} = \sqrt{-g}\, \eta_{\alpha\beta} \qquad (12.5.1.2)$$

The existence of such a coordinate system can easily be asserted locally [5]. Globally, the situation has been much more studied in the Riemannian (i.e., "Euclidean") case, where "moduli" need to be introduced for multiply connected surfaces. Although no complete study exists, it seems that no such complication arises in the singularity-free pseudo-Riemannian case considered here.† This is because one usually assumes that the causal structure of the space-time surface swept out by the string is trivial, as we now explain.

† The singularity-free assumption excludes changes of topology in the Lorentzian case, and appears appropriate when discussing free strings.

We consider for definiteness the open string, which raises more interesting questions. Our aim is to show that one can define global coordinates (τ, σ), with $0 \leq \sigma \leq \pi$, such that the condition (12.5.1.1) holds. As is clear from our above discussion of the boundary conditions, the space-time vector $\partial X^A/\partial \sigma$ necessarily vanishes at the end points in conformal coordinates. But apart from this endpoint degeneracy, the coordinates (τ, σ) should be everywhere regular.

The assumption of trivial causal structure is equivalent to the demand that any event on the string world sheet is in the causal past and in the causal future of the boundaries. More precisely, if θ is a regular "time parameter" along the world line of one of the string ends (e.g., it can be the Minkowskian time t), it is assumed that one can assign to each event P on the world sheet a pair of coordinates (θ, θ') where (1) θ is the time at which one must send a light signal from the selected end along the string world sheet to arrive at P, and (2) θ' is the time at which a light signal sent from P will reach the end (see Figure 12.3) [15].

The coordinates (θ, θ') are globally defined and obviously lightlike. They provide a regular parametrization of the string history in the sense that the vectors $\partial X^A/\partial \theta$ and $\partial X^A/\partial \theta'$ are assumed to be linearly independent, except at the end points where one has

$$\partial X^A/\partial \theta = \partial X^A/\partial \theta' \qquad \text{(end points)} \qquad (12.5.1.3)$$

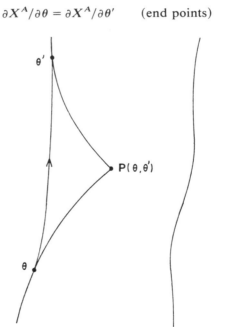

Figure 12.3. The lightlike coordinates (θ, θ') of a point on the string world sheet.

This last condition expresses the known behavior of the light rays in the vicinity of the boundaries, which arrive tangentially to the only lightlike direction there. Relation (12.5.1.3) is equivalent to $\partial X^A/\partial\sigma = 0$.

It is clear that more complicated causal structures with event horizons ("black holes") would not allow for a global definition of θ and θ'. These structures naturally arise in two-dimensional gravity [16], but their impact on flat-space free string theory (if any) has so far not been demonstrated. Hence they will not be considered here.

In terms of the above coordinates (θ, θ'), the metric reads

$$ds^2 = -F^2(\theta, \theta') \, d\theta \, d\theta' \qquad (12.5.1.4)$$

for some regular function $F(\theta, \theta')$. The minus sign is justified by the fact that ds^2 is negative for positive $d\theta$ and $d\theta'$, because then the event $(\theta + d\theta, \theta' + d\theta')$ is in the causal future of (θ, θ').

The range of θ is $[-\infty, +\infty]$, while θ' lies between θ and $\theta + 2l(\theta)$. The equation $\theta' = \theta$ describes the first end and the second end is given by $\theta' = \theta + 2l(\theta)$, where $2l(\theta) \neq 0$ is the time needed for a complete "round trip" starting from the first end at time θ.

At $\theta = \theta'$ and $\theta' = \theta + 2l(\theta)$, the function F vanishes, corresponding to the fact that the boundaries are lightlike. The metric (12.5.1.4) completely degenerates there, but this is partly due to the coordinate choice.

Once the metric has been expressed in the form (12.5.1.4), it is easy to construct conformal coordinates (τ, σ). One simply defines

$$\tau = \tfrac{1}{2}(\theta' + \theta) \qquad (12.5.1.5a)$$

and

$$\sigma = \tfrac{1}{2}(\theta' - \theta) \qquad (12.5.1.5b)$$

The relation (12.5.1.3) implies $\partial X^A/\partial\sigma = 0$ at the end points, as announced.

What is less obvious is that the range of σ, namely $[0, l(\theta)]$, is constant. This is not true for an arbitrary parametrization θ of the world line $\sigma = 0$, but can be made so by an appropriate change $\theta \to f(\theta)$: one uses the bouncing rays as a new clock. It will be proved below that the Minkowski time t possesses the property $l(t) = $ constant when the string obeys the equations of motion.

The interest of the conformal gauge is that the string dynamical equations become linear when the condition (12.5.1.1) holds. In explicit form

$$\eta^{\lambda\mu}\partial_\lambda\partial_\mu X^A = 0 \qquad (12.5.1.6)$$

while the constraints $T_{\alpha\beta} = 0$ become

$$\dot{X}^2 + X'^2 = 0, \qquad \dot{X} \cdot X' = 0 \qquad (12.5.1.7)$$

($\dot{X}^2 = \eta_{AB} \dot{X}^A \dot{X}^B$, and so on).

Exercises

1. Show that the conformal gauge is equivalent to $N = 1$, $N^1 = 0$.
2. Discuss the conformal gauge for the closed string.
3. **a.** Consider, in three dimensions, the two-dimensional surface defined by

$$X^0 = \tau$$

$$X^1 = \cos \alpha + \tau, \qquad 0 \leq \alpha \leq 2\pi$$

$$X^2 = \sin \alpha$$

Compute the induced metric on the world sheet.

   b. Show that it takes an infinite amount of time for a light ray to go from $\alpha = 0$ to $\alpha = \pi$. Can a light ray go in the direction of decreasing α?

   c. Discuss the possibility of setting up a conformal coordinate system on the whole world sheet.

   d. Consider the open string history described by the above equations, for $0 \leq \alpha \leq \pi$. Show that the end points move with the speed of light at right angles to the string, but that the condition (12.1.7.8b) is not fulfilled. Draw the lines orthogonal to the lines $\tau = $ const.

12.5.2. Light-Cone Gauge

The conformal-gauge condition does not completely fix the coordinate system. As already noted, the conformal group remains as the residual gauge group.

In order to fully specify the gauge, further conditions must be imposed. For instance, one can prescribe what the "time" τ is (in a way compatible with the conformal gauge).

Once τ is fixed, the only conformal transformations which remain are

$$\tau' = \tau \qquad (12.5.2.1)$$

and

$$\sigma' = \pm \sigma + b \qquad (12.5.2.2)$$

where b is a constant. If one requires in addition that the conformal transformation preserves the spatial orientation, one is left with the plus sign in relation (12.5.2.2).

In the open-string case, the range of the new coordinate σ' must also extend from 0 to π (the transformation must become the identity at the boundary). The residual gauge group then reduces to the identity.

In the closed case, it only makes sense to demand that σ' be periodic of period 2π. However, this is clearly fulfilled by transformation (12.5.2.2) no matter what value is taken by the constant b. Hence, one sees that giving τ and imposing the conformal gauge does not completely fix the coordinate system in the closed case. The possibility remains of performing zero-mode spatial translations

$$\tau' = \tau \tag{12.5.2.3}$$

and

$$\sigma' = \sigma + b \tag{12.5.2.4}$$

In the light-cone gauge, one takes τ proportional to X^+,

$$\tau = \beta X^+ \tag{12.5.2.5}$$

where

$$X^\pm = \frac{1}{\sqrt{2}} (X^0 \pm X^{d-1}) \tag{12.5.2.6}$$

are lightlike coordinates in Minkowski space. The reason why equation (12.5.2.5) is so convenient is that, in lightlike coordinates, the flat metric reads

$$ds^2_{\text{Mink}} = -2\, dX^+ \, dX^- + \sum_i (dX^i)^2, \quad i = 1, \ldots, d-2 \tag{12.5.2.7}$$

and is linear in dX^+.

The conditions (12.5.1.1) and (12.5.2.5) define the light-cone gauge.

In order to determine whether condition (12.5.2.5) is an acceptable parametrization, one must perform two checks: (1) that X^+ is a time coordinate compatible with the conformal gauge, and (2) that the hyperplanes $X^+ = \text{const}$ intersect the string history along one-dimensional lines which are spacelike. This is done in the next section, where it is shown that there is almost no problem for trajectories obeying the classical equations of motion.

We determine first the constant β in equation (12.5.2.5). The equations of motion for X^+ yield

$$\dot{X}^+ = 2\pi\alpha' N \mathcal{P}^+ + N^1 (X^+)' \tag{12.5.2.8a}$$

and

$$\dot{\mathcal{P}}^+ = (NX^{+\prime})' + (N^1\mathcal{P}^+)' \qquad (12.5.2.8b)$$

Since the conformal gauge is equivalent to $N = 1$ and $N^1 = 0$, the condition (12.5.2.5) implies, together with equation (12.5.2.8a), that $\mathcal{P}^+(\sigma)$ has no nonzero mode:

$$\mathcal{P}^+(\sigma) = p^+/\pi \qquad \text{(open string)} \qquad (12.5.2.9a)$$

or

$$\mathcal{P}^+(\sigma) = p^+/2\pi \qquad \text{(closed string)} \qquad (12.5.2.9b)$$

Furthermore, one obtains β from equation (12.5.2.8a):

$$\tau = X^+/2p^+\alpha' \qquad \text{(open string)} \qquad (12.5.2.10a)$$

or

$$\tau = X^+/p^+\alpha' \qquad \text{(closed string)} \qquad (12.5.2.10b)$$

Conversely, we assume that equations (12.5.2.9) and (12.5.2.10) hold. Then, one finds from relation (12.5.2.8a) that $N = 1$ [$\dot{p}^+ = 0$, as can be seen by integrating equation (12.5.2.8b) over σ], and from relation (12.5.2.8b) that $N^{1\prime} = 0$ (provided $p^+ \neq 0$). Accordingly, N^1 is a function of τ only. In the open-string case, that function has to vanish as a result of the boundary conditions. The coordinate system must be conformal, and one can adopt relations (12.5.2.9) and (12.5.2.10) as the definition of the light-cone gauge.

In the closed-string case $N^1(\tau)$ is not determined, so that conditions (12.5.2.9) and (12.5.2.10) are weaker than the original light-cone gauge conditions. However, because these conditions are more convenient to manipulate—they are expressed directly in terms of the canonical variables—we will also adopt them as coordinate conditions for the closed string and refer to them from now on as light-cone gauge conditions. What is important to realize is that conditions (12.5.2.9) and (12.5.2.10) still allow for arbitrary, time-dependent coordinate transformations given by

$$\tau' = \tau \qquad \text{and} \qquad \sigma' = \sigma + f(\tau) \qquad (12.5.2.11)$$

(time-dependent zero-mode translations). This residual gauge invariance plays a key role in the quantum theory. Its associated constraint implies that the first excited state in the spectrum has spin 2.

Exercise

a. Let (τ, σ) be conformal ("isothermal") coordinates. Show that a necessary condition for $\tau'(\tau, \sigma)$ to be an acceptable new time of another conformal coordinate system is that $\partial^2 \tau'/\partial \tau^2 - \partial^2 \tau'/\partial \sigma^2 = 0$.

b. Given $\tau'(\tau, \sigma)$ obeying (a), describe geometrically how to construct the new conformal coordinate system (τ', σ').

c. When τ' solves (a), one can write

$$\tau' = \tfrac{1}{2}[f(\tau + \sigma) + g(\tau - \sigma)]$$

and

$$\sigma' = \tfrac{1}{2}[f(\tau + \sigma) - g(\tau - \sigma)]$$

Show that the coordinate system (τ', σ') is acceptable if and only if f and g are invertible.

d. One sometimes takes not relation (12.5.2.10a), but $\tau = X^+$. What is the range of σ under this condition? [Note: equation (12.5.2.9a) should be read as $\mathcal{P}^+(\sigma) = p^+/\text{range of } \sigma$, so as to have $\int \mathcal{P}^+(\sigma)\,d\sigma = p^+$.]

12.5.3. General Solution of the String Classical Equations

In the conformal gauge, the string dynamical equations are readily solved as

$$X^A(\tau, \sigma) = \tfrac{1}{2}[f^A(\tau + \sigma) + g^A(\tau - \sigma)] \qquad (12.5.3.1)$$

or, in terms of the null coordinates θ and θ',

$$X^A(\theta, \theta') = \tfrac{1}{2}[f^A(\theta) + g^A(\theta')] \qquad (12.5.3.2)$$

The light-cone gauge is acceptable and will be globally well-defined if and only if $f^+(\theta)$ and $g^+(\theta')$ are invertible [see Exercise above]. This might fail to happen in some circumstances.

To investigate when this failure of the light-cone gauge occurs, we must first take into account the conditions $T_{\alpha\beta}(X) = 0$ as well as the boundary conditions at $\theta' = \theta$ and $\theta' = \theta + 2l(\theta)$. (We consider from now on only the open string. Remarks on the closed string will be grouped together in Section 12.5.7.)

The conditions $T_{\alpha\beta}(X) = 0$ will be fulfilled, i.e., the coordinate system θ, θ' will be null, if and only if the vectors $df^A/d\theta$ and $dg^A/d\theta'$ are lightlike,

$$\eta_{AB}\frac{df^A}{d\theta}\frac{df^B}{d\theta} = 0, \qquad \eta_{AB}\frac{dg^A}{d\theta'}\frac{dg^B}{d\theta'} = 0 \qquad (12.5.3.3)$$

Furthermore, the boundary condition (12.5.1.3), which summarizes the behavior of the string near its ends, will hold at the first end point $\theta = \theta'$ if the lightlike vectors $df^A/d\theta$ and $dg^A/d\theta'$ are equal at $\theta' = \theta$. But if $f'^A(\theta) = g'^A(\theta)$, one can assume without loss of generality that the functions f^A and g^A are identical,

$$f^A(\theta) = g^A(\theta) \qquad (12.5.3.4)$$

Condition (12.5.1.3) at the other end point will then hold if

$$f'^A(\theta) = f'^A(\theta + 2l(\theta))$$

By taking for θ one of the appropriate parametrizations that make σ range from 0 to π, i.e., $l(\theta) = \pi$, we obtain

$$f'^A(\theta) = f'^A(\theta + 2\pi) \qquad (12.5.3.5)$$

The periodicity of the derivative f'^A in θ implies

$$f^A(\theta + 2\pi) = f^A(\theta) + 2\pi\alpha' p^A \qquad (12.5.3.6)$$

where p^A is some constant "zero mode." Explicit computation shows that p^A is nothing but the string momentum.

Accordingly, one sees that the motion of the string is completely determined by a single function $f^A(\theta)$, i.e., by the motion of its end $\theta = \theta'$, which moves with the velocity of light along a helicoidal trajectory,

$$X^A(\theta, \theta') = \tfrac{1}{2}[f^A(\theta) + f^A(\theta')] \qquad (12.5.3.7a)$$

$$\frac{df^A}{d\theta}\frac{df^B}{d\theta}\eta_{AB} = 0 \qquad (12.5.3.7b)$$

$$f^A(\theta + 2\pi) = f^A(\theta) + 2\pi\alpha' p^A \qquad (12.5.3.7c)$$

If p^A is timelike, there exists a "rest frame" in which $p^A = (p, 0, 0, \ldots, 0)$ and in which the end $\theta = \theta'$ describes a periodic closed orbit with the velocity of light.

The result (12.5.3.7) has interesting consequences. First, one finds from relation (12.5.3.7c) that p^A can be written as a sum of future pointing lightlike vectors,

$$p^A = \frac{1}{2\pi\alpha'}\int_0^{2\pi}\frac{df^A}{d\theta}d\theta \qquad (12.5.3.8)$$

and hence is necessarily timelike or null,

$$p^2 \equiv p^A p_A \leq 0 \qquad (12.5.3.9)$$

The classical string has no tachyonic state of motion.

Furthermore, equality holds in relation (12.5.3.9) if and only if $df^A/d\theta$ is parallel, for all θ, to a given lightlike direction μ^A,

$$df^A/d\theta = k(\theta)\mu^A \qquad (12.5.3.10a)$$

This implies that

$$f^A(\theta) = \bar{k}(\theta)\mu^A + f_0^A \qquad (12.5.3.10b)$$

for an appropriate function $\bar{k}(\theta)$. Substitution of equations (12.5.3.10) into relation (12.5.3.7a) leads one to conclude that lightlight momenta p^A correspond to strings collapsed to a point, moving in uniform motion with the speed of light. These are the ground-state motions.

Equations (12.5.3.7) enable one to also discuss the validity of the light-cone gauge on shell. The whole question is whether $f^+(\theta)$ is invertible. From relation (12.5.3.7c) one sees that two cases need to be considered:

1. $p^+ = 0$. The expression $p^2 = -2p^+p^- + \sum (p^i)^2$ is nonpositive from relation (12.5.3.9), so this can only occur if $p^2 = 0$ and $p^i = 0$ ($|p^-| < \infty$). The string is in its collapsed ground state, and moves with the velocity of light along the "last" spatial direction. The function $f^+(\theta)$ is then constant and cannot be one-to-one. The light-cone gauge certainly fails for such motions.
2. $p^+ \neq 0$. In this case, the situation is much better. Indeed let us assume $f^+(\theta_1) = f^+(\theta_2)$ with $\theta_1 < \theta_2$. Because the hyperplane $X^+ = $ const is a null hyperplane, it has a degenerate metric and every direction on it is either spacelike or null,

$$ds^2\big|_{\text{induced on } X^+ = \text{const}} = \sum_i (dX^i)^2 \qquad (12.5.3.11)$$

The null directions are given by $X^i = $ const.

Now, if the end-point trajectory crosses twice a given hyperplane $X^+ = $ const, the points of intersection must possess the same values of X^i, namely

$$X^i(\theta_1) = X^i(\theta_2) \qquad (12.5.3.12)$$

for this is the only way the event θ_2 can be in the future of θ_1.

Furthermore, one must have

$$X^i(\theta) = X^i(\theta_1) \atop X^+(\theta) = X^+(\theta_1)\} \theta_1 \le \theta \le \theta_2 \quad (12.5.3.13)$$

for all θ between θ_1 and θ_2, because this is the only causal curve that can join the two events.

However, one then finds that the induced metric on the string world sheet,

$$ds^2|_{\text{string}} = [-X'^+(\theta)X'^-(\theta') - X'^+(\theta')X'^-(\theta)$$
$$+ X'^i(\theta)X'_i(\theta')] \, d\theta \, d\theta' \quad (12.5.3.14)$$

is degenerate for $\theta_1 \le \theta, \theta' \le \theta_2$, in contradiction to our "conformal-gauge assumption" that the coordinates θ, θ' provide a regular parametrization of the string.† This means $\theta_1 = \theta_2$ and $f^+(\theta)$ can be inverted.

In addition, relation (12.5.3.11) enables one to see that the intersections of the string history with the hyperplanes $X^+ = $ const are nontimelike; X^+ is therefore a time coordinate. The conclusion is that the light-cone gauge is a good gauge when $p^+ \ne 0$, for "on-shell" sheets with a trivial causal structure.

One may wonder whether the problems mentioned above when $p^+ = 0$ should be worried about. Since these only concern ground-state pointlike motions of the string, some light can be shed on that question by considering a free, massless, relativistic particle in the light-cone gauge. It is easily seen that no serious difficulty arises in the quantum theory. This seems to indicate that the problems with $p^+ = 0$ should not concern us, and is in agreement with the fact that the string light-cone gauge quantization makes predictions identical with those of other methods which do not need a gauge-fixing condition.

Exercises

1. a. Describe the classical ground state in terms of oscillator variables.

    b. Consider the first excited states and show from the classical "Virasoro conditions" that $p^2 \le 0$. Can one have $p^2 = 0$? What can be said about the "pure gauge" nature of these modes?

2. Show, using equation (12.5.3.7a), that the Minkowskian time needed for a light ray to go from one end to the other and back is time-independent.

3. The parametrization θ considered in relations (12.5.3.7) is not unique. Show that one can impose $f'^A(\theta)p_A = p^A p_A$.

† The points $X^A(\theta, \theta')$, $\theta_1 \le \theta, \theta' \le \theta_2$ actually all lie on the same lightlike line.

4. Consider a free relativistic massless particle.

  a. Write the reduced action in the gauge
$$\tau = mX^+/p^+$$

  b. Show that one can take as independent degrees of freedom $X^i, p_i, u^- \equiv X^- - p^-\tau/m$, and p^+. Show that these variables are canonically conjugate.

  c. Derive the light-cone gauge Hamiltonian.

  d. Solve the Schrödinger equation. Prove the existence of states which go backward in Minkowskian time (i.e., for which $p^0 < 0$). Do such states appear in the more conventional gauge $\tau = X^0$?

  e. Write the Poincaré generators. Order them so that they are formally self-adjoint. Show that they close in the Poincaré algebra even quantum mechanically.

12.5.4. Independent Degrees of Freedom—Dirac Brackets

We now apply the machinery of canonical gauge fixing for constrained Hamiltonian systems [13]. The first step is to identify the independent degrees of freedom, in terms of which the constraints can be solved for the "dependent" variables. We will work right away with the oscillator coordinates a_n^A.

The gauge conditions (12.5.2.9) and (12.5.2.10) are equivalent to†

$$a_n^+ = 0 \quad \text{and} \quad X_0^+ = 2\alpha' p^+ \tau \quad (12.5.4.1)$$

and show that a_n^+ and X_0^+ are dependent variables. If relations (12.5.4.1) fully fix the gauge (and we know this is so, since the lapse N and shift N^1 are completely determined by these relations), one should be able to solve the Virasoro constraints $L_n = 0$ for the conjugate variables a_n^- and p^-.

Indeed, one finds from constraints (12.4.1.8) that

$$a_n^- = iL_n^{tr}/\sqrt{2\alpha' n}\, p^+ \quad (n > 0) \quad (12.5.4.2)$$

and

$$p^- = L_0^{tr}/2\alpha' p^+ \quad (12.5.4.3)$$

where L_0^{tr} and L_n^{tr} are given by the same expressions as L_0 and L_n [expressions (12.4.1.8a-c)], but with summation only over transverse variables. The great virtue of the light-cone gauge is that the Virasoro constraints are linear in a_n^- and p^- and can be solved as simply as in relations (12.5.4.2) and (12.5.4.3).

The variables p^+, X_0^-, p^i, X_0^i, and a_n^i are completely unconstrained and are taken as independent variables. They correspond to the true degrees

† As mentioned previously, we consider from now on the open string.

of freedom of the string ($d-2$ transverse oscillator variables plus the center-of-mass coordinates).

The next step in the analysis consists of computing the Dirac brackets of the independent variables. (Those of the dependent variables follow from their expressions in terms of the independent ones since, by the definition of the Dirac brackets, the constraints and gauge conditions have vanishing Dirac brackets with everything else and can be used freely before computing the brackets.) To that end, one rewrites the gauge conditions and constraints (12.5.4.1)–(12.5.4.3) as

$$\chi_n \equiv a_n^+ = 0, \qquad \chi \equiv X_0^+ - 2\alpha' p^+ \tau = 0$$

$$\phi_n \equiv a_n^- - \frac{iL_n^{tr}}{\sqrt{2\alpha'n}p^+}, \qquad \phi \equiv p^- - \frac{L_0^{tr}}{2\alpha' p^+} = 0 \qquad (12.5.4.4)$$

The matrix C of the Poisson brackets of system (12.5.4.4) is easy to evaluate: the χ-gauge conditions have vanishing brackets among themselves, as do the first-class ϕ constraints. The only nonvanishing brackets are $[\chi_n, \phi_{-n'}] = i\delta_{n,n'}$ and $[\chi, \phi] = 1$, so that C reads formally

$$\begin{array}{cc} & \chi \quad \phi \\ C = \begin{pmatrix} 0 & I \\ -I & 0 \end{pmatrix} & \begin{array}{c} \chi \\ \phi \end{array} \end{array}$$

Its inverse matrix has the same form, so that the Dirac brackets are given by [13]

$$[F, G]_D = [F, G]_P + \text{``}[F, \phi]_P[\chi, G]_P - [F, \chi]_P[\phi, G]_P\text{''} \qquad (12.5.4.5)$$

The interesting thing about equation (12.5.4.5) is that $[F, G]_D = [F, G]_P$ wherever F and G commute with the χ-gauge conditions (i.e., $[F, \chi_n] = [G, \chi_n] = [F, \chi] = [G, \chi] = 0$). This is clearly the case for p^+, p^i, X_0^i, and a_n^i. Accordingly, these variables have identical Dirac and Poisson brackets.

The only change involves the variable X_0^-, since $[X_0^-, \chi] = 2\alpha' \tau \neq 0$. This suggests replacing X_0^- by u_0^-,

$$u_0^- = X_0^- - 2\alpha' p^- \tau \qquad (12.5.4.6)$$

which has the property $[u_0^-, \chi] = [u_0^-, \chi_n] = 0$. With u_0^- as new independent variables, the Dirac brackets have canonical values

$$[u_0^-, p^+]_D = -1, \qquad [X_0^i, p^j]_D = \delta^{ij}$$

$$[a_n^i, a_{n'}^{*j}]_D = -i\delta^{ij}\delta_{nn'} \qquad (12.5.4.7)$$

(all other Dirac brackets vanish). From now on, we will drop the symbol D in the Dirac brackets.

Exercise. Solve the constraints in the continuous representation $\mathcal{H} = \mathcal{H}_1 = 0$. Show that X_0^- appears as an integration constant in the general solution $X^-(\sigma)$ of $\mathcal{H}_1 = 0$.

12.5.5. Light-Cone Gauge Action—Light-Cone Gauge Hamiltonian

The gauge conditions (12.5.4.1) explicitly involve the time τ, so the time evolution of the dynamical variables in the light-cone gauge is not simply given by taking their Dirac bracket with the canonical Hamiltonian. Since this one vanishes,† the procedure would imply the meaningless statement that everything is τ-independent.

This recipe (use of the Dirac bracket in the dynamical equations) is only valid for time-independent gauge conditions [13], and clearly contradicts here the preservation of $\chi = 0$ by the time evolution (one would find instead $\dot{\chi} = [\chi, H]_D + \partial \chi / \partial \tau = 2\alpha' p^+$!).

In order to get the correct time evolution in the light-cone gauge, one must proceed along different directions. One possibility is to make (before writing the equations of motion) a time-dependent canonical transformation, which turns the gauge conditions into time-independent ones, and which generates a nonzero Hamiltonian through the explicit time derivative of the generating function S, $H \to H + \partial S / \partial \tau$. In our case, one finds (see Exercise 1 below) that

$$\partial S / \partial \tau = 2\alpha' p^+ p^- \qquad (12.5.5.1)$$

Hence, the light-cone gauge Hamiltonian is given by

$$H^{\text{L.C.}} = 2\alpha' p^+ p^- = L_0^{\text{tr}} / 2\alpha' \qquad (12.5.5.2)$$

Once the constraints are time-independent, one can replace the Poisson brackets by the Dirac brackets in the Hamiltonian equations and get the correct light-cone gauge time evolution.

Another method consists of computing the reduced light-cone gauge action, which is obtained by eliminating from the canonical action (12.2.2.3) the dependent variables X_0^+, a_n^+, p^-, and a_n^- by means of conditions (12.5.4.1)–(12.5.4.3). This is permissible, i.e., extremization of the light-cone

---

† It is seen from expression (12.2.2.4) that the Hamiltonian is a combination of the constraints.

gauge action leads to the correct dynamical equations. One finds that

$$S^{\text{L.C.}}[a_n^i, a_n^{*i}, p^i, p^+, X_0^i, u_0^-]$$

$$= \int d\tau \left[i \sum_n \int d\sigma \, a_{ni}^* \dot{a}_n^i + p_i \dot{x}_0^i - p^+ \dot{u}_0^- - H^{\text{L.C.}} \right] \qquad (12.5.5.3)$$

with $H^{\text{L.C.}}$ given by expression (12.5.5.2).

An examination of relation (12.5.5.3) shows that $H^{\text{L.C.}}$ is indeed the correct generator of the time evolution, and that the brackets of the independent variables are indeed given by system (12.5.4.7). The reduced action (12.5.5.3) plays a key role in the theory, e.g., it appears in the light-cone gauge path integral.

Exercises

1. Work out explicitly the generating function of the canonical transformation

$$u^- = X_0^- - 2\alpha' p^- \tau, \qquad u^+ = X_0^+ - 2\alpha' p^+ \tau$$

   (other variables fixed). Show that its explicit time derivative is given by equation (12.5.5.1).

2. Justify the use of the reduced light-cone gauge action. Show that the variational principle $\delta S^{\text{L.C.}} = 0$ summarizes all the light-cone gauge equations (for the independent variables). Find a general argument which proves that the brackets determined by action (12.5.5.3) are the standard Dirac brackets.

12.5.6. Poincaré Generators

Poincaré transformations in their original, linear form take us away from the light-cone gauge. In order to remain within the gauge conditions (12.5.2.9), (12.5.2.10) one must also perform a gauge transformation.

The "compensating" reparametrization is easily evaluated by noting that the correct (Poincaré + gauge) transformation is generated by the same generators as before, but this time acting through the *Dirac* bracket.

Indeed, the correct transformation must differ from the given Poincaré transformation by a reparametrization such that $\delta \chi_n = \delta \chi = 0$ ($\delta \phi_n = \delta \phi = 0$ is automatic, since Poincaré generators are gauge-invariant). But this is precisely what the use of the Dirac brackets guarantees, since $[\chi_n, \text{anything}]_{DB} = [\chi, \text{anything}]_{DB} = 0$.† The fact that χ depends explicitly on τ is of no relevance here, since Poincaré transformations are internal symmetries not involving τ.

---

† One also sees that the Dirac bracket $[F, \text{Poincaré generators}]_{DB}$ differs from the Poisson bracket $[F, \text{Poincaré generators}]_{PB}$ by a gauge transformation, because the Poincaré generators commute with the Virasoro constraints [and because of relation (12.5.4.5)].

Accordingly, the Poincaré generators in the light-cone gauge are still given by expressions (12.4.1.9), i.e., in terms of the light-cone gauge independent variables.

$$P^+ = p^+, \qquad P^i = p^i, \qquad P^- = p^- = L_0^{\text{tr}}/2\alpha'p^+ \qquad (12.5.6.1)$$

$$M^{+-} = \tfrac{1}{2}u_0^- p^+, \qquad M^{i+} = -\tfrac{1}{2}p^+ X_0^i \qquad (12.5.6.2\text{a})$$

$$M^{ij} = \tfrac{1}{2}(p^i X_0^j - p^j X_0^i) + \tfrac{1}{2}\sum_{n>0} i(a_n^{i*}a_n^j - a_n^{j*}a_n^i) \qquad (12.5.6.2\text{b})$$

$$M^{i-} = \frac{1}{2}\left(p^i u_0^- - \frac{L_0^{\text{tr}}}{2\alpha'p^+} X_0^i\right) - \frac{1}{2}\sum_{n>0} \frac{a_n^{*i}L_n^{\text{tr}} + L_{-n}^{\text{tr}}a_n^i}{\sqrt{2\alpha'n}p^+} \qquad (12.5.6.2\text{c})$$

We note that these generators are not linear or quadratic in the independent variables.

In order to arrive at expressions (12.5.6.2), we have set to zero the explicit time dependence—which appears through X^+—as can be done by an appropriate canonical transformation (that induced by the motion). This does not modify the Dirac-bracket algebra of the generators.

Exercises

1. Work out explicitly the transformation generated by system (12.5.6.1) and (12.5.6.2) through the Dirac bracket. Separate it into an ordinary Poincaré transformation plus a reparametrization.

2. Check explicitly that system (12.5.6.1) and (12.5.6.2) closes in the Dirac bracket according to the Poincaré algebra.

12.5.7 Peculiarities of the Closed String

The light-cone gauge treatment of the closed string closely parallels what we did for the open one. The only difference is that the light-cone gauge conditions

$$X^+ = \alpha'p^+\tau \qquad \text{and} \qquad \mathcal{P}^+ = p^+/2\pi \qquad (12.5.7.1)$$

equivalent to

$$c_n^+ = \bar{c}_n^+ = 0 \qquad (12.5.7.2)$$

no longer completely fix the gauge. There remains one residual gauge invariance (zero-mode translations along σ), and one associated constraint: the zero mode of $\mathcal{H}_1$, the constraint-generator of σ reparametrizations, must vanish.

One thus finds that the "independent" variables p_i, X_0^i, u_0^-, p^+, and c_n^i, $\bar{c}_n^i$ are constrained by

$$L_0^{\text{tr}} - \bar{L}_0^{\text{tr}} = 0 \tag{12.5.7.3}$$

($=\int \mathcal{H}_1 \, d\sigma$). They still possess canonical Dirac brackets

$$[X_0^i, p_j] = \delta_j^i, \qquad [p^+, u_0^-] = 1 \tag{12.5.7.4a}$$

$$[c_n^i, c_{n'}^{*j}] = -i\delta^{ij}\delta_{nn'} = [\bar{c}_n^i, \bar{c}_{n'}^{*j}] \tag{12.5.7.4b}$$

(other brackets vanish).

Quantities p^- and c_n^-, $\bar{c}_n^-$ are again given by

$$p^- = \frac{L_0^{\text{tr}} + \bar{L}_0^{\text{tr}}}{\alpha' p^+} \tag{12.5.7.5}$$

$$c_n^- = \frac{2i L_n^{\text{tr}}}{\sqrt{2\alpha' n} p^+}, \qquad \bar{c}_n^- = \frac{2i \bar{L}_n^{\text{tr}}}{\sqrt{2\alpha' n} p^+} \tag{12.5.7.6}$$

while the Poincaré generators are

$$P^i = p^i, \qquad P^+ = p^+, \qquad P^- = \frac{L_0^{\text{tr}} + \bar{L}_0^{\text{tr}}}{\alpha' p^+} \tag{12.5.7.7}$$

$$M^{+-} = \tfrac{1}{2} u_0^- p^+, \qquad M^{i+} = -\tfrac{1}{2} p^+ X_0^i \tag{12.5.7.8a}$$

$$M^{ij} = \tfrac{1}{2}(p^i X_0^j - p^j X_0^i)$$

$$+ \tfrac{1}{2} \sum_{n>0} i[c_n^{i*} c_n^j + \bar{c}_n^{i*} \bar{c}_n^j - (i \leftrightarrow j)] \tag{12.5.7.8b}$$

$$M^{i-} = \frac{1}{2}\left(p^i u_0^- - \frac{L_0^{\text{tr}} + \bar{L}_0^{\text{tr}}}{\alpha' p^+} X_0^i\right)$$

$$- \sum_{n>0} \frac{c_n^{*i} L_n^{\text{tr}} + L_{-n}^{\text{tr}} c_n^i}{\sqrt{2\alpha' n} p^+} - \sum_{n>0} \frac{\bar{c}_n^{*i} \bar{L}_n^{\text{tr}} + \bar{L}_{-n}^{\text{tr}} \bar{c}_n^i}{\sqrt{2\alpha' n} p^+} \tag{12.5.7.8c}$$

They close in the Dirac bracket according to the Poincaré algebra. A nontrivial point is that now the remaining constraint (12.5.7.3) appears in the right-hand side of $[M^{i-}, M^{j-}]$, given by

$$[M^{i-}, M^{j-}] = \frac{i}{2\alpha'(p^+)^2}[c_n^i c_n^{*j} - c_n^j c_n^{*i} - \bar{c}_n^i \bar{c}_n^{*j} + \bar{c}_n^j \bar{c}_n^{*i}](L_0^{\text{tr}} - \bar{L}_0^{\text{tr}})$$

$$\approx 0 \qquad (12.5.7.9)$$

Exercises

1. Show that constraint (12.5.7.3) appears as a condition for the constraint $\mathcal{H}_1(\sigma) = 0$ to be solvable for a periodic $X^-(\sigma)$ [no zero mode in $X'^-(\sigma)$].

2. Verify relation (12.5.7.9).

Chapter 13

Quantization of the Nambu–Goto String

13.1. GENERAL CONSIDERATIONS—VIRASORO ALGEBRA

13.1.1. Introduction

The most striking feature of quantum string models is perhaps their prediction of a critical space-time dimension outside of which the quantum theory is problematical. For the bosonic model, that space-time dimension is 26.

There are at least three different ways to derive the critical dimension. The first is based on the light-cone gauge formulation of the theory; one finds that, because of quantum noncommutativity of operators, the quantum Poincaré generators fail to close according to the Poincaré algebra, unless $d = 26$ [17]. The second, called the "covariant approach," uses no gauge condition and works with all (true and gauge) degrees of freedom. One shows that negative-norm states are absent from the physical subspace only if $d \leq 26$ [18]. Finally, the last approach, which is probably the deepest but also the least understood, is based on Becchi–Rouet–Stora–Tyutin symmetry. This symmetry is found to be realized quantum-mechanically only if $d = 26$ [19].

All of these approaches use in an essential way the property that the algebra of the Virasoro constraints acquires a nonvanishing central charge upon quantization (the central extension is known as the "Virasoro algebra"). Our first task is therefore to compute explicitly this dimension-dependent central charge, within the covariant formalism.

13.1.2. Fock Representation—Virasoro Operators

In order to compute the central charge, one must be precise as to in which (pseudo-)Hilbert space, and how, the basic operators of the theory are to be defined. It turns out that the actual central extension depends on the representation chosen for the basic commutation relations, something which is possibly not stressed enough in the literature.

The constraints are quadratic and L_0 is of harmonic-oscillator type, so it appears natural to choose a Fock representation. One therefore assumes the existence of a vacuum state $|0, 0\rangle$ annihilated by all "destruction operators" a_n^A (including a_n^0, in order to preserve Lorentz invariance), and possessing zero d-momentum,

$$a_n^A|0, 0\rangle = 0, \qquad p^A|0, 0\rangle = 0 \tag{13.1.2.1}$$

The entire space is generated by application of the "creation" operators a_n^{A*} and $\exp i k \cdot X_0$. This latter operator creates momentum according to

$$\exp i k \cdot X_0 |0, 0\rangle = |0, k\rangle \tag{13.1.2.2a}$$

and

$$p^A|0, k\rangle = k^A|0, k\rangle \tag{13.1.2.2b}$$

where the momentum eigenstates are normalized as

$$\langle 0, k'|0, k\rangle = (2\pi)^d \delta^{(d)}(k - k') \tag{13.1.2.2c}$$

A general state is a superposition of $|\lambda, k\rangle$ states, where λ is the oscillator occupation number (actually, one λ for each oscillator a_n^A) and k is the d-momentum.

These prescriptions give a definite meaning to the operators X^A, p_B, a_m^A, a_n^{B*}, and to their commutation relations, which are

$$[X_0^A, p_B] = i\delta_B^A \tag{13.1.2.3}$$

and

$$[a_m^A, a_n^{B*}] = \eta^{AB}\delta_{m,n} \tag{13.1.2.4}$$

13 • Quantization of Nambu–Goto String

The time component of relation (13.1.2.4) is $[a_m^0, a_n^{0*}] = -1$, so one generates negative-norm states when acting on the vacuum with an odd number of creation operators a_n^{0*}. This is unavoidable if explicit Lorentz invariance is required within our Fock-space representation.

Elimination of the negative-norm states will be the subject of Section 13.4. Our only purpose now is to examine how ordering problems affect the L_n algebra.

To that end, we note that the Virasoro operators with $n \neq 0$ are given by products of commuting operators and hence are unaffected by any ordering ambiguity,

$$L_n = -i\sqrt{2n\alpha'}p_A a_n^A + \sum_{m>0} \sqrt{(m+n)m}\, a_{Am}^* a_{m+n}^A$$

$$-\tfrac{1}{2} \sum_{m=1}^{n-1} \sqrt{(n-m)m}\, a_{Am} a_{n-m}^A \qquad (13.1.2.5a)$$

$$L_{-n} = L_n^* = i\sqrt{2n\alpha'}p_A a_n^{A*} + \sum_{m>0} \sqrt{(m+n)m}\, a_{Am+n}^* a_m^A$$

$$-\tfrac{1}{2} \sum_{m=1}^{n-1} \sqrt{(n-m)m}\, a_{An-m}^* a_m^{*A} \qquad (13.1.2.5b)$$

($n > 0$). We have chosen to normal order L_n and L_{-n} in expressions (13.1.2.5), but we could as well have taken the "antinormal" order without changing L_n or L_{-n}.

The only true ordering ambiguity lies in the zero mode L_0. Since we have adopted a Fock representation, the quantum analog of L_0 must be well defined within Fock space. This means that it can at most differ from the normal-ordered expression

$$\alpha'p^2 + \sum_n n a_{An}^* a_n^A \qquad (13.1.2.6)$$

by a finite constant α_0.

We henceforth reserve the symbol L_0 for the normal-ordered expression (13.1.2.6),

$$L_0 = \alpha'p^2 + \sum_n n a_{An}^* a_n^A \qquad (13.1.2.7)$$

and recollect in subsequent calculations that the quantum analog of the classical L_0^{cl} might be $L_0 - \alpha_0$,

$$L_0^{cl} \to L_0 - \alpha_0 \qquad (13.1.2.8)$$

with α_0 *finite*. We note that a different ordering prescription, with an infinite α_0 (e.g., antinormal ordering), clearly leads to an ill-defined Fock-space operator.

Exercise. Define a representation of the commutation relations (13.1.2.3) and (13.1.2.4) in a "different Hilbert space," in which antinormal ordering makes sense. Note: in that representation, a_n^0 (a_n^j) creates positive- (negative-) norm states.

13.1.3. Virasoro Algebra

Our strategy in computing the quantum algebra of the Virasoro operators is the following. The only reason why the quantum algebra differs from $i\hbar$ times its classical analog is the presence in $[L_m, L_n] - (m-n)L_{m+n}$ of terms which classically cancel each other, but which no longer do so quantum-mechanically because they are differently ordered.

Now, L_{m+n} is quadratic in the basic variables and is in normal-order form according to our Fock-space choice. Accordingly, quantum effects can occur only when the commutator $[L_m, L_n]$, which is also quadratic in the basic variables, is not normal-ordered. This means that the central charge is equal to the term acquired upon normal-ordering of $[L_m, L_n]$.

Since we know the classical form of the commutator, we simply need to compute the "anomalous" term. Moreover, we expect—as is easily checked—a nonvanishing central charge only when $L_{m+n} = L_0$, i.e., $n = -m$, since it is L_0 alone which is afflicted by an ordering ambiguity.

We therefore compute $[L_m, L_{-m}]$ with $m > 0$:

$$[L_m, L_{-m}] = \text{N.O.} + \frac{1}{4} \sum_{k,s=1}^{m-1} (m-k)k(m-s)s(a_k \cdot a_s^* \delta_{k,s} + a_k \cdot a_{m-s}^* \delta_{m-k,s})$$

$$(13.1.3.1)$$

where N.O. stands for a normal-ordered expression whose precise form need not concern us here, since it is irrelevant for the central charge.

We note that the second term in the right-hand side of expression (13.1.3.1), although not in normal-order form, is well defined in Fock space because the sum only contains a finite number of terms. (It is important to carry out the calculations so that the expressions which appear at any stage are well defined in the chosen representation space. This is guaranteed by the use of the normal order. Had we taken a different representation space, that appropriate to "antinormal order" say, we would conduct the computations differently and would find a different central charge.)

Normal ordering of equation (13.1.3.1) easily yields

$$[L_m, L_{-m}] = \text{N.O.} + \frac{d}{12}(m^3 - m) \qquad (13.1.3.2)$$

(we use the identity $6\sum_{k=1}^{m-1}(m-k)k = m(m-1)(m+1)$; d appears through the trace $\eta^A{}_A$, which counts the number of oscillators).

The conclusion of this computation is that one indeed acquires an "anomalous" c-number term in the quantum Virasoro algebra, proportional to the space-time dimension,

$$[L_m, L_n] = (m-n)L_{m+n} + \frac{d}{12}(m^3 - m)\delta_{m,-n} \qquad (13.1.3.3)$$

Exercises

1. Evaluate the central charge in the (silly) representation associated with antinormal order. Show that one gets the opposite value $(-d/12)(m^3 - m)\delta_{m,-n}$, even though the quantities L_n are unchanged ($n \neq 0$).
2. Can one remove the central charge in equation (13.1.3.3) by redefining L_m as $L_m \to L_m + \alpha_m$, where α_m are constants?
3. Prove that the central charge in expression (13.1.3.3) is the most general one (up to trivial redefinitions $L_m \to L_m + \alpha_m$).

   Hints.
   a. Show that one can choose α_m so that $k_{0m} = 0$ in $[L_m, L_n] = (m-n)L_{m+n} + k_{mn}$ (with $k_{mn} = -k_{nm}$); α_0 is not determined by that requirement.
   b. Write the constraints on k_{mn} which follow from the Jacobi identity.
   c. Set one index equal to zero in the Jacobi identity and conclude $k_{mn} = 0$ unless $n = -m$ (with the choice $k_{0m} = 0$), i.e., $k_{mn} = k(m)\delta_{m,-n}$ ($m > 0$).
   d. Show that $k(m)$ obeys

   $$(n-m)k(m+n) + (2n+m)k(m) - (2m+n)k(n) = 0$$

   Adjust α_0 so that $k(1) = 0$. Find then $k(n) = \frac{1}{6}(n^3 - n)k(2)$, as desired.

13.1.4. Virasoro Constraints versus the Wheeler–De Witt Equation

We have emphasized above the great similarities between the Virasoro conditions and the constraints of general relativity. They both have a common origin, reparametrization invariance.

In quantum gravity, the "super-Hamiltonian constraint" $\mathcal{H} = 0$ and the "supermomentum constraint" $\mathcal{H}_i = 0$ are imposed as quantum conditions on the physical states,

$$\mathcal{H}|\psi\rangle = 0, \qquad \mathcal{H}_i|\psi\rangle = 0 \qquad (13.1.4.1)$$

These conditions guarantee the gauge invariance of the quantum theory in the physical subspace. In the "Schrödinger representation," where the spatial metric $g_{ij}(x)$ is diagonal, $\psi = \psi[g_{ij}(x)]$, equations (13.1.4.1) are known as the "Wheeler–De Witt" equations [20a].

A necessary condition for these equations to make sense is that the quantum operators $\mathcal{H}$ and $\mathcal{H}_i$ are still "first class," $[\mathcal{H}_\alpha, \mathcal{H}_\beta] \sim C^\gamma_{\alpha\beta} \mathcal{H}_\gamma$, without an "anomalous" term. If such a term was present, one would get from equations (13.1.4.1) further conditions on the physical states that might completely kill the theory.

Therefore an important question in quantum gravity is to find a representation for the field operators and an ordering of the constraints so that they remain "first class." As we saw, the explicit computation of the possible anomalous terms cannot be made—apart from extremely formal considerations—without a definite choice of a Hilbert space in which to represent the basic commutation relations. To our knowledge, this problem has not been solved so far; recent developments are discussed elsewhere [20b].

If one attempts to quantize the string model along the Wheeler-De Witt lines, serious problems are immediately encountered. The equations analogous to conditions (13.1.4.1) are

$$L_n|\psi\rangle = 0 \tag{13.1.4.2}$$

for all n, and are clearly inconsistent owing to the unremovable central charge in the Virasoro algebra. Accordingly, one must weaken equations (13.1.4.2) by demanding that

$$L_n|\psi\rangle = 0, \qquad n > 0 \tag{13.1.4.3}$$

for positive n only [and $(L_0 - \alpha_0)|\psi\rangle = 0$, see below]. This weakened version of the classical constraints is now apparently consistent, since the quantities L_n with $n > 0$ form a true group without central charge: the central charge occurs in equation (13.1.3.3) only when one index is positive and the other negative. Full consistency of the theory for $d \leq 26$ is actually proved in Section 13.4.

Comparison of equations (13.1.4.1) and (13.1.4.3) raises a number of conceptual questions, which we try to answer here.

First question. Since we only impose half of the constraints in equations (13.1.4.3), is it clear that we have enforced gauge invariance in the quantum theory and that we are not retaining too many degrees of freedom? The answer is in general no. It turns out, however, that the critical value $d = 26$ below which solutions to equations (13.1.4.3) possess a nonnegative norm, is also such that there is a remaining quantum gauge invariance, in a sense to be made precise below, in the physical subspace. This gauge invariance eliminates further degrees of freedom, so that the classical and quantum strings both possess the same number of degrees of freedom. For $d < 26$ this is not true, and it is not clear that the theory based on equations

13 • Quantization of Nambu–Goto String

(13.1.4.3), although consistent, is indeed the quantum version of the classical string studied above (important features are lost).

Second question. Is it conceivable that one should somehow weaken the Wheeler–De Witt equation of quantum gravity, as it would be necessary if a (c- or q-number) "central charge" appears in the constraint "algebra"? Yes, it is, but no work along these lines has yet been done.

Third question. Conversely, should one not try to use a different representation of the string operators so as to avoid the central charge?

Again, it might very well be possible to construct such a representation and, if so, it is very likely that the resulting quantum theory would be very different from the one explained here. It could be that this yet-to-be-constructed theory would possess an intrinsic interest of its own (e.g., through the occurrence of infinite-dimensional representations of the Lorentz algebra). Moreover, because that theory would not be based on the use of oscillator variables, it might be more easily extendable to higher-dimensional objects, such as the membrane.

However, to the author's knowledge, this subject has not been investigated. Accordingly, the rest of the book will be devoted to the conventional quantum theory, which finds its roots in the dual models.

Exercise. Consider the closed string, to which the above considerations of course apply (one has two Virasoro algebras, with identical central charges).

a. The Hamilton–Jacobi equation is given by

$$\frac{\delta S}{\delta X^A(\sigma)} \frac{\delta S}{\delta X_A(\sigma)} + X'^A(\sigma) X'_A(\sigma) = 0 \tag{i}$$

$$X'^A(\sigma) \frac{\delta S}{\delta X^A(\sigma)} = 0 \tag{ii}$$

It is the classical analog of the Wheeler–De Witt equation.

From (i), it is tempting to try solutions for which $\delta S/\delta X^A(\sigma)$ is linear in $X'^A(\sigma)$. So one sets

$$\frac{\delta S}{\delta X^A(\sigma)} = k_{AB} X'^B(\sigma) \tag{iii}$$

with k_{AB} a constant tensor obeying

$$k_{AB} k^A{}_D = -\eta_{BD} \tag{iv}$$

Show that (ii) implies that k_{AB} is antisymmetric and that S is given by

$$S = \tfrac{1}{2} \int_0^{2\pi} k_{AB} X^A(\sigma) X'^B(\sigma) \, d\sigma \tag{v}$$

$$k_{AB} = -k_{BA} \qquad \text{(vi)}$$

Show also that (iv) and (vi) possess solutions in any space time of even dimensionality. Is k_{AB} real?

b. Compute the Poincaré generators for solutions $X^A(\sigma)$, $\mathcal{P}_A(\sigma)$ described by S. Discuss the formal classical solution generated, in four dimensions, by

$$S = \int X^0(\sigma) X^{1\prime}(\sigma)\,d\sigma + i \int X^2(\sigma) X^{3\prime}(\sigma)\,d\sigma$$

$X^0(\sigma, \tau = 0) = 0$, $\quad X^1(\sigma, \tau = 0) = a\cos\sigma$, $\quad X^2(\sigma, \tau = 0) = a\sin\sigma$, $\quad X^3(\sigma, \tau = 0) = 0$

c. Consider now the wave functional $\psi[X^A(\sigma)] = \exp iS[X^A(\sigma)]$. Prove that it solves the quantum Wheeler-De Witt equations†

$$-\frac{\delta^2 \psi}{\delta X^A(\sigma)\delta X_A(\sigma)} + X^{\prime A}(\sigma) X'_A(\sigma)\psi = 0 \qquad \text{(vii)}$$

$$X^{\prime A}(\sigma) \frac{\delta\psi}{\delta X^A(\sigma)} = 0 \qquad \text{(viii)}$$

[no matter what regularized value is assigned to $\delta'(0)$]. The existence of solutions to (vii) and (viii) indicates the absence of central charge in the representation in which the field $X^A(\sigma)$ is diagonal, with $\mathcal{P}_A$ to the right of $X^{\prime A}$ in (viii) (if one can make sense of this representation).

d. *Scalar product.* It seems natural to define formally an inner product in the space of $\psi[X^A(\sigma)]$ as

$$\langle \psi, \chi \rangle = \int \mathscr{D} X^A(\sigma) \psi^*[X^A(\sigma)] \chi[X^A(\sigma)] \qquad \text{(ix)}$$

[or $\langle \psi, \chi \rangle = \int \prod dX_n^A \psi^*(X_n^A)\chi(X_n^A)$ in terms of Fourier components]. This scalar product is positive definite and formally Lorentz invariant (negative norms are thus linked with the Lorentz-invariant Fock representation). Physical state solutions to $L_n|\psi\rangle = 0$ are likely to be not normalizable, because one also integrates over pure gauge degrees of freedom in (ix) and, indeed, the above $\psi[X^A(\sigma)]$ contains an oscillating factor and an unbounded one.

For physical states, (ix) should be modified by the insertion of a gauge condition (hopefully in a Lorentz-invariant way). Yet, this should not be enough to make $\psi[X^A(\sigma)] = \exp iS[X^A(\sigma)]$ normalized, as can be seen by considering how $\psi_{k_{AB}}[X^A(\sigma)] = \exp\frac{1}{2}i\int k_{AB} X^A(\sigma) X^{\prime B}(\sigma)\,d\sigma$ transforms under the Poincaré group. Show that

$$U(a, \Lambda)\psi_{k_{AB}}[X^A(\sigma)] = \psi_{k'_{AB}}[X^A(\sigma)]$$

with

$$k'_{AB} = \Lambda^C{}_A \Lambda^D{}_B k_{CD}$$

Hence, the quantities $\psi_{k_{AB}}$ belong to a finite-dimensional representation of the Lorentz group. Can $\psi_{k_{AB}}$ have finite norm (knowing that U is unitary—if the scalar product is Lorentz invariant!—and that the scalar product is positive definite)?

---

† This solution was communicated to the author by L. Mezincescu.

e. Compute P_A and $\langle M_{AB}\rangle$.

f. The above solution cannot be directly extended to the open string. Prove that $S = \frac{1}{2}\int_{-\pi}^{\pi} k_{AB} X^A(\sigma) X'^B(\sigma)\, d\sigma \equiv 0$ for the open string (and hence it does not solve the Hamilton–Jacobi equation).

Show that $S = \frac{1}{2}\int_0^\pi k_{AB} X^A(\sigma) X'^B(\sigma)\, d\sigma$ does not possess well-defined functional derivatives because of nonvanishing boundary terms.

13.1.5. Virasoro Algebra and Kac–Moody Algebras

The Virasoro algebra

$$[L_m, L_n] = (m-n)L_{m+n} + \frac{c}{12}(m^3 - m)\delta_{m,-n} \qquad (13.1.5.1)$$

which, as we have seen, is intimately related to conformal symmetry, has recently found many applications not only in string theory, but also in statistical mechanics [21]. For this reason, the study of its representations has attracted considerable interest [22].

There exist other infinite-dimensional algebras with physical applications. Among these, the Kac–Moody algebras [23] also play a role in string models. They are defined as follows. Consider a finite-dimensional Lie algebra G with structure constants $C^a{}_{bc}$ and (matrix) generators T_a. The generators M_{am} of the associated Kac–Moody affine algebra obey the commutation relations

$$[M_{am}, M_{bn}] = C^c{}_{ab} M_{c\,m+n} + cm\delta_{m,-n} g_{ab} \qquad (13.1.5.2)$$

where g_{ab} is a tensor invariant under the adjoint action; m and n are here positive and negative integers.

Kac–Moody algebras with vanishing central charge ($c = 0$) can easily be constructed as follows. Consider the matrices $M_{am} = T_a\, e^{im\theta}$. One finds that

$$[M_{am}, M_{bn}] = C^c{}_{ab} M_{c\,m+n} \qquad (13.1.5.3)$$

which is called a loop algebra because it is associated with mappings of the circle into the Lie algebra G: as θ ranges from 0 to 2π, M_{am} describes a loop in G.

We have already encountered a Kac–Moody algebra previously: that of the translation currents j_A^α. Indeed, the lightlike components of j_A^α, given by $j_A^\pm = j_A^0 \pm j_A^1$, are equal to†

$$j_A^\pm(\sigma) = -\frac{\sqrt{2\alpha'}}{2}\left(\mathcal{P}_A(\sigma) \pm \frac{1}{2\pi\alpha'} X'_A(\sigma)\right) \qquad (13.1.5.4)$$

† We have introduced appropriate normalization factors in equation (13.1.5.4) in order to simplify subsequent relations.

where we note that ± does not refer here to background space-time light-cone coordinates but rather to lightlike directions along the string.

Straightforward computations indicate that

$$[j_A^+(\sigma), j_B^+(\sigma')] = \frac{i}{2\pi} \eta_{AB} \delta'(\sigma, \sigma') \qquad (13.1.5.5a)$$

$$[j_A^-(\sigma), j_B^-(\sigma')] = -\frac{i}{2\pi} \eta_{AB} \delta'(\sigma, \sigma') \qquad (13.1.5.5b)$$

$$[j_A^+(\sigma), j_B^-(\sigma')] = 0 \qquad (13.1.5.5c)$$

In the case of the open string,† one defines

$$j_A(\sigma) = \begin{cases} j_A^+(\sigma), & 0 \le \sigma \le \pi \\ j_A^-(\sigma), & -\pi \le \sigma \le 0 \end{cases} \qquad (13.1.5.6)$$

and Fourier-expands $j_A(\sigma)$,

$$j_A(\sigma) = \frac{1}{2\pi} \sum_n \alpha_{An} \exp -in\sigma \qquad (13.1.5.7)$$

It follows from equations (13.1.5.5) that

$$[\alpha_{Am}, \alpha_{Bn}] = m \eta_{AB} \delta_{m,-n} \qquad (13.1.5.8)$$

The α quantities, namely, the Fourier components of the currents, close (in the commutator) according to the Kac-Moody algebra based on the translation group, with a nontrivial central charge.

Quantities α_{An} can be related to the oscillator variables by

$$\alpha_{A0} = -\sqrt{2\alpha'} p_A$$
$$\alpha_{An} = i\sqrt{n} a_{An}, \qquad \alpha_{A-n} = (\alpha_{An})^* \qquad (n > 0) \qquad (13.1.5.9)$$

They differ from the α_{An} of Exercise 2, Section 12.4.1, by the mere factor $\sqrt{2\alpha'}$.

There exists close ties between the Virasoro algebra of the energy-momentum components and the Kac-Moody algebra (13.1.5.8) of the

---

† Our general policy in the subsequent pages will be again to treat the open string first and then briefly comment on the peculiarities of the closed one when necessary.

translation currents. This is because one has a "Sugawara type" of theory, with an energy-momentum tensor which is quadratic in the currents [24]. Indeed, one sees from equations (12.3.2.1) and (12.3.2.5) that

$$L_n = \pi \int_{-\pi}^{+\pi} :j_A(\sigma)j^A(\sigma): e^{in\sigma} \, d\sigma \qquad (13.1.5.10)$$

where : : denotes normal ordering. Expansion (13.1.5.7) enables relation (13.1.5.10) to be expressed in the form

$$L_n = \tfrac{1}{2}\sum_m :\alpha_{A\,n-m}\alpha^A_m: \qquad (13.1.5.11)$$

(in agreement with Exercise 2, Section 12.4.1).

Exercises

1. Study the algebra of the Lorentz currents.
2. Study the algebra of the quantities L_n together with the α_m.

13.1.6. Virasoro Algebra in Curved Backgrounds

The exact quantum theory of strings in curved backgrounds has only been constructed in very particular cases and raises many difficult issues. Since this is an important subject, we indicate here how to proceed.

The types of background which have been considered are products of d-dimensional Minkowski space by an SO(N) or SU(N) group space [25],

$$\text{M} \times \text{SO}(N) \quad \text{or} \quad \text{M} \times \text{SU}(N) \qquad (13.1.6.1)$$

Furthermore, the compactified radius of the group manifold is taken to be quantized in units of the string tension,

$$R^2 = \frac{\alpha'}{2}|K| \qquad (13.1.6.2)$$

where K is an integer.

The string action is the quadratic form of the Nambu–Goto action, plus a Wess–Zumino term with a coefficient adjusted so that conformal invariance is restored (details can be found in Refs. 25 and 26; this is how the quantization condition (13.1.6.2) enters the analysis).[†]

[†] The string is treated as a σ model.

Relation (13.1.6.2) enables one to show [26] that the algebra of the SU(N) or SO(N) currents assumes a very simple form: the currents just comprise a Kac–Moody algebra based on SU(N) or SO(N).

Because the energy-momentum tensor components are still bilinear in the currents, one can evaluate the algebra of the normal-ordered L_n [27, 25], which is again the Virasoro algebra

$$[L_n, L_m] = (n - m)L_{n+m} + \frac{c}{12}(n^3 - n)\delta_{n,-m} \qquad (13.1.6.3)$$

with a central charge c given by

$$c = d + \frac{(N^2 - 1)|K|}{N + |K|} \qquad \text{for SU}(N) \qquad (13.1.6.4a)$$

and

$$c = d + \frac{\tfrac{1}{2}N(N - 1)|K|}{|K| + N - 2} \qquad \text{for SO}(N) \qquad (13.1.6.4b)$$

Although interesting, the present models possess the peculiar feature that they cannot be reached continuously from the flat-space models, since the curvature of the internal manifold only takes quantized values.

Exercise. Is c an integer?

Note added. Between the giving of the lectures on which this book is based and its going to press, important developments along different lines have taken place [28]. In the work cited in the relevant papers [28], one quantizes perturbatively the string in a general background. Consistency of the quantum theory imposes conditions on the background. It would be out of place to describe this interesting work here.

13.2. BECCHI–ROUET–STORA–TYUTIN (BRST) QUANTIZATION OF THE STRING

13.2.1. BRST Quantization—A Rapid Survey

The quickest—but less well understood—way of getting the critical dimension 26 is to adopt BRST methods, which possess the further advantage of being purely algebraic. The idea that BRST techniques might be useful in the quantum theory of string models originated from work by Polyakov [7a], and was applied explicitly in Ref. 19. The BRST approach also seems to play an important role in the second-quantized version of string theory [29], so we briefly review it here.

One of the key conceptual questions that one may ask in the quantum theory of gauge fields is whether the theory is actually gauge invariant. It is well known that gauge invariance in the quantum domain is a subtle problem, the meaning of which depends on the approach to quantization adopted.

We consider here that approach in which all field components, including those corresponding to pure gauge degrees of freedom, are treated as dynamical operators on an equal footing. Additional local fields, the "ghosts," are then introduced in order to compensate for the spurious quantum effects induced by the pure gauge operators.

This approach to the quantum theory has the great advantage of preserving one important feature of field theories, namely, their space-time locality. Furthermore, the basic fields all possess c-number as opposed to q-number commutation relations (which are well known to lead to complicated factor-ordering questions).

The dynamics of the pure gauge degrees of freedom is generated by adding to the classical gauge-invariant Lagrangian the "gauge-fixing" and "Fadde'ev-Popov" terms. Different gauge-fixing conditions imply different Lagrangians, and hence different Hamiltonian operators in the quantum linear space in which the gauge field and ghost operators are realized.

To prove the gauge invariance of the theory, one must show that these different Hamiltonians possess identical matrix elements between physical states.

An important observation was made by Fradkin and his collaborators [30], that different gauge choices in the Lagrangian lead formally to BRST invariant, Hermitian Hamiltonians which are related by†

$$H' = H + [K, \Omega] \qquad (13.2.1.1)$$

where K is a gauge-dependent operator and Ω is the (Hermitian) BRST charge.‡ Contrary to K, Ω is gauge independent. Various people rediscovered later that $H' - H$ is a "total Ω-derivative."

Furthermore, Fradkin and his collaborators have also given a general prescription for constructing Ω in *any* gauge theory (with or without closed algebra) and have indicated that the BRST generator is nilpotent if one neglects quantum factor-ordering effects,

$$\Omega^2 = \tfrac{1}{2}[\Omega, \Omega] = 0 \qquad (13.2.1.2)$$

† This is formal because ordering difficulties are neglected; relation (13.2.1.1) has been obtained by manipulation of the path integral.
‡ The quantum bracket [,] stands for the anticommutator for odd objects like K and Ω.

We note that equation (13.2.1.2) implies $[[K, \Omega], \Omega] = 0$, as required by the BRST invariance of both H and H'. The classical roots of the remarkable work [30], as well as a review of its main properties, may be found in Ref. 31, to which we refer the reader for details.

Now, it is clear that not any pair of quantum states can be such that

$$\langle \chi | H' | \psi \rangle = \langle \chi | H | \psi \rangle \tag{13.2.1.3}$$

since $\langle \chi | [K, \Omega] | \psi \rangle$ does not vanish in general. This leads to the definition of physical states as states annihilated by the BRST charge [32],

$$\Omega | \psi \rangle = 0 \tag{13.2.1.4}$$

The physical-state condition (13.2.1.4) implies equation (13.2.1.3) when $|\chi\rangle$ and $|\psi\rangle$ are physical states. The fact that only a subset of the full linear space of states is "physical" was of course expected, since we have included pure gauge and ghost degrees of freedom.

Owing to the nilpotency condition (13.2.1.2), any state of the form $\Omega|\chi\rangle$, for some $|\chi\rangle$, is physical. But such a state decouples from all other physical states by the hermiticity of Ω, so that one must identify $|\psi\rangle$ with $|\psi\rangle + \Omega|\chi\rangle$,

$$|\psi\rangle \sim |\psi\rangle + \Omega|\chi\rangle \tag{13.2.1.5}$$

States like $\Omega|\chi\rangle$ are called null states.

Similarly, BRST invariant operators (like H) map the physical subspace on itself and are called observables. Observables like $[K, \Omega]$ create null states and should be identified with zero,

$$H \sim H + [K, \Omega] \tag{13.2.1.6}$$

Transformations (13.2.1.5) and (13.2.1.6) are called quantum gauge transformations because, as we have already mentioned, different choices of gauge in the path integral yield Hamiltonians related as in equation (13.2.1.6). Furthermore, the equivalence relation (13.2.1.5) reduces the physical subspace (13.2.1.4) in such a way that the classical and quantum theories possess the same number of degrees of freedom. This property, namely adequate cancellation of pure gauge degrees of freedom, is of course another manifestation of the gauge-invariance principle. It must indeed hold if gauge invariance is realized quantum mechanically. (It is noteworthy that the eigenvalues of Ω all vanish and its canonical Jordan form involves one-dimensional and two-dimensional blocks. Conditions (13.2.1.4) and (13.2.1.6) "kill" the two-dimensional blocks.)

It should be clear that the quantum gauge invariance (13.2.1.5) and (13.2.1.6) would not exist if Ω was not nilpotent. For instance, $\Omega|\chi\rangle$ would no longer be a physical null state, and the number of true quantum degrees of freedom would differ from what is expected. Hence, one can say that the nilpotency of Ω is the quantum expression of the gauge-invariance principle. Without $\Omega^2 = 0$, one could not go from one gauge to another, and gauge invariance would not be realized at the quantum level.

We investigate in Section 13.2.4 under which condition $\Omega^2 = 0$ holds in string theory. It has been shown [19] that nilpotency requires $d = 26$.

13.2.2. Classical Expression of the BRST Charge

The recipe for writing the classical Ω from the gauge constraints is given in the literature [30, 31]. To each constraint G_a, one associates a canonically conjugate pair of ghosts η^a, $\mathcal{P}_a$ obeying

$$(\eta^a)^* = \eta^a, \qquad (\mathcal{P}_a)^* = -\mathcal{P}_a \qquad (13.2.2.1)$$

and

$$[\eta^a, \mathcal{P}_b] = -\delta^a_b \qquad (13.2.2.2)$$

The BRST generator reads

$$\Omega = G_a \eta^a - \tfrac{1}{2} \mathcal{P}_c C^c{}_{ab} \eta^b \eta^a \qquad (13.2.2.3)$$

(with $[G_a, G_b] = C^c{}_{ab} G_c$, and $C^c{}_{ab}$ are structure constants), and clearly obeys

$$[\Omega, \Omega] = 0 \qquad (13.2.2.4)$$

In equations (13.2.2.2) and (13.2.2.4), [,] denotes the symmetric Poisson bracket corresponding to classical fermionic variables.

In our case, the constraints are $\mathcal{H}(\sigma)$ and $\mathcal{H}_1(\sigma)$. The boundary conditions for the respective ghost variables are the same as those for the Lagrange multipliers, namely,

$$\eta^{\perp(2k+1)}(\sigma) = 0 = \eta^{1(2k)}(\sigma) \qquad \text{at } \sigma = 0, \pi \qquad (13.2.2.5a)$$
$$\mathcal{P}^{(2k+1)}_{\perp}(\sigma) = 0 = \mathcal{P}^{(2k)}_1(\sigma) \qquad (13.2.2.5b)$$

Here, $F^{(2n)}$ is the derivative of order $2n$ with respect to σ. Hence, only cosine modes will appear in the Fourier transforms of $\eta^\perp$, $\mathcal{P}_\perp$, and only sine modes will appear in those of η^1, $\mathcal{P}_1$. With conditions (13.2.2.5), Ω is a well-defined canonical generator since there is no unwanted boundary term in $\delta\Omega$.

In terms of the Fourier components L_m of the constraints and those η_m, $\mathcal{P}_m$ of the ghosts, one finds from equation (13.2.2.3) that

$$\Omega = \sum_m L_m \eta_{-m} + \sum_{n,m} m \mathcal{P}_n \eta_m \eta_{-n-m} \quad (13.2.2.6)$$

with

$$[\eta_m, \mathcal{P}_n] = -i\delta_{m,-n} \quad (13.2.2.7a)$$

$$\eta_m^* = \eta_{-m}, \quad \mathcal{P}_n^* = \mathcal{P}_{-n} \quad (13.2.2.7b)$$

where η_m is the mth Fourier component (in the $\exp im\sigma$ basis) of $\eta^\perp + \eta^1$, and so on. We have inserted an appropriate factor of i in $\mathcal{P}_n$.

One easily checks that Ω is real and classically nilpotent:

$$\Omega^* = \Omega \quad (13.2.2.8a)$$

and

$$[\Omega, \Omega] = 0 \quad (13.2.2.8b)$$

Exercises

1. Check relations (13.2.2.8a) and (13.2.2.8b) directly from equation (13.2.2.6).

2. Construct the BRST invariant extensions of the Poincaré generators [31]. Do the ghosts carry a Lorentz index?

13.2.3. Ghost Fock Space

We now go to the quantum theory. The ghosts η_m, $\mathcal{P}_m$ become operators and, from relations (13.2.2.7), obey

$$[\eta_m, \mathcal{P}_n] = \delta_{m,-n} \quad (13.2.3.1)$$

(where [,] now denotes the anticommutator, according to the rule: Poisson brackets of Fermi variables → anticommutator), and

$$\eta_m^* = \eta_{-m}, \quad \mathcal{P}_n^* = \mathcal{P}_{-n} \quad (13.2.3.2)$$

One sees from equations (13.2.3.1) and (13.2.3.2) that the zero modes η_0, $\mathcal{P}_0$ are different from the others, in that they are real and satisfy

$$[\eta_0, \mathcal{P}_0] = 1 \quad (13.2.3.3)$$

13 • Quantization of Nambu–Goto String

The irreducible representation space for equation (13.2.3.3) is well known [33]. It is two-dimensional and is isomorphic to the space of functions of one Grassmann variable η^0,†

$$f = a + b\eta^0 \qquad (13.2.3.4a)$$

$$\eta^0\text{: multiplication by } \eta^0 \qquad (13.2.3.4b)$$

$$\mathcal{P}_0 = \partial/\partial\eta^0 \qquad (13.2.3.4c)$$

$$(f, g) = \int f^* g \, d\eta^0 \qquad (13.2.3.4d)$$

The states $f_1 = 1$ and $f_2 = \eta^0$ have zero norm and scalar product $+1$. Alternatively, one can consider the states $(1 \pm \eta^0)/\sqrt{2}$, which are orthogonal to each other and have norm ± 1.

The ghost state space is the direct product of the space (13.2.3.4) with the Fock space of the higher modes η_m, $\mathcal{P}_m$, built as follows. One defines the operators

$$f_m = \frac{1}{\sqrt{2}}(\eta_m + \mathcal{P}_m), \qquad f_m^* = \frac{1}{\sqrt{2}}(\eta_m^* + \mathcal{P}_m^*)$$

$$g_m = \frac{1}{\sqrt{2}}(\eta_m - \mathcal{P}_m), \qquad g_m^* = \frac{1}{\sqrt{2}}(\eta_m^* - \mathcal{P}_m^*) \qquad (13.2.3.5)$$

$(m > 0)$, which obey

$$[f_m, f_m^*] = -[g_m, g_m^*] = 1 \qquad (13.2.3.6)$$

(other anticommutators vanish).

The ghost vacuum is annihilated by the "destruction operators" f_m, g_m,

$$f_m|0\rangle = g_m|0\rangle = 0 \qquad (13.2.3.7)$$

and is assumed to possess unit norm, $\langle 0|0\rangle = 1$. The creation operators f_m are of the usual type, while the g_m create negative-norm states (the existence of such states is expected in a ghost space!).

---

† In order to have $0 = \eta^0\eta_m + \eta_m\eta^0 = \eta^0\mathcal{P}_m + \mathcal{P}_m\eta^0 = \cdots = \mathcal{P}^0\mathcal{P}_m^* + \mathcal{P}_m^*\mathcal{P}^0$, the operators (13.2.3.4b) and (13.2.3.4c) should actually be multiplied by $(-)^{N_F}$, where N_F is the number of ghost fermions, but this factor will be omitted here.

The subsequent computations will be conducted so as to be meaningful, at each stage, in the above Fock space. This will be done by relying on the normal-ordered form of operators. We note that this amounts to writing all η_m^*, $\mathcal{P}_m^*$ to the left of η_m, $\mathcal{P}_m$, since η_m^* and $\mathcal{P}_m^*$ only involve creation operators while η_m and $\mathcal{P}_m$ are expressible solely in terms of destruction operators. As to η_0 and $\mathcal{P}_0$, the antisymmetric ordering will be adopted.

Exercise. What is the space of the zero modes for the closed string?

13.2.4. Nilpotency of the Quantum BRST Operator

The quantum version of the BRST generator (13.2.2.6) has the form

$$\Omega = \sum_{n=-\infty}^{+\infty} (L_n - \alpha_0 \delta_{n,0}) \eta_n^* - 2 \sum_{n>0} n \mathcal{P}_0 \eta_n^* \eta_n + \sum_{m>0} m(\mathcal{P}_m^* \eta_m + \eta_m^* \mathcal{P}_m) \eta_0$$

$$- \sum_{n>0} \sum_{m=1}^{n-1} (n \mathcal{P}_m^* \eta_{n-m}^* \eta_n + n \eta_n^* \eta_{n-m} \mathcal{P}_m + m \mathcal{P}_n^* \eta_{n-m} \eta_m$$

$$+ m \eta_m^* \eta_{n-m}^* \mathcal{P}_n)$$

$$+ \sum_{n>0} \sum_{m>0} m(\eta_{n+m}^* \mathcal{P}_n \eta_m - \mathcal{P}_n^* \eta_m^* \eta_{n+m})$$

$$= \sum_{n=-\infty}^{+\infty} (L_n - \alpha_0 \delta_{n,0}) \eta_n^* - 2 \sum_{n>0} n \mathcal{P}_0 \eta_n^* \eta_n + \sum_{m>0} m(\mathcal{P}_m^* \eta_m + \eta_m^* \mathcal{P}_m) \eta_0$$

$$- \sum_{n>0} \sum_{m=1}^{n-1} [(2n-m)(\mathcal{P}_m^* \eta_{n-m}^* \eta_n + \eta_n^* \eta_{n-m} \mathcal{P}_m) + m(\mathcal{P}_n^* \eta_{n-m} \eta_m$$

$$+ \eta_m^* \eta_{n-m}^* \mathcal{P}_n)] \quad (13.2.4.1)$$

The only ordering ambiguities appear in L_0 (hence the presence of α_0 in the first term of Ω) and in the η_0 terms. However, this latter ambiguity can be absorbed in α_0 and yields nothing new.

The first condition on Ω, namely its hermiticity, is obvious with equation (13.2.4.1). It remains to check its nilpotency.

The BRST generator fails to be nilpotent for two reasons. First, the L_n no longer form a true algebra, due to the central charge. Second, the ghosts also contribute anomalous terms. Indeed, the anticommutator $[\mathcal{P}_a^* \eta_b^* \eta_c, \eta_d^* \eta_e \mathcal{P}_f]$ is not in normal form.† When bringing it to that form,

---

† It is easy to see that the only anomalous ghost terms come from this anticommutator (where η, $\mathcal{P}$ also denote the zero modes). If zero modes occur in equation (13.2.4.2), then N.O. refers also to antisymmetric ordering (and there is a factor of $\frac{1}{2}$ in the anomalous term).

one generates classically absent contributions,

$$[\mathcal{P}_a^* \eta_b^* \eta_c, \eta_d^* \eta_e \mathcal{P}_f] = -\delta_{ae}\delta_{fb}\eta_d^* \eta_c + \text{N.O.} \qquad (13.2.4.2)$$

Equation (13.2.4.2) enables one, after some algebra, to find that $\Omega^2 = 0$ is given explicitly by

$$\Omega^2 = \frac{1}{2}\sum_{n>0}\left[n^3\left(\frac{d}{6}-\frac{13}{3}\right) + n\left(4\alpha_0 - \frac{d}{6}+\frac{1}{3}\right)\right]\eta_n^* \eta_n \qquad (13.2.4.3)$$

where the ghost anomalous contribution is dimension-independent and can be read from expression (13.2.4.3) by setting $d = \alpha_0 = 0$. The central charge of the Virasoro algebra generates the term proportional to d and there is also, of course, an α_0 contribution.

Demanding nilpotency of Ω yields

$$d = 26 \quad \text{and} \quad \alpha_0 = 1 \qquad (13.2.4.4)$$

Gauge invariance is realized quantum mechanically only in the critical dimension $d = 26$. For $d \neq 26$, this essential feature is lost. (The meaning of the second condition $\alpha_0 = 1$ will be discussed later. It unfortunately implies the presence of a tachyon in the spectrum.)

Exercises

1. Derive expression (13.2.4.3).

2. What would Ω^2 be if one had taken an "anti-Fock" representation for the ghosts, based on a vacuum annihilated by f_m^* and g_m^*? Is this alternative representation *a priori* as natural as that used above? Discuss.

3. Insert back in the theory the non-Weyl-invariant degrees of freedom g_{11}, their momenta, as well as the momenta conjugate to N and N^1. Write the corresponding BRST generator. Argue that the new piece in Ω does not contribute to the anomaly.

4. Derive the critical dimension for the closed string ($d = 26!$). What is α_0? ($\alpha_0 = 1$ for each L_0; alternatively $\alpha_0[L_0 + \bar{L}_0] = 2$, $\alpha_0[L_0 - \bar{L}_0] = 0$.)

13.2.5. Critical Dimension in Curved Backgrounds

BRST techniques can also be applied to the curved-background models of Section 13.2.6. Actually, the only thing needed in the computation of the critical dimension by BRST methods is the algebra of the generators L_n; their explicit form is not necessary.

The BRST operator is still given by expression (13.2.4.1), with L_n the appropriate Virasoro generators of the curved-background models.

The ghost anomalous contribution therefore remains the same. The only change in the final answer (13.2.4.3) is the replacement of d by the new value c of the central charge. The nilpotency of Ω implies accordingly

$$d + \frac{(N^2-1)|K|}{N+|K|} = 26 \qquad \text{for SU}(N) \qquad (13.2.5.1\text{a})$$

$$d + \frac{\tfrac{1}{2}N(N-1)|K|}{|K|+N-2} = 26 \qquad \text{for SO}(N) \qquad (13.2.5.1\text{b})$$

We note that the critical Minkowski space-time dimension d calculated from system (13.2.5.1) is not a positive integer for all N and K. Therefore relations (13.2.5.1) cannot be satisfied for those values of N and K. This makes the previous flat-space result (13.2.4.4) even more remarkable, since one can remove anomalous terms by adjusting a parameter which is forced to take positive-integer values.

13.2.6. Physical Subspace

The physical subspace is defined by

$$\Omega|\psi\rangle = 0 \qquad (13.2.6.1)$$

We wish to characterize the solutions of this equation. To that end, following Kato and Ogawa [19], we introduce two new Hermitian operators which play an important role, because they commute with Ω. The first is given by

$$L = \sum_{n\geq 0} n a_n^{*A} a_{An} + \sum_{n\geq 0} n(\mathcal{P}_n^*\eta_n + \eta_n^*\mathcal{P}_n) \qquad (13.2.6.2\text{a})$$

$$= \sum_{n\geq 0} n(a_n^{*A} a_{An} + f_n^* f_n - g_n^* g_n) \qquad (13.2.6.2\text{b})$$

and the second by

$$M = 2\sum_{n\geq 0} n\eta_n^*\eta_n \qquad (13.2.6.3\text{a})$$

$$= \sum_{n\geq 0} n(f_n^* f_n + g_n^* g_n + f_n^* g_n + g_n^* f_n) \qquad (13.2.6.3\text{b})$$

One easily checks that both L and M are BRST invariant,

$$[L, \Omega] = [M, \Omega] = 0 \qquad (13.2.6.4)$$

13 • Quantization of Nambu–Goto String

(e.g., $M \sim [\eta^0, \Omega])$. Furthermore

$$[L, M] = 0 \qquad (13.2.6.5)$$

Of particular interest is the operator L, which reduces in the bosonic sector to the "level operator" N given by

$$N = \sum_{n \geq 0} n a_n^{*A} a_{An} \qquad (13.2.6.6)$$

to be encountered again later on. The operator L is actually the BRST invariant extension of N and can be written as

$$L = N + N^f + N^g \qquad (13.2.6.7)$$

where N^f and N^g are the level operators for the f and g ghost modes, respectively. (There is a minus sign in $N^g = -\sum_{n \geq 0} n g_n^* g_n$ owing to the unusual anticommutation relations $[g_n, g_n^*] = -1$.) Operator L can be diagonalized and possesses positive eigenvalues.

The momentum operators p^A also commute with Ω, L, and M:

$$[p^A, \Omega] = [p^A, L] = [p^A, M] = 0 \qquad (13.2.6.8)$$

These properties enable us to search for solutions of equation (13.2.6.1) that simultaneously diagonalize L and p^A.† Their eigenvalues will be denoted by the same letters.

Theorem 1. The solutions of equation (13.2.6.1) with $L \neq -\alpha' p^2 + \alpha_0$ are pure gauge, i.e., they can be written in the form

$$|\psi\rangle = \Omega |\chi\rangle \qquad (13.2.6.9)$$

The proof of this theorem consists in noting that Ω can be decomposed as

$$\Omega = \bar{\Omega} + (\alpha' p^2 + L - \alpha_0) \eta^0 - M \mathcal{P}_0 \qquad (13.2.6.10)$$

where $\bar{\Omega}$ is an operator which does not involve the ghost zero modes η^0, $\mathcal{P}_0$ and whose explicit form can easily be read from expression (13.2.4.1). We note that $\bar{\Omega}$, η^0, and $\mathcal{P}_0$ also commute with L and p^A.

It is assumed that $L \neq -\alpha' p^2 + \alpha_0$. Let us expand $|\psi\rangle$ as $|\psi\rangle = |a\rangle + |b\rangle \eta^0$, where $|a\rangle$ and $|b\rangle$ are states which do not involve the ghost zero

† The Jordan form of M is not trivial and we do not assume M diagonal.

modes. Because $\alpha' p^2 + L - \alpha_0 \neq 0$, one can add to $|\psi\rangle$ a state $\Omega|\chi\rangle$ such that $|b\rangle = 0$. We take, e.g., $|\chi\rangle = -(\alpha' p^2 + L - \alpha_0)^{-1}|b\rangle$. With $|b\rangle = 0$, the BRST condition (13.2.6.1) on $|\psi\rangle$ implies that

$$\bar{\Omega}|a\rangle = 0 \quad \text{and} \quad (\alpha' p^2 + L - \alpha_0)|a\rangle = 0$$

from which it follows that $|a\rangle = 0$, as asserted.

This theorem enables one to consider only the case when

$$\alpha' p^2 + L - \alpha_0 = 0 \qquad (13.2.6.11)$$

to which we now turn.†

When relation (13.2.6.11) holds, $\bar{\Omega}$ is nilpotent and the BRST equation (13.2.6.1) with

$$|\psi\rangle = |a\rangle + |b\rangle \eta^0 \qquad (13.2.6.12)$$

becomes

$$\bar{\Omega}|b\rangle = 0, \qquad \bar{\Omega}|a\rangle - M|b\rangle = 0 \qquad (13.2.6.13)$$

From now on only states independent of the zero mode, like $|a\rangle$ and $|b\rangle$, will be considered, unless otherwise specified.

Theorem 2. a. The general solution to $\bar{\Omega}|b\rangle = 0$ can be written as

$$|b\rangle = |P\rangle|0\rangle_{\text{ghost}} + \bar{\Omega}|c\rangle \qquad (13.2.6.14)$$

where $|0\rangle_{\text{ghost}}$ denotes the ghost vacuum (for the nonzero modes),

$$\eta_n|0\rangle_{\text{ghost}} = \mathcal{P}_n|0\rangle_{\text{ghost}} = 0 \qquad (13.2.6.15)$$

and where $|P\rangle$ is a purely bosonic state (living in the Hilbert space of the a oscillators).

b. Furthermore, $|P\rangle$ obeys

$$L_n|P\rangle = 0, \qquad n > 0 \qquad (13.2.6.16)$$

and

$$(L_0 - \alpha_0)|P\rangle = 0 \qquad (13.2.6.17)$$

---

† By an appropriate Lorentz rotation, one can assume $p^i = 0$ (only p^+ and $p^- \neq 0$), but this will not be explicitly needed here.

We also take $p^A \neq 0$ throughout. The "infrared" case $p^A \neq 0$ is exceptional and easy to treat. For this reason, it will not be considered here.

Proof. See the paper of Kato and Ogawa [19], who actually proved a stronger version of relation (13.2.6.14), stating that $|P\rangle$ can in fact be taken to be purely transverse; see Section 13.4. The idea of their proof is sketched in Appendix A.

b. Once (a) is established the condition (13.2.6.17) is obvious, because one can express $|b\rangle$ in the form (13.2.6.14) by taking a vector $|c\rangle$ which lives in the $\alpha' p^2 + L - \alpha_0 = 0$ subspace ($\alpha' p^2 + L - \alpha_0$ commutes with $\bar{\Omega}$). Hence $|P\rangle|0\rangle_{\text{ghost}}$ is also annihilated by $\alpha' p^2 + L - \alpha_0$, which easily leads to equation (13.2.6.17) since $N_f = N_g = 0$ for the ghost vacuum.

As to conditions (13.2.6.16), they simply follow from the BRST condition $\bar{\Omega}|P\rangle|0\rangle_{\text{ghost}} = 0$. This one reduces to

$$\sum_{n>0} L_n \eta_n^* |P\rangle|0\rangle_{\text{ghost}} = 0$$

and hence is equivalent to relation (13.2.6.16) because the states $\eta_n^*|0\rangle_{\text{ghost}}$ are all independent.

Theorem 3. The general solution to $\Omega|\psi\rangle = 0$ can be written as

$$|\psi\rangle = |P_1\rangle|0\rangle_{\text{ghost}} + |P_2\rangle|0\rangle_{\text{ghost}} \eta^0 + \Omega|\chi\rangle \qquad (13.2.6.18)$$

where $|P_1\rangle$ and $|P_2\rangle$ are purely bosonic states obeying

$$L_n|P_1\rangle = (L_0 - \alpha_0)|P_1\rangle = 0 \qquad (n > 0) \qquad (13.2.6.19a)$$

$$L_n|P_2\rangle = (L_0 - \alpha_0)|P_2\rangle = 0 \qquad (n > 0) \qquad (13.2.6.19b)$$

Proof. By Theorem 2, it follows that one can bring the coefficient of η^0 in $|\psi\rangle$ to the required form $[(\bar{\Omega}|c\rangle)\eta^0 = \Omega|\chi\rangle +$ terms which modify $|a\rangle$ only, with $|\chi\rangle = |c\rangle\eta^0$]. But then $M|b\rangle = 0$, and relations (13.2.6.13) imply $\bar{\Omega}|a\rangle = 0$. A second application of Theorem 2 leads to

$$|a\rangle = |P_1\rangle|0\rangle_{\text{ghost}} + \bar{\Omega}|c'\rangle$$

$$= |P_1\rangle|0\rangle_{\text{ghost}} + \Omega|c'\rangle$$

($\bar{\Omega}|c'\rangle = \Omega|c'\rangle$ for $|c'\rangle$ has no component along η^0 and is annihilated by $\alpha' p^2 + L - \alpha_0$.) This demonstrates equation (13.2.6.18).

Upon acting with Ω on equation (13.2.6.18) and using $\alpha' p^2 + L - \alpha_0 = 0$, we then infer the conditions (13.2.6.19).

Theorem 3 is interesting, because it completely relates the BRST physical states to the physical states (13.2.6.19) of the covariant approach, considered in Section 13.4. One sees that owing to the zero ghost mode degeneracy, there are twice as many BRST physical states as there are physical states in the covariant approach.† A similar phenomenon (doubling of states in the BRST approach) was found recently in a different context [34].

As a final point, we note that the quantum "gauge condition" expressing that $|\psi\rangle$ should be annihilated by the ghost oscillators, does not completely fix the BRST gauge. One can still add to $|\psi\rangle$ states of the form $\Omega|\chi\rangle$ such that $\Omega|\chi\rangle$ does not involve the ghosts. These states are the famous null states encountered in the covariant approach. Their decoupling can hence be understood in terms of the BRST symmetry (see the exercises for an example of such a null state).

Exercises

1. **a.** Consider the state $|R\rangle\mathcal{P}_1^*|0\rangle_{\text{ghost}} \equiv |\chi\rangle$, with $L_n|R\rangle = 0$, $n \geq 0$. Compute $\Omega|\chi\rangle$. Infer that the state $L_{-1}|R\rangle|0\rangle_{\text{ghost}}$ is a null state.

    b. Prove that the state $L_{-1}|R\rangle|0\rangle_{\text{ghost}}\eta^0$ is annihilated by the BRST operator and has vanishing scalar products with all other physical states. This means that it is equal to $\Omega|\chi\rangle$, for some $|\chi\rangle$. *Hint*: This state can be written as in equation (13.2.6.18), with $|P_1\rangle$ and $|P_2\rangle$ purely *transverse* states which must be zero by the scalar-product properties of the state $L_{-1}|R\rangle|0\rangle_{\text{ghost}}\eta^0$.

    c. Show that the state $|R\rangle\eta_1^*$ is equal to $\Omega|\chi\rangle$, for some $|\chi\rangle$.

    d. Show that the state $|R^0\rangle\eta_1^*|0\rangle_{\text{ghost}}$ is not a solution of $\Omega|\psi\rangle = 0$ when $|R^0\rangle|0\rangle_{\text{ghost}}$ is a physical state (there is a problem with the η^0 term). (This is why $|R\rangle$ above must obey $L_0|R\rangle = 0$, not $L_0|R\rangle = |R\rangle$.)

2. What does equation (13.2.6.18) become for the closed string? Show that there is a "double doubling."

13.2.7. Remarks on the Doubling

One of the main motivations for the use of BRST methods is to have a (pseudo-)Hilbert space in which all variables, including the pure gauge degrees of freedom and the ghosts, are realized as Hermitian operators on an equal footing. One of the conditions for the consistency of the formalism is that the solutions of $\Omega|\psi\rangle = 0$ belong to that Hilbert space.

Now, as we have indicated, the appropriate Hilbert space is given by the direct product of the (bosonic and fermionic) oscillator Fock space by the Hilbert space for the zero modes X_0^A, p_A, η^0, $\mathcal{P}_0$. The scalar product

---

† Solution (13.2.6.18) is the best one can do when $|P_1\rangle$ and $|P_2\rangle$ are transverse, i.e., one cannot gauge away the $|P_1\rangle$ or $|P_2\rangle$ part, because in general these states are not orthogonal to all other physical states and hence are not what one calls "null" states. It will be argued in Section 13.2.7 that the doubling (13.2.6.18) is actually spurious in the open-string case and appears because of an inappropriate definition of the linear space of states $|\psi\rangle$.

is equal to

$$\langle\psi|\psi\rangle = \int d^{26}p \, d\eta^0 \, (\psi(p, \eta^0)|\psi(p, \eta^0)) \quad (13.2.7.1)$$

where $(\psi(p, \eta^0)|\psi(p, \eta^0))$ is the Fock-space inner product. [In the (p, η^0) representation, a state $|\psi\rangle$ is described by a function $\psi(p, \eta^0)$ with value in the Fock space.]

The states (13.2.6.18), having definite momentum, do not make the integral $\int d^{26}p$ finite. The way to cure this problem is well known: one simply considers appropriate wave packets with different p^A. However, owing to the zero-mode constraint $(L_0 - 1)|P\rangle = 0$ (on-the-mass-shell condition), one can only superpose plane waves with definite mass. Thus, one picks up an infinite factor of $\delta(0)$ in the scalar product of states belonging to the same mass level through the integration over the mass, which appears in equation (13.2.7.1).

Therefore, something must be done in order to obtain a satisfactory theory: equation (13.2.7.1) must be given a meaning for state solutions to $\Omega|\psi\rangle = 0$. Also, we must handle the doubling of BRST states found in the previous section. This question is linked to the regularization of equation (13.2.7.1).

There exists various ways of giving a meaning to equation (13.2.7.1) which, although superficially different, seem to be equivalent. None of these possesses complete rigor.

13.2.7a. Regularization of the Momentum Integral Only

One replaces the integral $\int d^{26}p$ by the appropriate Klein-Gordon scalar product, which differs from it by the factorization of an infinite $\delta(0)$ [33]. This scalar product is only defined for on-the-mass-shell states. Furthermore, Ω remains Hermitian for it does not involve the center-of-mass coordinates X_0^A.

Wave packets with definite mass are then acceptable (they now all have finite norm), and one must face the doubling

$$|\psi\rangle = |P_1\rangle|0\rangle_{\text{ghost}} + |P_2\rangle|0\rangle_{\text{ghost}} \eta^0 + \Omega|\chi\rangle \quad (13.2.7.2)$$

The attitude commonly held is that one should impose a "selection rule," which restricts the BRST states (13.2.7.2) to only one "sector" isomorphic to the space of states of the covariant approach ("truncation"). The imposition of such a selection rule is consistent with switching-on of interactions provided there is no transition from one sector to another, as will be the case if the interactions can be completely described within the covariant, ghost-free formalism. If this condition holds, the choice of selection rule is to a large extent a matter of convenience.

There exist at least two different ways to truncate the theory. These appear to be equivalent (provided the interactions indeed vanish in the ghost zero-mode sector).

1. As mentioned previously, the space of the ghost zero mode possesses negative-norm states. It is accordingly natural to retain a sector with positive-norm vectors only, such as that generated by $1 + \eta^0$ [34]. These vectors do not diagonalize the ghost-number operator (in what concerns the zero mode only). With this selection rule, the allowed physical states are

$$|\psi\rangle = \frac{1}{\sqrt{2}} |P\rangle |0\rangle_{\text{ghost}} (1 + \eta^0) + \Omega |\chi\rangle \qquad (13.2.7.3)$$

Their norm is given by

$$\langle \psi | \psi \rangle = (P|P)_{\text{Klein-Gordon-Fock}} \qquad (13.2.7.4)$$

2. In order to diagonalize the ghost-number operator, one imposes $|P_2\rangle = 0$. One also changes the scalar product in the space of the ghost zero mode so as to avoid zero-norm problems [19]. This method has been adopted in recent papers [35]. We simply point out that by changing the scalar product, either η^0, or $\mathscr{P}_0$, or both, are made non-Hermitian operators in the complete space of states. This makes Ω also non-Hermitian and no longer implies the decoupling of all states like $\Omega |\chi\rangle$.

13.2.7b. "BRST Supersymmetric" Regularization

It has been argued [31] that one should give a meaning to scalar products such as (13.2.7.1) by treating the pure gauge degrees of freedom—responsible for the occurrence of $\delta(0)$—and the ghosts on the same footing. Gauge degrees of freedom and ghosts are indeed "BRST supersymmetric partners."

One then inserts an appropriate regularizing function $f_\varepsilon(p, \eta^0)$ into equation (13.2.7.1) such that (1) the integral (13.2.7.1) becomes finite, and (2) $f_\varepsilon(p, \eta^0) \to 1$ as $\varepsilon \to 0$. With this factor, the norm of the state (13.2.7.2), where $|P_1\rangle$ and $|P_2\rangle$ are normalized wave packets with definite mass, is given in the limit $\varepsilon \to 0$ by

$$\langle \psi | \psi \rangle = \delta(0)[(P_1|P_2) + (P_2|P_1)] + (P_1|P_1) \qquad (13.2.7.5)$$

where $(P_i|P_j)$ is the Klein–Gordon–Fock-space inner product. The finite term $(P_1|P_1)$ remains because (loosely speaking) it is multiplied in expression (13.2.7.1) by $\delta(0)$ (integral over the mass) times 0 (integral over η^0). This ill-defined product $\delta(0) \times 0$ of bosonic and fermionic delta functions is made equal to one by the above regularization prescription.

In order for equation (13.2.7.5) to be finite and nontrivial, one must set $|P_2\rangle = 0$. This eliminates the doubling. Technically, this regularization is close to the previous one. The vanishing scalar product that we found was only one "leg" of the product $\delta(0) \times 0$. The idea here is that the integral $\int d\eta^0$ cannot be separated from the integral over the mass.

13.2.7c. Is Doubling Spurious?

It appears from this discussion that one should not attach too much importance to the doubling phenomenon. Of course, it would be nice to have a formulation, rigorous from the beginning, in which this phenomenon never appears. To the author's knowledge, such a formulation is still missing.

In order to reinforce the opinion that the doubling is spurious in the case of the open string, we note that $|P_2\rangle$ can be eliminated by adding to $|\psi\rangle$ a term of the form $\Omega|\chi\rangle$. Indeed, in the X_0^A representation, this amounts to solving

$$(\Box - m^2)\chi(X_0^A) = P_2(X_0^A)$$

(we have omitted the oscillator dependence). Here, $P_2(X_0^A)$ and $\chi(X_0^A)$ are the wave functions of the states $|P\rangle$, $|\chi\rangle$. Since $P_2(X_0^A)$ is a solution of the free wave equation $(\Box - m^2)P_2(X_0^A) = 0$, the function $\chi(X_0^A)$ blows up at timelike infinity, as expected (resonance).

For this reason, the associated state $|\chi\rangle$ is not normalizable, even after the scalar product is regularized. (The state $\Omega|\chi\rangle$ is accordingly not really a null state which decouples. As we saw, $|P_2\rangle\eta^0$ does have nonvanishing scalar products with other physical states.) Strictly speaking, one cannot consider the state $|\chi\rangle$. Yet, it is suggestive that there is no differentiable or topological obstruction in writing $|P_2\rangle\eta^0$ as $\Omega|\chi\rangle$ (only scalar-product arguments—which do not rest on a sound basis anyway—prevent one from doing it).

The present considerations suggest that one should view the zero-mode operators X_0^A, p_A, η^0, $\mathcal{P}_0$ as linear operators acting on some "bigger" linear space without scalar product. This space should include functions of X_0^A which blow up as $X_0^0 \to \infty$. In this space, the general solution of $\Omega|\psi\rangle = 0$ is $|\psi\rangle = |P\rangle|0\rangle_{\text{ghost}} + \Omega|\chi\rangle$, without a $|P_2\rangle$ term. A scalar product is then defined only for BRST invariant states, as $\langle\psi|\psi\rangle = (P|P)_{\text{Klein-Gordon-Fock}}$.

Exercise

a. Study the BRST quantization of the free, relativistic particle. Show that the above problems already appear in this simpler case.

b. Show that one can only remove half of the "double doubling" in the closed-string case by the above argument.

13.2.8. Miscellanea

13.2.8a. No-Ghost Theorem in the BRST Approach

The theorem of Kato and Ogawa indicates that in the decomposition (13.2.6.18), the bosonic states $|P_1\rangle$ and $|P_2\rangle$ can in fact be assumed to be "transverse": their "longitudinal" part (if any) can be removed by adding to $|\psi\rangle$ an appropriate state of the form $\Omega|\chi\rangle$.

Transverse states are, by definition, states created from the vacuum by the DDF operators. These operators, to be defined precisely below, generalize in an appropriate Virasoro invariant way the transverse oscillators a_n^i, a_n^{i*} to which they reduce in the light-cone gauge. They obey the same commutation relations. As a result, they only create positive-norm states.

This means that the norms of $|P_1\rangle$ and $|P_2\rangle$ are positive. Hence, the norm of $|\psi\rangle$ may be negative only because of the ghost zero mode. When this mode is handled as in the previous section, $\langle\psi|\psi\rangle$ is positive: all negative-norm states have been removed from the physical subspace ("no-ghost theorem"). A probabilistic interpretation can be given to the quantum theory.

This BRST-based proof of the no-ghost theorem is sketched in Appendix A. It is shown to be a consequence of Kugo and Ojima's quartet mechanism: the fermionic ghosts cancel the lightlike oscillators; only the transverse oscillators remain.

The original proof of the no-ghost theorem is reproduced in Section 13.4.

13.2.8b. Ghost-Number Operator

The ghost-number operator Q_c is defined so that

$$[A, Q_c] = \text{gh}(A)A \qquad (13.2.8.1)$$

Here, $\text{gh}(A)$ is the ghost number of the operator A such that

$$\text{gh}(X^A) = \text{gh}(\mathcal{P}_A) = 0 \qquad (13.2.8.2a)$$

$$\text{gh}(\eta^\perp) = \text{gh}(\eta^1) = 1 \qquad (13.2.8.2b)$$

$$gh(\mathcal{P}_\perp) = gh(\mathcal{P}_1) = -1 \qquad (13.2.8.2c)$$

$$gh(AB) = gh(A) + gh(B) \qquad (13.2.8.2d)$$

$$gh(A+B) = gh(A) \quad \text{if } gh(A) = gh(B) \qquad (13.2.8.2e)$$

Only the ghosts have a nonzero ghost number. With these conventions, $gh(\Omega) = +1$,

$$[\Omega, Q_c] = \Omega \qquad (13.2.8.3)$$

The commutation relations (13.2.8.1) determine the operator Q_c up to an arbitrary constant,

$$Q_c = -i \int (\mathcal{P}_\perp(\sigma)\eta^\perp(\sigma) + \mathcal{P}_1(\sigma)\eta^1(\sigma)) \, d\sigma + \text{const} \qquad (13.2.8.4)$$

which can be chosen so that Q_c is anti-Hermitian,

$$Q_c = \frac{-i}{2} \int [\mathcal{P}_\perp(\sigma)\eta^\perp(\sigma) - \eta^\perp(\sigma)\mathcal{P}_\perp(\sigma) + \mathcal{P}_1(\sigma)\eta^1(\sigma) - \eta^1(\sigma)\mathcal{P}_1(\sigma)] \, d\sigma$$

$$(13.2.8.5)$$

(In Ref. 31, the constant of operator (13.2.8.4) was set equal to zero. This choice is more adapted to a representation in which the qs are diagonal, while the form (13.2.8.5) is more suited to a Fock-space representation.)

The interest in the ghost-number operator is that physical observables are zero-ghost-number operators, i.e., they commute with Q_c,

$$[A, Q_c] = 0 \quad (\text{and } [A, \Omega] = 0) \qquad (13.2.8.6)$$

In terms of oscillator variables, expression (13.2.8.5) becomes

$$Q_c = -\tfrac{1}{2}(\eta^0 \mathcal{P}_0 - \mathcal{P}_0 \eta^0) - \sum_{n>0} (\eta_n^* \mathcal{P}_n - \mathcal{P}_n^* \eta_n) \qquad (13.2.8.7)$$

Although anti-Hermitian, the ghost-number operator possesses real eigenvalues. This means that all its eigenstates, with the possible exception

† It has been shown [31] that any gauge-invariant classical observable possesses a BRST-invariant extension such that relation (13.2.8.6) holds. But the converse might not be true: there exist "BRST observables," solutions to equation (13.2.8.6), which have no classical analog when a degeneracy like (13.2.6.18) occurs [34].

of those with eigenvalue zero (if any), must have zero—or ill-defined—norm.

Owing to the presence of the zero mode in equation (13.2.8.7), all the eigenvalues are half-integer.[†] This phenomenon is known as "fractionalization of the ghost number" [19].

13.2.8c. Conformal Invariance in the Quantum Theory

In the critical dimension $d = 26$, the BRST invariant extensions L_n^Ω of the constraints,

$$L_n^\Omega = [\Omega, \mathcal{P}_n] = L_n + \text{ghost contributions} \qquad (13.2.8.8)$$

close according to the conformal algebra without central charge [35],

$$[L_n^\Omega, L_m^\Omega] = (n - m)L_{n+m}^\Omega \qquad (13.2.8.9)$$

This means that the transformations generated by L_m^Ω through the commutator form a representation of the conformal group.

This corresponds well to our intuition, since we know that nilpotency of Ω, which only holds when $d = 26$, is equivalent to gauge invariance at the quantum level. Furthermore, one sees from equation (13.2.8.8) that L_n^Ω creates null states when acting on physical states,

$$|\psi\rangle \to |\psi\rangle + \varepsilon^n L_n^\Omega |\psi\rangle = |\psi\rangle + \varepsilon^n \Omega \mathcal{P}_n |\psi\rangle \approx |\psi\rangle \qquad (13.2.8.10)$$

so that the "gauge group" acts trivially on equivalence classes of physical states. The quantum theory is thus also gauge invariant in this more traditional sense.

One can take advantage of relation (13.2.8.10) by imposing a "gauge condition" in the quantum theory, for instance, that the representative $|\psi\rangle$ in each equivalence class should obey the light-cone gauge conditions $\langle\psi|a_n^+|\psi\rangle = 0$ (see below).

It should be kept in mind, however, that equation (13.2.8.10) appears to be only a subset of the full gauge invariance of the quantum theory

$$|\psi\rangle \to |\psi\rangle + \Omega|\chi\rangle \qquad (13.2.8.11)$$

(with $|\chi\rangle$ an arbitrary $|\psi\rangle$-dependent state). From that point of view, equation (13.2.8.11) is more fundamental than equation (13.2.8.10).

One can easily see that for physical states obeying the "BRST gauge condition"

$$\eta_n|\psi\rangle = \mathcal{P}_n|\psi\rangle = 0 \qquad (n > 0) \qquad (13.2.8.12)$$

[†] With the choice of constant in expression (13.2.8.4) as in relation (13.2.8.5).

half of the transformations (13.2.8.10) reduce to the identity: those generated by L_n^Ω, $n > 0$. As to the others, they act nontrivially on $|\psi\rangle$—although they create null states, of course—and they do not preserve condition (13.2.8.12) in general.

In some other gauge theories, all the gauge transformations analogous to equation (13.2.8.10) reduce to the identity. However, equation (13.2.8.11) still plays an essential, nontrivial role. That is why we feel it is more appropriate to focus on relation (13.2.8.11) and the whole apparatus of the BRST formalism based on the nilpotency condition, rather than on equation (13.2.8.10) alone.

Exercise

a. Compute explicitly L_n^Ω.
b. Evaluate $L_n^\Omega |\psi\rangle$ for the physical states (13.2.6.18).
c. Do the L_n^Ω commute with the ghost-number charge?

13.3. LIGHT-CONE GAUGE QUANTIZATION

13.3.1. Poincaré Invariance of the Quantum Theory

In the light-cone gauge quantization, the Hilbert space of the quantum states is a true Hilbert space without negative-norm states. The gauge has been completely fixed prior to quantization and every state is physical (except in the case of the closed string; see Section 13.3.3).

What is not obvious—and turns out to hold only in the critical dimension—is Poincaré invariance of the quantum theory.†

The light-cone gauge Poincaré generators are nonlinear, so their algebra develops a q-number anomaly which disappears under the following conditions:

$$d = 26 \quad \text{and} \quad \alpha_0 = 1 \qquad (13.3.1.1)$$

Hence, one finds the same conditions as in the other approaches to quantization. This is reassuring because the light-cone gauge method, although very practical, is in the author's opinion less fundamental than the others. For instance, it is not clear what would replace the light-cone gauge requirement of Poincaré invariance in a curved background, with no isometry. Furthermore, the light-cone gauge condition is a canonical gauge for a noninternal gauge symmetry (reparametrization invariance). It is a

---

† One trades manifest positiveness of the physical subspace against manifest Poincaré invariance. The two are equivalent in the critical dimension, where one can impose the light-cone gauge even quantum-mechanically.

priori unclear that it can be implemented quantum-mechanically, although in the cases of the bosonic and fermionic strings it can be justified *a posteriori*, as already mentioned.

The light-cone gauge classical Poincaré generators were found in Section 12.5.6, equations (12.5.6.1) and (12.5.6.2). We define the associated quantum operators by adopting the normal-ordering prescription for the oscillator variables, and the symmetric ordering for the (X, p) pairs. We also allow for an as yet undetermined constant α_0 in L_0.

Two remarks should be made at this point:

1. There is actually no true ordering ambiguity in the Poincaré generators—apart from the α_0 term—if one demands that these operators be Hermitian.
2. One must take the same α_0 in P^- and M^{i-} in order to preserve $[P^i, M^{i-}] = \frac{1}{2}i\, P^-$.

One easily checks that the Poincaré algebra is obeyed by the quantum Poincaré operators, with the exception of, possibly,

$$[M^{i-}, M^{j-}] = 0 \quad ? \qquad (13.3.1.2)$$

Everything else is in normal-ordered form. So the key question is: does relation (13.3.1.2) still hold?

Again, in order to compute $[M^{i-}, M^{j-}]$, one can rely on the classical computations (which indicate that $[M^{i-}, M^{j-}]$ vanishes classically) and only focus on the anomalous terms.

The potentially dangerous terms are:

1. The contribution proportional to α_0, classically absent.
2. The terms resulting from normal ordering of $L_0[X_0^i, \sum_{n>0} a_n^{*j} L_n^{\text{tr}} + L_{-n}^{\text{tr}} a_m^j]$.
3. The contributions resulting from normal ordering of the commutators of the cubic terms $[a_{-n}L_n^{\text{tr}}, a_{-m}L_{+m}^{\text{tr}}]$, $[a_{-n}L_n^{\text{tr}}, L_{-m}^{\text{tr}}a_m]$, and so on.

Upon explicit evaluation, one finds that the total anomaly in $[M^{i-}, M^{j-}]$ reads

$$\frac{1}{4(p^+)^2 \alpha'} \sum_{n>0} \left[\alpha_0 + \frac{d-2}{24}(n^2 - 1) - n^2 \right] (a_n^{*i} a_n^j - a_n^{*j} a_n^i) \qquad (13.3.1.3)$$

This expression vanishes if and only if the conditions (13.3.1.1) are fulfilled,

13 • Quantization of Nambu–Goto String

as announced. If these conditions are not satisfied, the quantum theory is not Poincaré invariant.

13.3.2. Description of the Spectrum

The spectrum of the string consists of an infinite number of states lying on linearly rising "Regge trajectories." These trajectories yield the "spin" of the states versus their mass squared,

$$J = \alpha' M^2 + \text{"intercept"} \qquad (13.3.2.1)$$

The first trajectory, at intercept +1, is called the "leading trajectory" (Figure 13.1). The others are the "daughter" trajectories.

This situation can be understood as follows. Because the quantum theory is Poincaré invariant, states fall into irreducible representations of the Poincaré group. These are characterized by

1. The mass.
2. The representation of the little group to which they belong; the relevant little group is $SO(d-1) = SO(25)$ for massive states, and $SO(24)$ for massless ones.

The mass is given by

$$\alpha' M^2 = -\alpha' P_A P^A$$
$$= N - \alpha_0 \qquad (13.3.2.2)$$

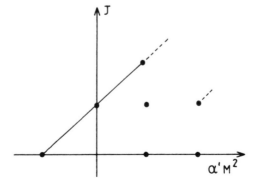

Figure 13.1. Using four-dimensional representation and terminology for "spin," one finds that the states of the bosonic string lie on linearly rising Regge trajectories.

where the "level number" N is given by

$$N = \sum_{n>0} n a_n^{i*} a_{in} \qquad (13.3.2.3)$$

One easily checks that N commutes with all Poincaré generators. States belonging to the same representation of the Poincaré group have identical level number.

The determination of the "spin" of a state is a little more intricate. The transverse indices are clearly SO(24) vector indices. For the massless state at the first excited level $N = 1$, the determination of the SO(24) representation is immediate. For massive states, one finds that different SO(24) tensors combine to form irreducible representations of SO(25).

The first state is the ground state, annihilated by all oscillators:

$$N = 0, \qquad |0\rangle, \qquad \alpha'M^2 = -1 \qquad (13.3.2.4)$$

The intercept parameter α_0 is one, so it is a scalar tachyon. This is the bad feature of the bosonic model.

The first excited state, at $N = 1$, is a massless transverse vector (spin 1, "photon").

$$N = 1, \qquad a_1^{i*}|0\rangle, \qquad \alpha'M^2 = 0, \qquad 24 \text{ states} \qquad (13.3.2.5)$$

Because we have only 24 states (and not 25 as would require a massive little group) at $N = 1$, we could actually have guessed that these states were massless if Lorentz invariance was to be realized at the quantum level. However, if the first excited states are massless, the ground state can only be a tachyon. This explains why we found $\alpha_0 = 1$.

The next states are at $N = 2$; we have

$$N = 2, \qquad a_2^{i*}|0\rangle, \qquad \alpha'M^2 = 1, \qquad 24 \text{ states}$$
$$a_1^{i*}a_1^{j*}|0\rangle, \qquad \alpha'M^2 = 1, \qquad 300 \text{ states} \qquad (13.3.2.6)$$

They combine to form a massive spin-2 representation of the Poincaré group (traceless symmetric 25×25 tensor = 324 states).

The next states are at $N = 3$:

$$N = 3, \qquad a_3^{i*}|0\rangle, \qquad \alpha'M^2 = 2$$
$$a_2^{*i}a_1^{*j}|0\rangle, \qquad \alpha'M^2 = 2 \qquad (13.3.2.7)$$
$$a_1^{*i}a_1^{*j}a_1^{*k}|0\rangle, \qquad \alpha'M^2 = 2$$

13 • Quantization of Nambu–Goto String

The associated massive representation of the Poincaré group is no longer reducible. One gets one "spin 3" and two "spin 1" representations: new, "daughter" trajectories start at $N = 3$ (we note that there are "holes" in the picture in Figure 13.1).

The analysis proceeds along the same lines for the higher levels. Two points are worthy of mention. First, the number of states increases exponentially with the level number N [9]. Second, a state is completely specified not only by its oscillator occupation numbers but also, of course, by its momentum p^i, p^+. Knowing p^i and p^+, one can compute p^-. The resulting p^0 can be positive or negative, so that one has simultaneously states moving forward (particles) and backward (antiparticles) in Minkowskian time X^0. For a string without further quantum numbers, one usually identifies those states.

13.3.3. Closed String—Poincaré Invariance

The closed string is analyzed similarly, the only new feature being the presence of the (single) constraint

$$(L_0^{\text{tr}} - \bar{L}_0^{\text{tr}})|\psi\rangle = (N - \bar{N})|\psi\rangle = 0 \tag{13.3.3.1}$$

on the physical states.† Here N ($\bar{N}$) is the level number for the right (left) movers, where

$$N = \sum_n n C_n^{*i} C_{ni} \tag{13.3.3.2a}$$

$$\bar{N} = \sum_n n \bar{C}_n^{*i} \bar{C}_{ni} \tag{13.3.3.2b}$$

We discussed at length previously that constraint (13.3.3.1) reflects the residual gauge invariance of the light-cone gauge (constant translations in σ).

The checking of the Poincaré algebra is again straightforward, except for $[M^{i-}, M^{j-}]$. One finds that $[M^{i-}, M^{j-}]$ differs from a "weakly vanishing" operator (i.e., an operator with $L_0 - \bar{L}_0$ on the right, which annihilates the physical states) by an anomalous term. This term is zero if and only if

$$d = 26 \tag{13.3.3.3a}$$

---

† We assume that the normal-ordering ambiguity in L_0 is the same as in $\bar{L}_0$. This is actually not an independent assumption, for it follows from the demand that $[M^{i-}, M^{j-}]$ be weakly zero.

and

$$\alpha_0 = 2 \quad \text{(closed string)} \tag{13.3.3.3b}$$

where α_0 denotes the ordering ambiguity in the sum $L_0 + \bar{L}_0$.

13.3.4. Spectrum (Closed String)

The mass-shell equation for the closed string is

$$\frac{\alpha'}{2} M^2 = N + \bar{N} - 2 \tag{13.3.4.1}$$

(see Chapter 12). The slope of the Regge trajectories is half the open-string value. In addition, physical states must be such that

$$N = \bar{N} \tag{13.3.4.2}$$

The ground state is the vacuum and is again a tachyon. The first excited states are at $N = \bar{N} = 1$:

$$N = \bar{N} = 1, \quad c_1^{i*} \bar{c}_1^{j*} |0\rangle \tag{13.3.4.3}$$

They are massless as they should be, since otherwise we would not have enough states to fill in representations of $SO(25)$.

The states (13.3.4.3) decompose into a traceless symmetric part (spin two, "graviton" g_{AB}), an antisymmetric part (2-form field B_{AB}), and a scalar (spin 0, dilaton ϕ). It is noteworthy that without the condition $N = \bar{N} = 1$, one could create vector states as first excited states. It is accordingly the reparametrization-invariance condition (13.3.4.2) which implies that the first excited state is a massless spin-two "graviton."

The next states are massive and are analyzed along the same lines as for the open string.

The whole spectrum decomposes into a part which is symmetric under the exchange $c \leftrightarrow \bar{c}$ of right and left movers, and an antisymmetric part. The exchange $c \leftrightarrow \bar{c}$ can be achieved by changing σ into $-\sigma$.

The symmetric truncation ("type I closed strings") corresponds to the "Shapiro–Virasoro" model and carries no σ orientation, since it is unchanged when σ goes into $-\sigma$. The full spectrum ("type II closed strings") corresponds to the "extended Shapiro–Virasoro" model and describes oriented strings, since now the transformation $\sigma \to -\sigma$ is no longer trivial on the states (for more on this, see Schwarz [36]). It appears that there is no truncation to the antisymmetric sector alone which is consistent with the switching-on of interactions.

13.4. COVARIANT QUANTIZATION

13.4.1. Elimination of Ghosts as the Central Issue in the Covariant Approach

The covariant quantization of the string possesses interesting mathematical ramifications. Furthermore, it appears to be more flexible than the other methods because it yields a consistent theory, at the free level at least, for more values of d and α_0 than just $d = 26$ and $\alpha_0 = 1$. (However, $d = 26$ and $\alpha_0 = 1$ stand on a special footing even in the covariant approach.)

We pointed out above that one cannot impose the Virasoro conditions

$$L_n|\psi\rangle = 0, \quad n \in z \qquad (13.4.1.1)$$

for all (positive and negative) values of n owing to the nontrivial central charge in the Virasoro algebra. Rather, one weakens conditions (13.4.1.1):

$$L_n|\psi\rangle = 0, \quad n > 0 \qquad (13.4.1.2a)$$

and

$$(L_0 - \alpha_0)|\psi\rangle = 0 \qquad (13.4.1.2b)$$

where α_0 denotes the ordering ambiguity in L_0. The state $|\psi\rangle$ is now a purely bosonic state involving only the a_n^A oscillators (no η_n or $\mathcal{P}_n$). (It was denoted by $|P\rangle$ in the BRST section.)

The weakening (13.4.1.2) is important, both technically and conceptually. It departs strongly from the usual prescription for handling constraints of the "Gauss law" type resulting from the gauge invariance of the action alone, in that it only enforces "half" of the two-dimensional reparametrization-invariance conditions in the quantum theory. For this reason, it is not clear that the theory based on system (13.4.1.2) is the quantum theory of the geometric, reparametrization-invariant string.

It turns out that also in the covariant approach, complete reparametrization invariance is regained, in a subtle way, for the critical values $d = 26$ and $\alpha_0 = 1$, again through the presence of null physical states, which can be factored out.

But apart from these conceptual questions, one can wonder under which conditions the quantum theory defined by system (13.4.1.2) make sense. That quantum theory will make sense if conditions (13.4.1.2) suffice to exclude all negative-norm states from the physical subspace. Such states are generated by a_n^{0*}.

Hence, the central issue in the covariant approach is the problem of ghost elimination. Here, "ghost" specifically refers to negative-norm states and, of course, not to the ghosts of the BRST approach.

The structure of the subspace defined by system (13.4.1.2) has been studied in Refs. 18 and 37 and the following conclusions were reached.

1. For $d > 26$ or $\alpha_0 > 1$, ghosts are not completely eliminated from the physical subspace.
2. For $d = 26$ and $\alpha_0 = 1$, the no-ghost theorem holds. If $\alpha_0 \neq 1$, ghosts are not eliminated.
3. Finally, for $d \leq 25$ and $\alpha_0 \leq 1$, there are no ghosts in the physical subspace.

The complete proof of these results will not be reproduced here. It can be found in Ref. 37. We only give below the proof of the no-ghost theorem for $d = 26$ and $\alpha_0 = 1$, where special features ("quantum gauge invariance") occur.

One can roughly understand why $d = 26$ and $\alpha_0 = 1$ arise as critical values from the following:

Why $\alpha_0 = 1$. Suppose $\alpha_0 > 1$. We consider states at the first level $N = 1$,

$$k_A a_1^{*A}|0, p\rangle \quad (13.4.1.3)$$

where k_A is an arbitrary d vector. The mass-shell condition

$$(L_0 - \alpha_0) k_A a_1^{*A}|0, p\rangle = 0 \quad (13.4.1.4)$$

yields the result that the d-momentum p^A is spacelike. Hence the on-shell states (13.4.1.3) are tachyonic when $\alpha_0 > 1$. One then observes that the only nontrivial equation (13.4.1.2a) is for $n = 1$. It reduces to

$$k_A p^A = 0 \quad (13.4.1.5)$$

The vector k_A is orthogonal to p^A, so it can be timelike.

On the other hand, the states (13.4.1.3) have a norm equal to[†]

$$k_A k^A \quad (13.4.1.6)$$

[†] As we have seen, the norm is actually infinite because one has a definite value of d momentum. To get finite-norm states one must consider wave packets [and factor out a $\delta(p_A p^A + m^2)$]. Subsequently, we will systematically forget the integration over the momenta in the scalar product.

13 • Quantization of Nambu–Goto String

Hence, for timelike quantities k_A the physical states (13.4.1.3) have a negative norm. This means that for $\alpha_0 > 1$, the Virasoro conditions fail to remove all negative-norm states from the physical subspace.

When $\alpha_0 = 1$, p^A is null, and k^A is necessarily spacelike or lightlike. If $k_A = \lambda p_A$, the corresponding physical state decouples from all other physical states, so that only $d - 2$ independent values of k_A are effectively relevant. This is in agreement with the fact that the states at level $N = 1$ are massless; the little group is $O(d - 2)$.

When $\alpha_0 < 1$, p^A is timelike. The vector k^A is then spacelike by equation (13.4.1.5). All $d - 1$ directions of the k_A vector are relevant for these massive states; $\alpha_0 = 1$ therefore appears as a critical (admissible) value with particular features.

Why $d = 26$. In order to see what goes wrong with $d > 26$, we consider the states at the second level $N = 2$ (see, e.g., Schwarz [38]),

$$|k\rangle = k_A a_2^{A^*}|0, p\rangle + \tfrac{1}{2} m_{AB} a_1^{A^*} a_1^{B^*}|0, p\rangle \tag{13.4.1.7}$$

with $m_{AB} = m_{BA}$ a symmetric $d \times d$ matrix. The norm of these states is given by

$$\langle k|k\rangle = k_A^* k^A + \tfrac{1}{2} m_{AB}^* m^{AB} \tag{13.4.1.8}$$

The Virasoro conditions imply that p^A obeys

$$\alpha' p_A p^A + 2 - \alpha_0 = 0 \tag{13.4.1.9}$$

and, furthermore, that one has

$$k_A = i m_{AB} p^B \sqrt{\alpha'} \tag{13.4.1.10a}$$

and

$$-i\sqrt{\alpha'} p^A k_A - \tfrac{1}{4} m_A{}^A = 0 \tag{13.4.1.10b}$$

Since we assume $\alpha_0 \leq 1$, p^A is timelike and we can take $p^A = (p, 0, 0, \ldots, 0)$ with

$$\alpha' p^2 = 2 - \alpha_0 \tag{13.4.1.11}$$

Equations (13.4.1.10) can be solved for k_A and m_{00} as functions of the independent components m_{0a} and m_{ab} of m_{AB},

$$k_0 = i m_{00} p \sqrt{\alpha'} = \frac{i m^a{}_a p \sqrt{\alpha'}}{4p^2 \alpha' + 1} \tag{13.4.1.12a}$$

$$k_a = im_{a0}p\sqrt{\alpha'} \qquad (13.4.1.12b)$$

$$m_{00} = \frac{m^a{}_a}{4p^2\alpha' + 1} \qquad (13.4.1.12c)$$

These latter relations enable the scalar product $\langle k|k\rangle$ to be expressed in the form

$$\langle k|k\rangle = \frac{|m^a{}_a|^2}{(9-4\alpha_0)^2}(\alpha_0 - \tfrac{3}{2}) + (1-\alpha_0)|m_{0a}|^2 + \tfrac{1}{2}m^*{}_{ab}m^{ab} \qquad (13.4.1.13)$$

The m_{0a} term is nonnegative provided $\alpha_0 \leq 1$, which we assume. The only negative contributions can then come from a nonvanishing trace $m^a{}_a$. If we take

$$m_{ab} = \frac{1}{d-1} m^c{}_c \delta_{ab} \qquad (13.4.1.14)$$

then equation (13.4.1.13) gives

$$\langle k|k\rangle = \frac{|m^a{}_a|^2}{(9-4\alpha_0)^2}\left[(\alpha_0 - \tfrac{3}{2}) + \frac{1}{2(d-1)}(9-4\alpha_0)^2\right] \qquad (13.4.1.15)$$

This term is positive provided

$$\alpha_0 - \tfrac{3}{2} + \frac{1}{2(d-1)}(9-4\alpha_0)^2 \geq 0 \qquad (13.4.1.16)$$

i.e.,

$$\alpha_0 \leq \frac{37 - d - \sqrt{(d-25)(d-1)}}{16} \quad (<1) \qquad (13.4.1.17a)$$

or

$$\alpha_0 \geq \frac{37 - d + \sqrt{(d-25)(d-1)}}{16} \qquad (13.4.1.17b)$$

if $d > 25$. [For $d \leq 25$, expression (13.4.1.16) is always positive.]

A detailed analysis shows that the first possibility (13.4.1.17a) is unacceptable, for it ultimately leads to negative-norm states at the higher levels $N > 2$ [37]. One is therefore left only with possibility (13.4.1.17b).

Now, function (13.4.1.17b) is an increasing function of d, which takes the maximum allowed value $\alpha_0 = 1$ for the critical dimension $d = 26$. When $d > 26$, α_0 given by expression (13.4.1.17b) is greater than 1.

In other words, if one takes $\alpha_0 = 1$, the expression (13.4.1.16) is negative for $d > 26$ and there are negative-norm states in the physical subspace. These negative-norm states become null states, which decouple from all other physical states in $d = 26$ dimensions [expression (13.4.1.16) is zero],† and positive-norm states in $d < 26$ dimensions. This is how 26 arises as the critical dimension.

The maximum number of null states which decouple appear for the critical values $\alpha_0 = 1$ and $d = 26$. One then finds that the dimension of the effective physical subspace at the second mass level is given by $25 + 26 \times 25/2$ (number of independent components of m_{AB}) $- 25 - 1$ (number of states which decouple). This number is also equal to $24 + 25 \times 24/2$, i.e., the number of light-cone gauge states at the second mass level ($a_2^{*i}|0\rangle$ and $a_1^{*i} a_1^{j*}|0\rangle$).

Exercise. Study the physical subspace at the third mass level.

13.4.2. Vertex Operators

Two operator functions which play a fundamental role in the covariant formalism are [39, 10, 38]

$$Q^A(\theta) = X_0^A + 2\alpha' p^A \theta + \sum_{n \geq 1} \sqrt{\frac{2\alpha'}{n}} (a_n^A \exp -in\theta + a_n^{A*} \exp in\theta)$$

(13.4.2.1)

and

$$P^A(\theta) = \frac{1}{2\alpha'} \frac{dQ^A}{d\theta}$$

$$= p^A + \sum_{n \geq 1} \sqrt{n}(-ia_n^A \exp -in\theta + ia_n^{A*} \exp in\theta)$$

(13.4.2.2)

These operators are often expressed as complex plane functions of $z = \exp i\theta$.

Function $Q^A(\theta)$ is just the position of the end point $\sigma = 0$ at "time" θ in a conformal gauge; $P^A(\theta)$ is the two-dimensional lightlike component

---

† In the absence of negative-norm states, a zero-norm state necessarily decouples, i.e., it has vanishing scalar product with all other physical states (otherwise, one could construct negative-norm states).

$j^A(\theta)$ of the space-time translation current. We have already encountered these functions previously (e.g., it was noted that the motion of the end point $\sigma = 0$ completely determines the classical string history; $P^A(\theta)$ was related to a Kac-Moody algebra).

Operators $Q^A(\theta)$ and $P^A(\theta)$ obey

$$[L_m, Q^A(\theta)] = -i\, e^{im\theta} \frac{dQ^A}{d\theta} \tag{13.4.2.3a}$$

$$[L_m, P^A(\theta)] = e^{im\theta}\left(-i\frac{d}{d\theta} + m\right) P^A(\theta) \tag{13.4.2.3b}$$

One says that $Q^A(\theta)$ has conformal spin 0, while $P^A(\theta)$ has conformal spin one. More generally, any operator function $X(\theta)$ such that

$$[L_m, X(\theta)] = e^{im\theta}\left(-i\frac{d}{d\theta} + mJ\right) X(\theta) \tag{13.4.2.4}$$

is said to be of conformal spin J.

Exercise. Check system of equations (13.4.2.3).

The (scalar) vertex operator is defined by

$$V_0(k, \theta) = :\exp ik_A Q^A(\theta): \tag{13.4.2.5}$$

where : : denotes normal ordering, which is necessary because $\exp ik_A Q^A(\theta)$ is ill-defined unless $k^2 = 0$, in which case the normal order is superfluous.

One easily finds

$$V_0(k, \theta) = \exp(i\theta L_0)\, V_0(k)\, \exp(-i\theta L_0) \tag{13.4.2.6a}$$

with

$$V_0(k) = \exp\left(ik_A \sum_{n=1}^{\infty} \sqrt{2n^{-1}\alpha'}\, a_n^{*A}\right) \exp(ik_A X_0^A) \exp\left(ik_A \sum_{n=1}^{\infty} \sqrt{2n^{-1}\alpha'}\, a_n^A\right) \tag{13.4.2.6b}$$

Although $Q^A(\theta)$ has conformal spin 0, $V_0(k, \theta)$ does not possess a vanishing conformal spin owing to the normal ordering in expression

(13.4.2.5). Rather, one gets

$$[L_n, V_0(k, \theta)] = e^{in\theta}\left(-i\frac{d}{d\theta} + n\alpha'k^2\right) V_0(k, \theta) \qquad (13.4.2.7)$$

i.e., $V_0(k, \theta)$ has conformal spin $\alpha'k^2$.

The transverse vector vertex operator $V_A(k, \theta)$, where k is a lightlike vector, is given by

$$\varepsilon^A V_A(k, \theta) = \varepsilon^A P_A(Q) \exp ik_B Q^B(\theta), \qquad k^2 = 0 \qquad (13.4.2.8)$$

Here, the polarization vector ε^A is transverse ($\varepsilon^A k_A = 0$).

Transversality implies that the factors $\varepsilon^A P_A(\theta)$ and $k_B Q^B(\theta)$ commute,

$$[k_B Q^B(\theta), \varepsilon^A P_A(\theta)] = 4\pi i\alpha' k_A \varepsilon^A \delta(\theta, \theta') \qquad (13.4.2.9a)$$

$$= 0 \qquad (13.4.2.9b)$$

while $k^2 = 0$ makes the normal ordering in the exponential unnecessary. Hence, the conformal spin of the transverse vertex operators is 1,

$$[L_m, \varepsilon^A V_A(k, \theta)] = e^{im\theta}\left(-i\frac{d}{d\theta} + m\right) \varepsilon^A V_A(k, \theta) \qquad (13.4.2.10)$$

$\varepsilon^A V_A(k, \theta)$ can be extended to a full vector vertex operator, but this will not be needed here (see, e.g., Schwarz [38]).

The vertex operators are essential for the discussion of interactions in string theory. However, since the study of the interactions is outside the scope of this review, our interest in the vertex operators lies in a somewhat different context: the transverse vector vertex operator can be used to construct the manifestly positive-norm DDF states, which completely span the physical subspace up to null states.

13.4.3. DDF States

We expect that when $d = 26$ and $\alpha_0 = 1$, the physical space is isomorphic to the space of transverse excitations, apart from null states that can be factored out. We have checked explicitly this property at the first two levels in Section 13.4.1.

In order to prove the property for all levels we need to explain precisely what is meant by "transverse states" in the covariant formalism. In the light-cone gauge, a transverse state was simply a state created by the transverse oscillators a_n^i. This definition cannot be taken over as such in the covariant approach, for a_n^i does not commute with the Virasoro operators,

$$[a_n^{i*}, L_m] \neq 0 \qquad (13.4.3.1)$$

and hence does not even create physical states out of the vacuum. In particular, a_n^{*i} increases the level number by n units, but does not modify the d momentum so as to remain on the mass shell ($[a_n^{*i}, p^A] = 0$).

One needs to define operators A_n^i which commute with L_m,

$$[A_n^i, L_m] = 0 \qquad (13.4.3.2)$$

and which reduce classically to a_n^i in the light-cone gauge,

$$A_n^i \approx a_n^i + k_n^{im} \chi_m \qquad (13.4.3.3)$$

so that they can be called "transverse." Here, χ_m are all the light cone-gauge conditions.

It is a general property of first-class constrained Hamiltonian systems that this problem always possesses a solution, at least when equality in relation (13.4.3.2) is replaced by "weak" equality (equality up to $L_m = 0$ constraints).

Rather than applying the standard recipes for finding that solution, Del Giudice et al. [40] used the transverse vector vertex operator to obtain A_n^i. Their construction is based on the fact that the operator function $\varepsilon^A V_A(k, \theta)$ is periodic in θ with period 2π, provided one considers states with momentum p^A such that

$$2\alpha' k_A \cdot p^A = \text{integer} \qquad (13.4.3.4)$$

When this condition holds, one finds that

$$\left[L_m, \int_0^{2\pi} \varepsilon^A V_A(k, \theta) \, d\theta \right] = -i \int_0^{2\pi} \frac{d}{d\theta} (e^{im\theta} \varepsilon^A V_A)$$

$$= 0 \qquad (13.4.3.5)$$

since V_A has conformal spin 1.

We select a ground state $|0, p_0\rangle$ whose d-momentum p_0^A satisfies the mass-shell condition

$$\alpha' p_0^A p_{0A} = 1 \qquad (13.4.3.6)$$

with p_0^A fixed once and for all.

In order to satisfy condition (13.4.3.4) we choose furthermore, again once and for all, a null vector k_0 which obeys

$$k_0^A k_{0A} = 0, \qquad 2\alpha' k_0^A p_{0A} = 1 \tag{13.4.3.7}$$

The only allowed values of k for given p_0 and k_0 are multiples of k_0,

$$k^A = n k_0^A \tag{13.4.3.8}$$

If we take, for instance, $p_0^A = (0, 0, \ldots, 0, 1/\sqrt{\alpha'})$ and $k^A = (-1/2\sqrt{\alpha'}, 0, 0, \ldots, 1/2\sqrt{\alpha'})$, then a transverse vector ε_A assumes the form $(0, \varepsilon_i, 0)$. We note that a definite choice of k breaks manifest Lorentz invariance.

It follows from condition (13.4.3.5) that the operators

$$A_{ni} = \int_0^{2\pi} V_i(nk_0, \theta) \, d\theta \tag{13.4.3.9}$$

$$= \int_0^{2\pi} P_i(\theta) \exp \frac{in}{\sqrt{2\alpha'}} Q^+(\theta) \, d\theta \tag{13.4.3.10}$$

commute with L_m for all n. These are the DDF operators.

If one sets $a_n^+ = 0$ in the classical analog of operator (13.4.3.10) (light-cone gauge conditions),† then

$$Q^+(\theta) = \sqrt{2\alpha'}\,\theta \tag{13.4.3.11a}$$

and thus

$$A_{ni} = \sqrt{n}\, a_{ni} \tag{13.4.3.11b}$$

Quantities A_{ni} accordingly satisfy the condition (13.4.3.3) (up to the factor $\sqrt{n}$).

It can easily be checked that

$$A_{ni}^* = A_{-ni} \tag{13.4.3.12}$$

One can also prove the commutation relations (see, e.g., Schwarz [38])

$$[A_m^i, A_n^j] = m \delta^{ij} \delta_{m,-n} \tag{13.4.3.13}$$

† The conditions $a_n^+ = 0$ cannot be viewed as operator equations in the covariant formalism. That is why we revert to the classical theory.

which imply that the 24 transverse "creation" operators A_n^{i*} ($n > 0$) span a positive-definite Hilbert subspace. This subspace is isomorphic to the Hilbert space of the light-cone gauge quantization.

Exercise. Compute $[P^A, A_n^i]$ and check explicitly that A_n^i creates on-the-mass-shell states (when acting on such states).

13.4.4. No-Ghost Theorem for $d = 26$, $\alpha_0 = 1$

The DDF states are obtained by acting with A_n^{i*} on the vacuum. They obey the equations

$$K_n|\psi\rangle = 0, \quad n > 0 \qquad (13.4.4.1)$$

together with the physical-state conditions

$$L_n|\psi\rangle = 0, \quad n > 0 \qquad (13.4.4.2)$$

and

$$(L_0 - 1)|\psi\rangle = 0 \qquad (13.4.4.3)$$

In equation (13.4.4.1), K_n is just the lightlike operator

$$K_n = k_A \alpha_n^A = \alpha_n^+ \qquad (\alpha_n^A = \sqrt{n} a_n^A, \text{ etc.}) \qquad (13.4.4.4)$$

The property (13.4.4.1) simply follows from the fact that the DDF states do not involve the oscillators a_n^-, a_n^{-*}.

The K_n operators possess the properties

$$K_n = K_{-n}^*, \quad [K_m, K_n] = 0, \quad [L_m, K_n] = -nK_{m+n} \qquad (13.4.4.5)$$

It can be shown that, conversely, equations (13.4.4.1)–(13.4.4.3) completely characterize the DDF states: any solution to these equations is a linear combination of states of the form $(A_{n_1}^{i_1}*)^{\nu_1}(A_{n_2}^{i_2}*)^{\nu_2} \cdots (A_{n_n}^{i_n}*)^{\nu_n}|0\rangle$ [18].

It is sometimes convenient to consider not only on-the-mass-shell states, but states which fail to be on the mass shell by an integer number,

$$(L_0 + l - 1)|\phi\rangle = 0, \quad l = \text{integer} \qquad (13.4.4.6)$$

Such states can be obtained by acting with DDF operators on a ground state which does not fulfill $(L_0 - 1)|0\rangle = 0$, but rather has momentum equal to

$$p_0^A + nk_0^A \qquad (13.4.4.7)$$

where n is an integer.

In the sequel, we will only consider states with momentum given by expression (13.4.4.7) for some n. These states obey condition (13.4.4.6). It is easy to see that when acting on a state (13.4.4.7) with DDF operators, Virasoro operators, or K_n operators, then one remains within the class defined by system (13.4.4.7) and (13.4.4.6). Furthermore, the momentum of any solution of (13.4.4.6) can be written as in equation (13.4.4.7) by an appropriate Lorentz transformation, with $n < 0$ if $l \geq 1$.

The DDF operators are appropriate extensions of the transverse oscillators a_n^i. The K_n operators are just the lightlike oscillators a_n^+. Furthermore, because $p \cdot k \neq 0$, p^A possesses a component along the minus lightlike direction, which means that the first term in L_n contains a_n^-.

It thus comes as no surprise that the states created from the vacuum by A_p^{i*}, K_{-n}, and $L_{-m}(n, m, p > 0)$ completely span the Hilbert space.

Lemma [18]. Any state with momentum (13.4.4.7) can be written as a linear combination of states of the form

$$L_{-1}^{\lambda_1} L_{-2}^{\lambda_2} \cdots L_{-n}^{\lambda_n} K_{-1}^{\mu_1} \cdots K_{-m}^{\mu_m} |d\rangle \qquad (13.4.4.8)$$

where $|d\rangle$ refers to a state created from a vacuum with momentum (13.4.4.7) (with possibly a different n) by the DDF operators.

Proof [18]. By counting arguments, the proof amounts to establishing the linear independence of the states (13.4.4.8).

A null spurious state is, by definition, a physical state orthogonal to all other physical states and, in particular, to itself. We found such states at the first two levels (for $d = 26$, $\alpha_0 = 1$). As the no-ghost theorem indicates, they actually appear at all levels.

No-ghost theorem ($\alpha_0 = 1$, $d = 26$) [18]. Any physical state $|\psi\rangle$ solution to $L_n|\psi\rangle = 0$ ($n > 0$) and $(L_0 - 1)|\psi\rangle = 0$ can be written as

$$|\psi\rangle = |f\rangle + |ns\rangle \qquad (13.4.4.9)$$

where $|f\rangle$ is in the space spanned by the DDF states and $|ns\rangle$ is a null spurious state.

Proof. The previous lemma enables one to write

$$|\psi\rangle = |\phi\rangle + |s\rangle \qquad (13.4.4.10)$$

where $|\phi\rangle$ is a state (13.4.4.8) with $\lambda_1 = \lambda_2 = \cdots = \lambda_n = 0$ (only K_{-n} operators, if any), and $|s\rangle$ is a state (13.4.4.8) with at least one L_{-n}. Since $(L_0 - 1)|\psi\rangle = 0$, both $|\phi\rangle$ and $|s\rangle$, which are independent, are on the mass shell.

The key step in the proof is the demonstration that $|\phi\rangle$ and $|s\rangle$ are physical states,

$$L_n|\phi\rangle = 0 = L_n|s\rangle, \quad n > 0 \tag{13.4.4.11}$$

The L_n with positive n are generated by L_1 and L_2, so it suffices to establish that $L_1|\phi\rangle = L_2|\phi\rangle = L_1|s\rangle = L_2|s\rangle = 0$, or, what is the same, that $L_1|\phi\rangle = \tilde{L}_2|\phi\rangle = L_1|s\rangle = \tilde{L}_2|s\rangle = 0$. Here, $\tilde{L}_2$ is given by

$$\tilde{L}_2 = L_2 + \tfrac{3}{2}L_1^2 \tag{13.4.4.12}$$

By construction, one has

$$|s\rangle = L_{-1}|s_0\rangle + \tilde{L}_{-2}|s_{-1}\rangle \tag{13.4.4.13}$$

because the L_{-n} ($n \geq 1$) are generated by L_{-1} and L_{-2}. Furthermore, it follows from the L_n algebra that

$$L_0|s_0\rangle = 0 \quad \text{and} \quad L_0|s_{-1}\rangle = -|s_{-1}\rangle \tag{13.4.4.14}$$

This implies that

$$L_1|s\rangle = |s'\rangle \tag{13.4.4.15a}$$

and

$$\tilde{L}_2|s\rangle = |s''\rangle + (\tfrac{1}{2}d - 13)|s_{-1}\rangle \tag{13.4.4.15b}$$

where $|s'\rangle$ and $|s''\rangle$ are states of the form (13.4.4.8) with at least one L_{-n}. (For instance, $|s'\rangle = L_{-1}L_1|s_0\rangle + \tilde{L}_{-2}L_1|s_{-1}\rangle$.) For $d = 26$, the second term in the right-hand side of equation (13.4.4.15b) disappears and $\tilde{L}_2|s\rangle$ reduces to $|s''\rangle$.

Now, $L_1|\phi\rangle$ and $\tilde{L}_2|\phi\rangle$ are states of the form (13.4.4.8) *without* any L_{-n}. Indeed, one can pass L_1 and $\tilde{L}_2$ to the right of the K_{-m} operators in $|\phi\rangle$ without generating any L_{-n} ($[L_n, K_m] \sim K_{n+m}$), until L_1 and $\tilde{L}_2$ reach the state $|d\rangle$ which they annihilate.

It follows from this remark and the linear independence of the states (13.4.4.8) with different exponents that $L_1|s\rangle$ and $L_1|\phi\rangle$ are linearly independent, as are $\tilde{L}_2|s\rangle$ and $\tilde{L}_2|\phi\rangle$. However, since $L_1|\psi\rangle = \tilde{L}_2|\psi\rangle = 0$, one infers from equation (13.4.4.10) that $|\phi\rangle$ and $|s\rangle$ obey the same equations. This implies relation (13.4.4.11).

13 • Quantization of Nambu–Goto String

Having established that $|\phi\rangle$ and $|s\rangle$ are both physical states, it is straightforward to complete the proof of the theorem. The state $|\phi\rangle$ obeys $K_n|\phi\rangle = 0$ because the K_n commute. Hence, it is a DDF state since these are fully characterized by equations (13.4.4.1), (13.4.4.2), and (13.4.4.3) [18]. [In other words, in the expansion (13.4.4.8) of $|\phi\rangle$, every μ_i—and every λ_i—is zero.] Besides, owing to equation (13.4.4.13), the state $|s\rangle$ is a physical state which decouples from every physical state,

$$(\langle s_0|L_{-1})|\psi\rangle = \langle s_0|L_1|\psi\rangle = 0 \quad \text{etc.}$$

and hence it is indeed a null spurious state.

Since the DDF subspace has a positive-definite inner product, the theorem implies $\langle\psi|\psi\rangle \geq 0$.

Exercise. Demonstrate relations (13.4.4.15a) and (13.4.4.15b).

13.4.5. Quantum Gauge Invariance

The no-ghost theorem establishes that the quotient space of the physical subspace by the null spurious states is isomorphic to the Hilbert space containing only transverse excitations. Hence, the light-cone gauge and covariant methods of quantization are equivalent.

The addition of null spurious states to a physical state is the quantum reflection of the gauge invariance of the classical theory. It implies that one can completely formulate the quantum theory in terms of DDF states alone, or, what is the same in view of the isomorphism between DDF states and light-cone gauge states, that one can implement the light-cone gauge quantum mechanically.

This result only holds for $d = 26$ and $\alpha_0 = 1$. For $d \leq 25$ and $\alpha_0 \leq 1$, the physical subspace is still positive definite but it does not contain enough null spurious states. The DDF states no longer completely characterize the physical subspace. As a result, there are more degrees of freedom in the quantum theory than in the classical theory. The conditions $L_n|\psi\rangle = 0$ ($n > 0$), $(L_0 - 1)|\psi\rangle = 0$ are not enough to insure full reparametrization invariance at the quantum level.

Thus one sees that, even though there may be no ghosts for other values of the space-time dimensionality and the intercept parameter, the critical values $d = 26$ and $\alpha_0 = 1$ rest on a very special basis.

It should be noted that the null spurious states $|ns\rangle$ of the covariant approach are actually nothing but particular null states $\Omega|\chi\rangle$ of the BRST approach. We indeed pointed out above that the BRST gauge conditions $c_n|0\rangle = \mathcal{P}_n|0\rangle = 0$ (BRST ghost vacuum) admitted a residual "gauge group" $|\psi\rangle \to |\psi\rangle + \Omega|\chi\rangle$ with appropriate vectors $|\chi\rangle$. These residual transformations correspond to the addition of null spurious states. As we mentioned, in order to completely fix the BRST gauge, it was necessary to demand that $|P_1\rangle$ and $|P_2\rangle$ in solution (13.2.6.18) be transverse, namely DDF, states.

Finally, we remark that Lorentz invariance is manifest in the covariant formalism since the covariant Lorentz generators obey the Lorentz algebra and commute with L_n.

However, the DDF states are not manifestly Lorentz invariant for one needs to select a particular d-vector, k_0^A, to define them. For this reason, the covariant Lorentz generators do not map the DDF subspace onto itself, but rather also introduce nonvanishing null states. One can define "improved" Lorentz generators whose action does not involve these null states. These generators, completely defined within the DDF subspace, are just the Lorentz generators of the light-cone gauge approach (with $a_n^i \to A_n^i$).

Exercise. No-ghost theorem for $d \leq 25$, $\alpha_0 = 1$.

a. Show that the $d \leq 25$, $\alpha_0 = 1$ physical subspace can be realized as a subspace of the $d = 26$, $\alpha_0 = 1$ physical subspace.

b. Conclude that there is no negative-norm state solution to $L_n|\psi\rangle = 0$ $(n > 0)$, $(L_0 - 1)|\psi\rangle = 0$ when $d \leq 25$ and $\alpha_0 = 1$.

Chapter 14

The Fermionic String: Classical Analysis

14.1. LOCAL SUPERSYMMETRY IN TWO DIMENSIONS

The bad feature of the bosonic model is the presence of a tachyon in the spectrum, which violates causality.

A tachyon appears in the spectrum because α_0, the intercept parameter, is positive. The intercept parameter is related to the ordering ambiguity in L_0, i.e., to the "zero-point energy" [41], so it is natural to seek supersymmetric models with an equal number of bosonic and fermionic variables. It is indeed a general rule of supersymmetry that supersymmetric multiplets tend to cancel unwanted effects, so that one might achieve $\alpha_0 = 0$ upon appropriate inclusion of anticommuting degrees of freedom.

Supersymmetry is now a rather old subject in field theory and requires no further motivation.

There are *a priori* two different ways one can implement supersymmetry in the string models:

1. The bosonic string can be viewed as a two-dimensional field theory describing d scalar fields coupled to gravity. This theory is invariant under two-dimensional changes of coordinates. One can enlarge the symmetry to two-dimensional local supersymmetry by introducing the supersymmetric partners ψ_λ and Γ^A of the metric $g_{\alpha\beta}$ and of the scalar fields X^A, respectively. Here, ψ_λ is a two-dimensional "spin-$\frac{3}{2}$"

field, while Γ^A is a d-dimensional vector/2-dimensional spinor. It turns out that the only degrees of freedom are carried by the "matter supermultiplet" (X^A, Γ^A). The gauge field $(g_{\alpha\beta}, \psi_\lambda)$ of supergravity is pure gauge in two dimensions. This approach leads to the Neveu-Schwarz-Ramond model [42, 43].
2. The bosonic string is invariant under global Poincaré transformations in d dimensions. One can enlarge this symmetry into global invariance under the Poincaré supergroup (i.e., under the graded extension of the Poincaré group). This leads to the superstring of Green and Schwarz [44].

Surprisingly enough, the first approach yields (when appropriately truncated) the same theory as the second one. However, what is evident in the new formalism of Green and Schwarz [44], namely global supersymmetry, does not appear in an obvious way in the old formalism. Conversely, two-dimensional local supersymmetry, which also plays a key role, is not clear in the superstring. It would be nice to have a formalism in which both types of supersymmetry are obviously present.

In this chapter and the next, we study the fermionic string of Neveu, Schwarz, and Ramond. The superstring will be examined in Chapter 16.

The Neveu-Schwarz-Ramond model is an $N = 1$ supergravity theory, which can be derived in two different ways:

1. One can adopt the techniques of supergravity [45]. These naturally lead to a consideration of the extended models with $N = 2$ or $N = 4$ [46] [these models will not be discussed here because their critical dimensions are $d = 2$ ($N = 2$) or $d = -2$ ($N = 4$)].
2. Alternatively, one can focus on the Virasoro constraints, which have been shown to play an important role in the bosonic case, and try to take "their square root" so as to obtain a model with constraints that form a graded extension of the Virasoro algebra ("super-Virasoro" algebra).

The latter approach is adopted here [47]. We recall that on taking the "square root" of the mass-shell condition $p^2 + m^2 = 0$ for a free relativistic particle, one derives the Dirac equation.

14.2. SUPERCONFORMAL ALGEBRA

14.2.1. Square Root of the Bosonic Constraints and Fermionic Constraints

We have shown that the conformal algebra is the direct product of twice the diffeomorphism group in one dimension. This direct product structure appears explicitly in terms of the generators $Q^+(\sigma)$ and $Q^-(\sigma)$:

$$Q^+(\sigma) = P_A P^A = 2\pi(\mathcal{H} + \mathcal{H}_1) \qquad (14.2.1.1)$$

and

$$Q^-(\sigma) = S_A S^A = 2\pi(\mathcal{H} - \mathcal{H}_1) \qquad (14.2.1.2)$$

Here, we have set

$$P_A(\sigma) = \pi\sqrt{2\alpha'}\mathcal{P}_A(\sigma) + \frac{1}{\sqrt{2\alpha'}} X'_A(\sigma) \qquad (14.2.1.3)$$

$$S_A(\sigma) = \pi\sqrt{2\alpha'}\mathcal{P}_A(\sigma) - \frac{1}{\sqrt{2\alpha'}} X'_A(\sigma) \qquad (14.2.1.4)$$

Hence

$$[P_A(\sigma), P_B(\sigma')] = 2\pi\eta_{AB}\delta'(\sigma, \sigma') \qquad (14.2.1.5a)$$

$$[P_A(\sigma), S_B(\sigma')] = 0 \qquad (14.2.1.5b)$$

$$[S_A(\sigma), S_B(\sigma')] = -2\pi\eta_{AB}\delta'(\sigma, \sigma') \qquad (14.2.1.5c)$$

[The function $P_A(\sigma)$ defined here differs from $P_A(\sigma)$ in Section 13.4.2 by a mere numerical factor.]

We now introduce new, real, fermionic (i.e., anticommuting) variables $\Gamma^A_i(\sigma)$ and new fermionic constraint generators $\mathcal{S}_i(\sigma)$ ($i = 1, 2$ is a spinor index in two dimensions) in such a way that, upon anticommutation, $\mathcal{S}_i(\sigma)$ yield the Hamiltonian constraints $Q^\pm(\sigma)$.

The simplest choice is to assume†

$$[\Gamma^A_i(\sigma), \Gamma^B_j(\sigma')] = -4\pi i \eta^{AB}\delta_{ij}\delta(\sigma, \sigma') \qquad (14.2.1.6)$$

and define

$$\mathcal{S}_1(\sigma) = \Gamma^A_1(\sigma) P_A(\sigma) \qquad (14.2.1.7a)$$

$$\mathcal{S}_2(\sigma) = \Gamma^A_2(\sigma) S_A(\sigma) \qquad (14.2.1.7b)$$

† The Poisson bracket (14.2.1.6) is symmetric, as it should be for fermionic variables. It is imaginary, because $\Gamma^A_i(\sigma)$ are real.

Straightforward computation gives

$$[\mathcal{S}_1(\sigma), \mathcal{S}_1(\sigma')] = -4\pi i\left(P_A P^A + \frac{i}{2}\Gamma_1^A \frac{d\Gamma_{1A}}{d\sigma}\right)\delta(\sigma, \sigma') \quad (14.2.1.8a)$$

and

$$[\mathcal{S}_2(\sigma), \mathcal{S}_2(\sigma')] = -4\pi i\left(S_A S^A - \frac{i}{2}\Gamma_2^A \frac{d\Gamma_{2A}}{d\sigma}\right)\delta(\sigma, \sigma') \quad (14.2.1.8b)$$

$$[\mathcal{S}_1(\sigma), \mathcal{S}_2(\sigma')] = 0 \quad (14.2.1.8c)$$

The right-hand side of system (14.2.1.8) does not assume the desired form, unless one redefines $Q^+(\sigma)$ and $Q^-(\sigma)$ as

$$Q^+(\sigma) = P_A P^A + \frac{i}{2}\Gamma_1^A \frac{d\Gamma_{1A}}{d\sigma} \quad (14.2.1.9a)$$

$$Q^-(\sigma) = S_A S^A - \frac{i}{2}\Gamma_2^A \frac{d\Gamma_{2A}}{d\sigma} \quad (14.2.1.9b)$$

The generators $Q^+(\sigma)$ and $Q^-(\sigma)$ receive correction terms from the fermionic variables. The modified form (4.2.1.9) looks physically reasonable, if one recalls that $Q^+(\sigma)$ and $Q^-(\sigma)$ are the T^{++} and T^{--} lightlike components of the energy-momentum tensor. It is then natural that the fermionic degrees of freedom contribute to $T^{\alpha\beta}$.

The real test as to whether equations (14.2.1.9) are acceptable consists in checking that $Q^\pm(\sigma)$ and $\mathcal{S}_i(\sigma)$ form a closed (graded) algebra. [From now on, $Q^\pm(\sigma)$ denotes system (14.2.1.9).] This demand is easily seen to be fulfilled, since one finds not only

$$[\mathcal{S}_1(\sigma), \mathcal{S}_1(\sigma')] = -4\pi i Q^+(\sigma)\delta(\sigma, \sigma') \quad (14.2.1.10a)$$

[relation (14.2.1.8a) above], but also

$$[Q^+(\sigma), \mathcal{S}_1(\sigma')] = 2\pi(2\mathcal{S}_1(\sigma) + \mathcal{S}_1(\sigma'))\delta'(\sigma, \sigma') \quad (14.2.1.10b)$$

and

$$[Q^+(\sigma), Q^+(\sigma')] = 4\pi(Q^+(\sigma) + Q^+(\sigma'))\delta'(\sigma, \sigma') \quad (14.2.1.10c)$$

Hence, the constraint generators $\mathcal{S}_1(\sigma)$ and $Q^+(\sigma)$ form a "first-class" system. Similarly, one finds that $\mathcal{S}_1(\sigma)$ and $Q^+(\sigma)$ have vanishing bracket with $\mathscr{S}_2(\sigma)$ and $Q^-(\sigma)$, and

$$[\mathscr{S}_2(\sigma), \mathscr{S}_2(\sigma')] = -4\pi i Q^-(\sigma)\delta(\sigma, \sigma') \quad (14.2.1.11a)$$

$$[Q^-(\sigma), \mathscr{S}_2(\sigma')] = -2\pi(2\mathscr{S}_2(\sigma) + \mathscr{S}_2(\sigma'))\delta'(\sigma, \sigma') \quad (14.2.1.11b)$$

$$[Q^-(\sigma), Q^-(\sigma')] = -4\pi(Q^-(\sigma) + Q^-(\sigma'))\delta'(\sigma, \sigma') \quad (14.2.1.11c)$$

We note that $[Q^+(\sigma), Q^+(\sigma')]$ remains unchanged.

It follows from this analysis that one can consistently impose the conditions

$$Q^\pm(\sigma) = 0 = \mathscr{S}_i(\sigma) \quad (14.2.1.12)$$

Relations (14.2.1.10) and (14.2.1.11) define the ($N = 1$) graded extension of the conformal algebra, also known as the superconformal algebra. The constraint generators $\mathscr{S}_i(\sigma)$ and $Q^\pm(\sigma)$, and the superalgebra (14.2.1.10) and (14.2.1.11) completely characterize the fermionic string model.

The square-root approach appears to be very efficient for obtaining the Hamiltonian constraints that characterize the fermionic string model (they were originally given in the above "continuous" representation by Iwasaki and Kikkawa [48]).

One could have obtained $Q^\pm(\sigma)$ and $\mathscr{S}_i(\sigma)$ by supergravity methods, as follows. The $N = 1$ supergravity Lagrangian coupled to the supersymmetric matter multiplet (X^A, Γ^A) is given by [45]

$$\mathscr{L} = -\frac{1}{4\pi\alpha'}\eta_{\alpha\beta}e\{g^{\alpha\beta}\partial_\alpha X^A \partial_\beta X^B - e_a^\alpha \bar{\Gamma}^A \rho^a \partial_\alpha \Gamma^B$$

$$- 2e_a^\alpha e_b^\beta \bar{\psi}_\alpha \rho^b \rho^a \Gamma^A (\partial_\beta X^B + \tfrac{1}{2}\bar{\Gamma}^B \psi_\beta)\} \quad (14.2.1.13)$$

where ρ^a are the two-dimensional γ matrices, e_a^α are the two-dimensional vielbeins, e is the determinant of e_a^α ($=\sqrt{-g}$), and γ_α is the "spin-$\tfrac{3}{2}$" supersymmetric partner of $g_{\alpha\beta}$. The d matter multiplets contain "spin" 0 and $\tfrac{1}{2}$ fields (in two dimensions).

The action (14.2.1.13) possesses much invariance:

1. Local supersymmetry and invariance under changes of coordinates in two dimensions.

2. Local Weyl and "super-Weyl" invariance†

$$X^A \to X^A, \qquad \Gamma^A \to \Lambda^{-1/2}\Gamma^A$$
$$e_\alpha^a \to \Lambda e_\alpha^a, \qquad \psi_\alpha \to \Lambda^{1/2}\psi_\alpha + \rho_\alpha \varphi \qquad (14.2.1.14)$$

3. Local two-dimensional Lorentz rotations of the vielbeins.
4. Global Poincaré invariance (for which Γ^A is a d vector). The action is not globally supersymmetric (in d dimensions).

As a result of the gauge invariances and the absence of kinetic terms for the "graviton" and "gravitino," the Legendre transformation is singular. One needs to apply the Dirac method in order to obtain the Hamiltonian. The procedure follows that for the bosonic string.

After introducing the appropriate simplifications of the formalism, the Hamiltonian (on using the approach of Ref. 49) is derived in the form

$$H = \int d\sigma \, (N\mathcal{H} + N^1 \mathcal{H}_1 + \bar{M}\mathcal{S}) \qquad (14.2.1.15)$$

where the constraints $\mathcal{H} = 0$, $\mathcal{H}_1 = 0$, and $\mathcal{S}_i = 0$ generate, respectively, reparametrizations ($\mathcal{H}$ and $\mathcal{H}_1$) and local supersymmetry transformations ($\mathcal{S}_i$). In equation (14.2.1.15), the local Lorentz gauge has been fixed by taking one leg of the vielbein orthogonal to the lines $\tau = $ const.

The remaining canonical variables are $X^A(\sigma)$, $\mathcal{P}_A(\sigma)$, and the two-dimensional spinors $\Gamma^A(\sigma) \equiv (\Gamma_i^A(\sigma))$, which obey the brackets (14.2.1.6).‡ The constraints $\mathcal{H}(\sigma)$ and $\mathcal{H}_1(\sigma)$ are related to $Q^\pm(\sigma)$:

$$\mathcal{H} = \frac{1}{4\pi}(Q^+ + Q^-) \qquad (14.2.1.16a)$$

$$\mathcal{H}_1 = \frac{1}{4\pi}(Q^+ - Q^-) \qquad (14.2.1.16b)$$

while the fermionic constraint $\mathcal{S}$ denotes $(\mathcal{S}_1, \mathcal{S}_2)$. Finally, the fermionic Lagrange multiplier M is equal to ψ_0.

This completes what we have to say about the derivation of the Hamiltonian from the Lagrangian (14.2.1.13). Indeed, we have already obtained the Hamiltonian above by applying the square-root approach,

---

† As a result of invariance (14.2.1.14), the "constraints" $T_{\alpha\beta} = 0$ and $J_\alpha = 0$ (zero supercurrent) are not independent. One finds indeed $T_\alpha{}^\alpha + \frac{1}{2}J_\alpha \psi^\alpha = 0$ (up to the Γ^A equations of motion) and $J_\alpha \rho^\alpha = 0$.
‡ We note that, in the canonical formalism, $\Gamma^A(\sigma)$ carries a factor $g_{11}^{1/4}$ ($\Gamma^A_{\text{can}} = g_{11}^{1/4} \Gamma^A_{\text{Lagr}}$).

which directly yields the canonical generators in terms of the dynamical variables alone. We also stress that there is no need to fix the local supersymmetries or reparametrizations in order to arrive at the form (14.2.1.7) and (14.2.1.9) of the constraint generators $\mathcal{H}$, $\mathcal{H}_1$, and $\mathcal{S}_i$.

Exercises

1. Derive the constraints $\mathcal{H}$, $\mathcal{H}_1$, and $\mathcal{S}$ from the Lagrangian.
2. Compare the spinor terms in $\mathcal{H}$ and $\mathcal{H}_1$ with the energy-momentum components $T_{\alpha\beta}(\Gamma)$.
3. Compute $[\Gamma_i^A(\sigma), \int d\sigma' \mathcal{H}_1(\sigma')N^1(\sigma')]$ and show that indeed $\Gamma_i^A(\sigma)$ carries weight $\tfrac{1}{2}$.
4. Compute the kinetic term in the canonical action which reproduces the brackets (14.2.1.6).

14.2.2. Boundary Conditions

We assume that the bosonic variables $X^A(\sigma)$, $\mathcal{P}_A(\sigma)$, $N(\sigma)$, and $N^1(\sigma)$ still obey the same boundary conditions as before. We wish to find the behavior of $\Gamma_i^A(\sigma)$ at the end points.

14.2.2a. Open String

The boundary conditions on $\Gamma_i^A(\sigma)$ should be such that the functionals $\int_0^\pi N\mathcal{H}\, d\sigma$, $\int_0^\pi N^1\mathcal{H}_1\, d\sigma$, and $\int_0^\pi \bar{M}\mathcal{S}\, d\sigma$ are well defined as generators, i.e., one does not pick up "surface terms" at $\sigma = 0, \pi$ in their variations.

We compute $\delta \int_0^\pi N\mathcal{H}\, d\sigma$. The boundary term

$$N(\Gamma_{1A}\delta\Gamma_1^A - \Gamma_{2A}\delta\Gamma_2^A)\big|_0^\pi \qquad (14.2.2.1)$$

is obtained. In order that the term (14.2.2.1) should vanish, one might be tempted to set $\Gamma_{iA}(\sigma) = 0$ at the end points. However, this condition is too strong. Indeed, one would infer that, in order to maintain (in time) Γ_{iA}(end points) $= 0$, all derivatives of Γ_{iA} should vanish at the end points (use the Γ equations of motion). Hence something else must be done.

Since $N(0)$ and $N(\pi)$ are independent, the only possibility is to relate Γ_1^A and Γ_2^A at the boundaries. Two different choices have been considered:

$$\begin{aligned}\Gamma_1^A(0) &= \Gamma_2^A(0) \\ \Gamma_1^A(\pi) &= \Gamma_2^A(\pi)\end{aligned} \qquad \text{(Ramond)} \qquad (14.2.2.2)$$

or

$$\begin{aligned}\Gamma_1^A(0) &= \Gamma_2^A(0) \\ \Gamma_1^A(\pi) &= -\Gamma_2^A(\pi)\end{aligned} \qquad \text{(Neveu-Schwarz)} \qquad (14.2.2.3)$$

(One easily checks that only the relative sign at 0, π is relevant.) The first choice corresponds to the Ramond model, the second to the Neveu-Schwarz model.

Subject to conditions (14.2.2.2), the boundary term (14.2.2.1) vanishes. Furthermore, there is no problem with the generator $\int_0^\pi N^1 \mathcal{H}_1 \, d\sigma$ since N^1 is zero at the end points. Finally, one sees from $\delta \int_0^\pi \bar{M} \mathcal{S} \, d\sigma$ that the supergauge parameter must be restricted by

$$M^1 = M^2 \qquad \text{at } \sigma = 0$$
$$M^1 = M^2 \text{ (R)} \quad \text{or} \quad M^1 = -M^2 \text{ (NS)} \qquad \text{at } \sigma = \pi \tag{14.2.2.4}$$

One can again summarize conditions (14.2.2.2) and (14.2.2.3) by extending the interval $[0, \pi]$ to $[-\pi, \pi]$. We define

$$\Gamma^A(\sigma) = \begin{cases} \Gamma_1^A(\sigma), & 0 \leq \sigma \leq \pi \\ \Gamma_2^A(-\sigma), & -\pi \leq \sigma \leq 0 \end{cases} \tag{14.2.2.5}$$

Because $\Gamma_1^A(0) = \Gamma_2^A(0)$, $\Gamma^A(\sigma)$ is continuous at the origin. [Actually, the preservation in time of the boundary conditions by the equations of motion implies $\Gamma_1^{A(n)}(0) = (-)^n \Gamma_2^{A(n)}(0)$ for all n, so that all derivatives match.] It is then seen that $\Gamma^A(\sigma)$ as given by equation (14.2.2.5) is periodic, of period 2π, in the case of the Ramond boundary conditions. In the Neveu-Schwarz case, $\Gamma^A(\sigma)$ is antiperiodic $[\Gamma^A(\sigma + 2\pi) = -\Gamma^A(\sigma)]$.

If the functions

$$Q^+(\sigma) = P^2 + \tfrac{1}{2}i\Gamma^A \frac{d\Gamma^A}{d\sigma}, \qquad -\pi \leq \sigma \leq \pi \tag{14.2.2.6a}$$

and

$$\mathcal{S}(\sigma) = \Gamma^A(\sigma) P_A(\sigma), \qquad -\pi \leq \sigma \leq \pi \tag{14.2.2.6b}$$

are introduced on the whole interval $[-\pi, +\pi]$, one can replace $Q^+(\sigma) = Q^-(\sigma) = 0 = \mathcal{S}_i(\sigma)$ on $[0, \pi]$ by

$$Q^+(\sigma) = 0 = \mathcal{S}(\sigma), \qquad -\pi \leq \sigma \leq \pi \tag{14.2.2.7}$$

on $[-\pi, +\pi]$.

14.2.2b. Closed String

The open-string analysis suggests that one should consider both antiperiodic and periodic $\Gamma_i^A(\sigma)$. Such an approach leads to three possibilities

[38, 50]:

$$\left.\begin{array}{l}\Gamma_1^A(\sigma=0)=\Gamma_1^A(\sigma=2\pi)\\ \Gamma_2^A(\sigma=0)=\Gamma_2^A(\sigma=2\pi)\end{array}\right\} \quad (14.2.2.8a)$$

$$\left.\begin{array}{l}\Gamma_1^A(\sigma=0)=-\Gamma_1^A(\sigma=2\pi)\\ \Gamma_2^A(\sigma=0)=\Gamma_2^A(\sigma=2\pi)\end{array}\right\} \quad (14.2.2.8b)$$

$$\left.\begin{array}{l}\Gamma_1^A(\sigma=0)=-\Gamma_1^A(\sigma=2\pi)\\ \Gamma_2^A(\sigma=0)=-\Gamma_2^A(\sigma=2\pi)\end{array}\right\} \quad (14.2.2.8c)$$

The supergauge parameters $\varepsilon_1(\sigma)$ and $\varepsilon_2(\sigma)$ must be restricted in the same way. Under these conditions, the Hamiltonian is a consistent generator (the "end-point terms" at $\sigma = 0$ and 2π cancel).

14.2.3. Supergauge Transformations—Light-Cone Gauge Conditions

The constraints $\mathcal{S}(\sigma) = 0$ generate local supersymmetry transformations, or, as they are also called, "supergauge" transformations. These are:

$$\delta\Gamma_1^A(\sigma) = \left[\Gamma_1^A(\sigma), \int \varepsilon^1(\sigma')\Gamma_1^B(\sigma')P_B(\sigma')\,d\sigma'\right]$$

$$= 4\pi i \varepsilon^1(\sigma) P^A(\sigma) \quad (14.2.3.1)$$

$$\delta\Gamma_2^A(\sigma) = 4\pi i \varepsilon^2(\sigma) S^A(\sigma) \quad (14.2.3.2)$$

$$\delta X^A(\sigma) = \pi\sqrt{2}\alpha'\varepsilon^i(\sigma)\Gamma_i^A(\sigma) \quad (14.2.3.3)$$

$$\delta P_A(\sigma) = 2\pi(-\varepsilon^1\Gamma_{A1} + \varepsilon^2\Gamma_{A2})' \quad (14.2.3.4)$$

The light-cone gauge is defined as above by $X^+ \sim p^+\tau$, $\mathcal{P}^{+\prime} = 0$. One must also fix the supergauge, for which one adopts as gauge conditions

$$\Gamma_i^+(\sigma) = 0 \quad (14.2.3.5)$$

There are as many conditions in equation (14.2.3.5) as there are supergauge parameters. Furthermore, once $\Gamma_i^+ = 0$, one can no longer perform supergauge transformations without violating equation (14.2.3.5) (assuming $p^+ \neq 0$). The supergauge is completely fixed.

We recall that the conditions $X^+ \sim p^+\tau$ and $\mathcal{P}^{+\prime} = 0$ still allow for zero-mode translations in σ in the case of the closed string.

Exercise. Show that there is enough freedom at the end points to satisfy equation (14.2.3.5).

14.2.4. Poincaré Generators

The variables X^A and Γ_i^A are Lorentz vectors, so one can write straightforwardly the Lorentz generators,

$$M_{AB} = \tfrac{1}{2} \int_0^{\pi \text{ or } 2\pi} (\mathcal{P}_A X_B - \mathcal{P}_B X_A) \, d\sigma$$

$$- \frac{1}{8\pi i} \int_0^{\pi \text{ or } 2\pi} \sum_i \Gamma_{Ai} \Gamma_{Bi} \, d\sigma \quad (14.2.4.1)$$

It is easy to check that $[\Gamma_{Ai}(\sigma), M_{BC}]$ yields the correct transformation law for $\Gamma_{Ai}(\sigma)$. Furthermore, the translations generators are unchanged:

$$P^A = \int_0^{\pi \text{ or } 2\pi} \mathcal{P}^A(\sigma) \, d\sigma \quad (14.2.4.2)$$

because only $X^A(\sigma)$ is modified in a space-time translation ($\delta \Gamma_i^A = 0$).

14.3. FOURIER MODES (OPEN STRING)

14.3.1. Fourier Expansion of the Fields

The bosonic fields obey the same boundary conditions as in the bosonic model. They therefore possess the same Fourier decomposition.

The Fourier expansion of the Fermionic field $\Gamma^A(\sigma)$ depends on whether $\Gamma^A(\sigma)$ is periodic or antiperiodic.

14.3.1a. Ramond Model

The function $\Gamma^A(\sigma)$ is periodic and one sets

$$\Gamma^A(\sigma) = \sum_m \Gamma_m^A \exp -im\sigma, \quad m = 0, \pm 1, \pm 2, \ldots \quad (14.3.1.1)$$

The field $\Gamma^A(\sigma)$ is real, hence

$$\Gamma_m^{A*} = \Gamma_{-m}^A \qquad (14.3.1.2)$$

Therefore the zero mode is self-conjugate. Furthermore, the Poisson brackets satisfy

$$[\Gamma_m^A, \Gamma_{m'}^B] = -2i\eta^{AB}\delta_{m,-m'} \qquad (14.3.1.3)$$

14.3.1b. Neveu–Schwarz Model

The function $\Gamma^A(\sigma)$ is now antiperiodic, so its Fourier expansion only contains half-integer modes:

$$\Gamma^A(\sigma) = \sqrt{2}\sum_s b_s^A \exp -is\sigma, \qquad s = \pm\tfrac{1}{2}, \pm\tfrac{3}{2}, \ldots \qquad (14.3.1.4)$$

It does not include a zero mode.

One easily obtains the Poisson-bracket relations:

$$[b_r^A, b_s^B] = -i\eta^{AB}\delta_{r,-s} \qquad (14.3.1.5)$$

and

$$b_s^{A*} = b_{-s}^A \qquad (14.3.1.6)$$

14.3.2. Super-Virasoro Generators

We define

$$L_n = \frac{1}{4\pi}\int_{-\pi}^{+\pi} d\sigma\, e^{in\sigma} Q^+(\sigma) \qquad (14.3.2.1)$$

$$G_s = \frac{1}{4\pi\sqrt{2}}\int_{-\pi}^{+\pi} d\sigma\, e^{is\sigma}\mathcal{S}(\sigma) \qquad \text{(Neveu–Schwarz)} \qquad (14.3.2.2)$$

$$F_n = \frac{1}{4\pi\sqrt{2}}\int_{-\pi}^{+\pi} d\sigma\, e^{in\sigma}\mathcal{S}(\sigma) \qquad \text{(Ramond)} \qquad (14.3.2.3)$$

In terms of Fourier modes, the super-Virasoro generators L_n, G_r, and F_n are given by

$$L_n = L_n^{\text{bosonic}} + \tfrac{1}{2}\sum_{s\geq 1/2}(2s+n)b_s^{*A}b_{An+s}$$

$$+ \tfrac{1}{2}\sum_{s=1/2}^{n-1/2} sb_{n-s}^A b_{AS} \qquad (n>0) \qquad (14.3.2.4a)$$

$$L_{-n} = L_n^* \qquad (14.3.2.4b)$$

$$L_0 = L_0^{\text{bosonic}} + \sum_{s>0} s b_s^{*A} b_{As} \qquad (14.3.2.4c)$$

and

$$G_r = \sqrt{2\alpha'} b_r^A p_A - i \sum_{0<n<r} \sqrt{n} b_{r-n}^A a_{An}$$

$$- i \sum_{n>r} \sqrt{n} b_{n-r}^{*A} a_{An} + i \sum_{n>0} \sqrt{n} b_{n+r}^A a_{An}^* \qquad (r>0) \qquad (14.3.2.5a)$$

$$G_{-r} = G_r^* \qquad (14.3.2.5b)$$

for the Neveu–Schwarz model, and

$$L_n = L_n^{\text{bosonic}} + \tfrac{1}{4} \sum_{k>0} (2k+n) \Gamma_k^{*A} \Gamma_{An+k}$$

$$+ \tfrac{1}{4} \sum_{k<n} k \Gamma_{n-k}^A \Gamma_{Ak} + \tfrac{1}{4} n \Gamma_0^A \Gamma_{An} \qquad (n>0) \qquad (14.3.2.6a)$$

$$L_{-n} = L_n^* \qquad (14.3.2.6b)$$

$$L_0 = L_0^{\text{bosonic}} + \tfrac{1}{2} \sum_{k>0} k \Gamma_k^{*A} \Gamma_{Ak} \qquad (14.3.2.6c)$$

and

$$F_n = \frac{1}{\sqrt{2}} \Bigg[\sqrt{2\alpha'} \Gamma_n^A p_A - i \sum_{0<k\le n} \sqrt{k} \Gamma_{n-k}^A a_{Ak}$$

$$- i \sum_{k>n} \sqrt{k} \Gamma_{k-n}^{A*} a_{Ak} + i \sum_{k>0} \sqrt{k} a_{Ak}^* \Gamma_{k+n}^A \Bigg] \qquad (n>0) \qquad (14.3.2.7a)$$

$$F_{-n} = F_n^* \qquad (14.3.2.7b)$$

$$F_0 = \frac{1}{\sqrt{2}} \left(\sqrt{2\alpha'} \Gamma_0^A p_A - i \sum_{k>0} \sqrt{k} \Gamma_k^{A*} a_{Ak} + i \sum_{k>0} a_{Ak}^* \Gamma_{Ak} \right) \qquad (14.3.2.7c)$$

for the Ramond model.

The super-Virasoro generators close classically according to the following (graded) Poisson-bracket algebra:

$$[G_r, G_s] = -2i L_{r+s} \qquad (14.3.2.8a)$$

$$[L_m, G_r] = i(r - m/2)G_{m+r} \qquad (14.3.2.8b)$$

$$[F_m, F_n] = -2iL_{m+n} \qquad (14.3.2.8c)$$

$$[L_m, F_n] = i(n - m/2)F_{m+n} \qquad (14.3.2.8d)$$

$$[L_m, L_n] = i(n - m)L_{m+n} \qquad (14.3.2.8e)$$

It goes without saying that in all the above equations, r and s stand for half-integers while m, n, and k are integers.

14.3.3. Poincaré Generators

In terms of Fourier modes, the Poincaré generators are given by

$$P^A = p^A \qquad (14.3.3.1)$$

$$M^{AB} = \tfrac{1}{2}(p^A X_0^B - p^B X_0^A)$$
$$+ \tfrac{1}{2} \sum_{n>0} i(a_n^{A*} a_n^B - a_n^{B*} a_n^A) + I^{AB} \qquad (14.3.3.2a)$$

with

$$I^{AB} = \frac{i}{2} \sum_{s>0} (b_s^{A*} b_s^B - b_s^{B*} b_s^A) \qquad \text{(Neveu-Schwarz)} \qquad (14.3.3.2b)$$

or

$$I^{AB} = \frac{i}{4} \sum_{m>0} (\Gamma_m^{A*}\Gamma_m^B - \Gamma_m^{B*}\Gamma_m^A) + \frac{i}{4}\Gamma_0^A \Gamma_0^B \qquad \text{(Ramond)} \qquad (14.3.3.2c)$$

We note that the last term in expression (14.3.3.2c) is antisymmetric in A and B—as it should be—since $\Gamma_0^A \Gamma_0^B = -\Gamma_0^B \Gamma_0^A$.

14.3.4. Remarks on the Closed String

The closed string is again essentially the "direct product" of two open strings. One finds two sets of "open-string constraints," which are L_n, $\bar{L}_n$, F_m, $\bar{F}_m$ [if both $\Gamma_1^A(\sigma)$ and $\Gamma_2^A(\sigma)$ are periodic], L_n, $\bar{L}_n$, G_r, $\bar{F}_m$ [if $\Gamma_1^A(\sigma)$ is antiperiodic and $\Gamma_2^A(\sigma)$ periodic], or L_n, $\bar{L}_n$, G_r, $\bar{G}_r$ [if both $\Gamma_1^A(\sigma)$ and $\Gamma_2^A(\sigma)$ are antiperiodic].

In the closed-string expressions for the constraints the momentum p^A is replaced by $(\tfrac{1}{2})p^A$, while the oscillators a_n^A and b_n^A, Γ_n^A are replaced by the right and left movers c_n^A, $\bar{c}_n^A$, b_n^A, $\bar{b}_n^A$, Γ_n^A, $\bar{\Gamma}_n^A$.

The right and left sectors are related through the zero-mode constraints only.

14.3.5. Super-Virasoro Algebra

In the covariant approach to quantization, all dynamical variables are treated as operators in a pseudo-Hilbert space. Besides the former bosonic operators, one has fermionic creation and destruction operators b_r^{A*} and b_r^A obeying†

$$[b_r^A, b_s^{B*}] = \eta^{AB}\delta_{r,s} \qquad (r, s > 0) \qquad (14.3.5.1)$$

in the Neveu–Schwarz model. The vacuum is annihilated by all b_r^A ($r > 0$)

$$b_r^A |0, p\rangle = 0 \qquad (r > 0) \qquad (14.3.5.2)$$

In the Ramond model, the nonzero modes Γ_m^A play a similar role,

$$[\Gamma_m^A, \Gamma_n^{B*}] = 2\eta^{AB}\delta_{m,n} \qquad (m, n > 0) \qquad (14.3.5.3)$$

$$\Gamma_m^A |0, p\rangle = 0 \qquad (m > 0) \qquad (14.3.5.4)$$

One must also represent the zero modes Γ_0^A. These form a Clifford algebra,

$$\Gamma_0^A \Gamma_0^B + \Gamma_0^B \Gamma_0^A = 2\eta^{AB} \qquad (14.3.5.5)$$

They can therefore be identified with Γ matrices in d dimensions.

Hence, one sees that the ground state of the Ramond model (and all the excited states) must belong to the representation space of the Γ matrices. It accordingly carries spin $\frac{1}{2}$ (ground-state spinor). In d dimensions, the ground state is $2^{d/2}$ degenerate.‡ The excited states also possess half-integer spin.

The Hilbert space constructed here contains many negative-norm states, which are generated by the temporal components of the oscillators. Also, relation (14.3.5.5) implies $(\Gamma_0^0)^2 = -1$, which is compatible with $(\Gamma_0^0)^* = \Gamma_0^0$ only if the inner product is nondefinite positive.

---

† In relation (14.3.5.1) the bracket denotes the anticommutator, corresponding to Fermi variables.

‡ The Fermi oscillators anticommute with Γ_0^A. When d is even, one can redefine new Fermi oscillators given by $\Gamma_0^{d+1}\Gamma_m^A$, which commute with the Γ matrices. (Here, $\Gamma_0^{d+1} = \Gamma_0^0 \Gamma_0^1 \cdots \Gamma_0^{d-1}$.) It is then seen that the pseudo-Hilbert space under consideration is the direct product of the Fock space generated by the bosonic and fermionic oscillators, of the Hilbert space for the zero modes X_0^A, p^A, and of the Clifford-algebra representation space.

Elimination of the negative-norm states is the central issue in the covariant approach to quantization. It will not be studied in this section. Our purpose here is rather to point out that the classical (super-)algebra of the constraints is modified quantum mechanically by c-number anomalous terms.

The super-Virasoro operators are not afflicted by ordering ambiguities, except L_0, for which different orders may lead to two quantities L_0 which differ by a c number. Hence, we allow for an as yet arbitrary constant α_0 in the quantum L_0,

$$L_0^{cl} \to L_0 - \alpha_0 \qquad (14.3.5.6)$$

where L_0 denotes the normal-ordered expression.

The computation of the central charge in the super-Virasoro algebra proceeds exactly as in the bosonic case. For this reason we only reproduce the result. The modified (graded) commutation relations are

$$[L_n, L_m] = (n-m)L_{n+m} + \frac{d}{8}(n^3 - n)\delta_{n,-m} \qquad (14.3.5.7a)$$

$$[G_r, G_s] = 2L_{r+s} + \frac{d}{2}(r^2 - \tfrac{1}{4})\delta_{r,-s} \qquad (14.3.5.7b)$$

for the Neveu–Schwarz model, and

$$[L_n, L_m] = (n-m)L_{n+m} + \frac{d}{8}n^3 \delta_{n,-m} \qquad (14.3.5.8a)$$

$$[F_n, F_m] = 2L_{n+m} + \frac{d}{2}n^2 \delta_{n,-m} \qquad (14.3.5.8b)$$

for the Ramond model. The commutators $[L_n, G_r]$ or $[L_n, F_m]$ do not pick up anomalous terms.

It is noteworthy that in the Ramond model, the zero mode Γ_0^A also contributes to the anomaly (one no longer has $\Gamma_0^A \cdot \Gamma_0^A = 0$ quantum mechanically). That zero-mode contribution can be controlled by adopting the antisymmetric ordering for products of matrices Γ_0^A.

Owing to the central charge, one can only impose "half" of the constraints in the quantum theory,

$$(L_0 - \alpha_0)|\psi\rangle = 0, \qquad L_n|\psi\rangle = 0 \quad (n > 0) \qquad (14.3.5.9a)$$

and

$$G_s|\psi\rangle = 0 \quad (s > 0) \quad \text{or} \quad F_n|\psi\rangle = 0 \quad (n \geq 0) \quad (14.3.5.9b)$$

Exercises

1. Compute explicitly the central charge in relations (14.3.5.7) and (14.3.5.8).
2. Show that by adding to L_0 an appropriate constant, one can transform relation (14.3.5.8) to the form (14.3.5.7).

Chapter 15

The Fermionic String: Quantum Analysis

It is shown in this chapter that the fermionic-string critical dimension is equal to 10. Furthermore, the intercept parameter is $\frac{1}{2}$ for the Neveu–Schwarz model and vanishes with the Ramond boundary conditions. These critical values are derived through both BRST and light-cone gauge methods. They coincide with those given by the covariant approach, not studied here.

As a rule, we only consider the open string in actual computations, unless otherwise specified.

15.1. BECCHI–ROUET–STORA–TYUTIN (BRST) QUANTIZATION OF THE NEVEU–SCHWARZ MODEL

15.1.1. Ghost Fock Space

Besides the fermionic ghosts encountered previously, the ghost modes now include commuting degrees of freedom, which we denote by q_s and π_s. These are half-integer moded, since they are associated with the fermionic constraints $G_s = 0$. They obey

$$q_s^* = q_{-s}, \qquad \pi_s^* = \pi_{-s} \qquad (15.1.1.1)$$

and

$$[q_s, \pi_{s'}^*] = i\delta_{ss'} = [q_s^*, \pi_{s'}] \qquad (15.1.1.2)$$

All other commutators vanish.

One can define

$$\sqrt{2}\mu_s = q_s + i\pi_s \quad \text{and} \quad \sqrt{2}\nu_s = q_s - i\pi_s \tag{15.1.1.3}$$

These new variables obey oscillatorlike commutation relations. We note that ν_s^* create negative-norm states from the vacuum, and also that there is no zero-mode commuting ghost since there is no zero-mode fermionic constraint.

15.1.2. BRST Operator

According to the general rules given in Refs. 30 and 31, the BRST operator is given by

$$\Omega = \Omega^B + \sum_{r=-\infty}^{+\infty} G_r q_{-r} - \sum_{r,s} \mathcal{P}_{-r-s} q^r q^s$$
$$- i \sum_{s,m} \pi_{-m-s} \eta^m q^s (s - m/2) \tag{15.1.2.1}$$

where Ω^B is the bosonic string BRST generator.

Normal ordering of this equation yields the expression

$$\Omega = \Omega^B + \sum_{r>0} (q_r^* G_r + G_{-r} q_r) - \sum_{r,s>0} (\mathcal{P}_{r+s}^* q^r q^s + q^{*r} q^{*s} \mathcal{P}_{r+s})$$

$$- 2 \sum_{r>0} \sum_{0<s<r} (\mathcal{P}_{r-s}^* q_s^* q_r + q_r^* \mathcal{P}_{r-s} q_s)$$

$$- i \sum_{r,m>0} (r - m/2)(\pi_{m+r}^* \eta^m q^r - \eta^{*m} q^{*r} \pi_{m+r})$$

$$- i \sum_{r>0} \sum_{0<m<r} (r + m/2)(\pi_{r-m}^* \eta_m^* q_r - q_r^* \eta_m \pi_{r-m})$$

$$- i \sum_{m>0} \sum_{0<r<m} (r + m/2)(\eta_m^* \pi_{m-r} q_r - \pi_{m-r}^* q_r^* \eta_m)$$

$$- 2 \sum_{r>0} q_r^* q_r \mathcal{P}_0 - i \sum_{r>0} r(\pi_r^* q_r - q_r^* \pi_r) \eta^0 \tag{15.1.2.2}$$

There is actually no ordering ambiguity, except in the η^0 term. Two admissible orderings yield coefficients of η^0 which differ by a real c number, and which is absorbed in α_0.

Exercise. Check the classical nilpotency of Ω, $[\Omega, \Omega] = 0$.

15.1.3. Critical Dimension

Nilpotency of the quantum BRST generator is again not obvious, owing to the nonexact commutativity (or anticommutativity) of quantum operators.

Apart from the α_0 term and the central charge in the super-Virasoro algebra, the only other nonvanishing contributions to Ω^2 are those due to the ghosts. These latter contributions arise because the anticommutators analogous to the anticommutators (13.2.4.2) are not normal-ordered.

One finds explicitly

$$\Omega^2 = \sum_{n>0} \left[n^3 \left(\frac{d}{8} - \frac{5}{4} \right) + n \left(2\alpha_0 - \frac{d}{8} + \frac{1}{4} \right) \right] \eta_n^* \eta_n$$

$$+ \sum_{s>0} \left[s^2 \left(\frac{d-10}{2} \right) + \tfrac{1}{8}(16\alpha_0 - d + 2) \right] q_s^* q_s \quad (15.1.3.1)$$

If we set $\Omega^2 = 0$, then we obtain the announced conditions,

$$d = 10 \quad \text{and} \quad \alpha_0 = \tfrac{1}{2} \quad (15.1.3.2)$$

The intercept parameter is still positive and there is a tachyon in the spectrum.

Exercise. Check relation (5.1.3.1).

The fermionic-string critical dimension was recently computed by the same BRST methods as in Refs. 35, 51, and 52.

15.1.4. Structure of the Physical Subspace

We now characterize all solutions to $\Omega|\psi\rangle = 0$. For definiteness, the state $|\psi\rangle$ is regarded as a vector given by

$$|\psi\rangle = \sum \lambda_k |\psi_k\rangle \quad (15.1.4.1)$$

where λ_k are functions of the zero modes only,

$$\lambda_k = \lambda_k(X_0^A, \eta^0) \quad (15.1.4.2)$$

and where $|\psi_k\rangle$ are Fock-space vectors. The state (15.1.4.1) can also be written as

$$|\psi\rangle = |a\rangle + |b\rangle \eta^0 \quad (15.1.4.3)$$

if one wishes to single out the fermionic ghost zero mode.

The BRST operator (15.1.2.2) is

$$\Omega = (\alpha' p^2 + L - \alpha_0)\eta^0 - M\mathcal{P}_0 + \bar{\Omega} \qquad (15.1.4.4)$$

where, as in the case of the bosonic string, L is the BRST-invariant extension of the level-number operator, which also counts the ghosts:

$$L = N^a + N^b + N^\eta + N^q \qquad (15.1.4.5)$$

with

$$N^a = \sum_{n>0} n a_n^{*A} a_{An}, \qquad N^b = \sum_{s>0} s b_s^{*A} b_{As} \qquad (15.1.4.6)$$

$$N^\eta = \sum_{n>0} n(\mathcal{P}_n^* \eta_n + \eta_n^* \mathcal{P}_n) \qquad (15.1.4.7a)$$

$$N^q = -i \sum_{r>0} r(\pi_r^* q_r - q_r^* \pi_r) = \sum_{r>0} r(\mu_r^* \mu_r - \nu_r^* \nu_r) \qquad (15.1.4.7b)$$

M is the operator given by

$$M = 2 \sum_{n>0} n \eta_n^* \eta_n + 2 \sum_{r>0} q_r^* q_r \qquad (15.1.4.8)$$

and $\bar{\Omega}$ is the remaining part of the BRST generator. It does not involve the ghost zero modes.

One easily checks that the following commutation relations hold:

$$[L, M] = 0 = [L, \Omega] = [M, \Omega] \qquad (15.1.4.9)$$

and, furthermore, that $\bar{\Omega}$ is nilpotent on the subspace $\alpha' p^2 + L - \alpha_0 = 0$.

When acting on state (15.1.4.3), the BRST operator yields

$$\bar{\Omega}|a\rangle - M|b\rangle + [(\alpha' p^2 + L - \alpha_0)|a\rangle + \bar{\Omega}|b\rangle]\eta^0 \qquad (15.1.4.10)$$

This observation leads to

Theorem. The η^0 part of any state $|\psi\rangle$ can be absorbed in a state of the form $\Omega|\chi\rangle$.

Proof. We take $|\chi\rangle$ equal to $|a'\rangle$ (i.e., no η^0 contribution to $|\chi\rangle$), with $|a'\rangle$ the solution of

$$(-\alpha'\Box + L - \alpha_0)|a'\rangle = -|b\rangle \qquad (15.1.4.11)$$

This equation always admits solutions, as can be seen by expanding $|b\rangle$ as in equation (15.1.4.1),

$$|b\rangle = \sum_k \lambda_k(X_0^A)|b_k\rangle \qquad (15.1.4.12)$$

Here, one can assume the Fock-space vectors $|b_k\rangle$ to have definite level number L_k. Hence equation (15.1.4.11) becomes

$$(-\alpha'\Box + L_k - \alpha_0)\mu_k = -\lambda_k \qquad (15.1.4.13)$$

for the unknown coefficients $\mu_k(X_0^A)$ of $|a'\rangle$ in the expansion $|a'\rangle = \sum_k \mu_k|a'_k\rangle$. Clearly one can always solve equation (15.1.4.13), although μ_k, and hence also $|a'\rangle$, might blow up at infinity if $|b\rangle$ has a nonvanishing component along $\alpha'p^2 + L - \alpha_0 = 0$.

The $|a'\rangle$ solution of equation (15.1.4.11) yields $|b\rangle = 0$ by adding $\Omega|\chi\rangle$ to $|\psi\rangle$. This proves the theorem.

From now on we assume $|b\rangle = 0$ or, what is the same, we allow for unbounded functions such as the above $|\chi\rangle$ in $\Omega|\chi\rangle$, so that $|b\rangle$ can be removed.

At the same time, we regularize the ill-defined scalar products of solutions to $\Omega|\psi\rangle = 0$—which contain (as mentioned previously) the factor $\delta(0) \cdot 0$ arising from integrals over the mass and over η^0—by setting $\delta(0) \cdot 0$ equal to one, in order to recover the usual normalization of Klein–Gordon solutions. It was explained in Section 13.2.7 that there is so far no rigorous formulation which avoids this scalar-product difficulty.

Allowing for unbounded functions $|\chi\rangle$ in $\Omega|\chi\rangle$ has the great virtue of eliminating the doubling of states, which does not appear to have any great physical significance.

When $|b\rangle = 0$, the BRST-invariance physical condition reduces to

$$\bar{\Omega}|a\rangle = 0 \qquad (15.1.4.14a)$$

and

$$(\alpha'p^2 + L - \alpha_0)|a\rangle = 0 \qquad (15.1.4.14b)$$

The ghost zero mode has disappeared and $|a\rangle$ must be on the mass shell.

Equation (15.1.4.14a) is analyzed as in the bosonic model, by relying on the Kugo and Ojima quartet mechanism. One can prove that any state solution to equations (15.1.4.14) is of the form

$$|a\rangle = |P\rangle|0\rangle_{\text{ghost}} + \bar{\Omega}|c\rangle \qquad (15.1.4.15)$$

where $|0\rangle_{\text{ghost}}$ is the ghost vacuum (for the non-zero modes) and $|P\rangle$ is a physical state of the covariant approach,

$$(L_0 - \alpha_0)|P\rangle = 0$$

$$L_n|P\rangle = 0 \quad (n > 0) \tag{15.1.4.16}$$

$$G_r|P\rangle = 0 \quad (r > 0)$$

Actually, $|P\rangle$ can be assumed to be "transverse" in solution (15.1.4.15) [51–53]. This yields a BRST-based proof of the "no-ghost theorem." A discussion of "transverse," i.e., generalized DDF states in the fermionic-string model can be found in the papers quoted in Ref. 54.

Relations (15.1.4.16) and (15.1.4.16), and the previous theorem, together give.

Theorem. Any state solution of $\Omega|\psi\rangle = 0$ can be written as

$$|\psi\rangle = |P\rangle|0\rangle_{\text{ghost}} + \Omega|\chi\rangle \tag{15.1.4.17}$$

where $|P\rangle$ is a ghost-free, physical state,

$$(L_0 - 1/2)|P\rangle = 0$$

$$L_n|P\rangle = 0 \quad (n > 0) \tag{15.1.4.18}$$

$$G_r|P\rangle = 0 \quad (r > 0)$$

We finally note that if one restricts the asymptotic behavior of $|\chi\rangle$ at infinity, then one finds a doubling. Equation (15.1.4.17) must be replaced by

$$|\psi\rangle = |P_1\rangle|0\rangle_{\text{ghost}} + |P_2\rangle|0\rangle_{\text{ghost}}\eta^0 + \Omega|\chi\rangle \tag{15.1.4.19}$$

Also, in the closed-string case, there is a more serious "doubling" associated with the zero mode $L_0 - \bar{L}_0$.

15.2. BECCHI–ROUET–STORA–TYUTIN (BRST) QUANTIZATION OF THE RAMOND MODEL

15.2.1. Ghost Fock Space

The main difference with respect to the Neveu–Schwarz model is that now the commuting ghosts associated with the fermionic constraints F_n are

15 • Fermionic String: Quantum Analysis

integer moded and include a zero mode. One has

$$q_n^* = q_{-n}, \quad \pi_n^* = \pi_{-n} \tag{15.2.1.1}$$

and

$$[q_n, \pi_{n'}] = i\delta_{n,-n'} \tag{15.2.1.2}$$

One can define a Fock space for the nonzero modes q_n, π_n ($n \neq 0$),

$$q_n|0\rangle = \pi_n|0\rangle = 0 \quad (n > 0) \tag{15.2.1.3}$$

The pseudo-Hilbert space for the commuting ghosts is then obtained by taking the direct product of this Fock space with the space corresponding to the zero mode q^0. This space is tentatively realized in the usual way, as the space of functions of q^0:

$$f(q^0) \tag{15.2.1.4a}$$

$$\langle f, g \rangle = \int dq^0 f^*(q^0) g(q^0) \tag{15.2.1.4b}$$

$$q^0: \text{multiplication by } q^0 \tag{15.2.1.4c}$$

$$\pi_0 = \frac{1}{i} \frac{\partial}{\partial q^0} \tag{15.2.1.4d}$$

15.2.2. BRST Operator

The normal-ordered form of the BRST operator is given by

$$\Omega = \Omega^B + \sum_{n=-\infty}^{+\infty} F_n q_{-n} - \sum_{m,n>0} (\mathcal{P}^*_{m+n} q_m q_n + q_m^* q_n^* \mathcal{P}_{m+n})$$

$$- 2 \sum_{m>0} \sum_{0<n<m} (\overset{①}{\mathcal{P}^*_{m-n} q_n^* q_m} + \overset{③}{q_m^* \mathcal{P}_{m-n} q_n})$$

$$- i \sum_{n,m>0} (n - m/2)(\pi^*_{m+n} \eta^m q^n - \eta^{*m} q^{*n} \pi_{m+n})$$

$$- i \sum_{n>0} \sum_{0<m<n} (n + m/2)(\pi^*_{n-m} \overset{④}{\eta_m^*} q_n - q_n^* \overset{②}{\eta_m} \pi_{n-m})$$

$$- i \sum_{m>0} \sum_{0<n<m} (n + m/2)(\eta_m^* \overset{⑤}{\pi_{m-n}} q_n - \pi^*_{m-n} \overset{⑥}{q_n^*} \eta_m)$$

$$-2 \sum_{n>0} q_n^* q_n \mathcal{P}_0 \overset{⑦}{} - i \sum_{n>0} n(\pi_n^* q_n \overset{⑧}{-} q_n^* \pi_n) \eta^0$$

$$- \mathcal{P}_0 (q^0)^2 - 2 \sum_{n>0} (\mathcal{P}_n^* q_n \overset{⑨}{+} q_n^* \mathcal{P}_n) q^0 - \frac{3i}{2} \sum_{n>0} n(\eta_n^* q_n \overset{⑩}{-} q_n^* \eta_n) \pi_0$$

$$+ \frac{i}{2} \sum_{n>0} n(\pi_n^* \eta_n \overset{⑪}{-} \eta_n^* \pi_n) q^0 \tag{15.2.2.1}$$

This expression differs in structure from the Neveu–Schwarz BRST operator by the presence of the zero modes q^0 and π_0.

15.2.3. Critical Dimension

The critical dimension and the intercept parameter follow again from the requirement that Ω be nilpotent at the quantum level. The BRST operator fails to be nilpotent in general due to three types of terms:

1. Terms proportional to α_0. These are given (in Ω^2) by

$$2\alpha_0 \sum_{n>0} n \eta_n^* \eta_n + 2\alpha_0 \sum_{n>0} q_n^* q_n + \alpha_0 (q^0)^2 \tag{15.2.3.1}$$

2. Terms proportional to the central charge, and hence to the space-time dimension. These are equal to

$$\sum_{n>0} n^3 \frac{d}{8} \eta_n^* \eta_n + \sum_{n>0} n^2 \frac{d}{2} q_n^* q_n \tag{15.2.3.2}$$

3. Contributions resulting from the reordering of quartic ghosts terms. Those in $(\Omega^B)^2$ are given by

$$-\tfrac{13}{6} \sum_{n>0} n^3 \eta_n^* \eta_n + \tfrac{1}{6} \sum_{n>0} n \eta_n^* \eta_n \tag{15.2.3.3}$$

Those in the cross terms $\Omega^B \Omega^F + \Omega^F \Omega^B$ vanish, where Ω^F is the "fermionic part" of Ω, given explicitly in equation (15.2.2.1).

We now compute in detail the nonvanishing terms in $(\Omega^F)^2$, for the purpose of illustrating the calculations.

The "anomalous" terms come from the anticommutators analogous to equation (13.2.4.2) (possibly containing zero modes). The first such anomalous anticommutator comes from [①, ②] and is (in $\Omega^{F2} = \frac{1}{2}[\Omega^F, \Omega^F]$)

$$-2i \sum_{\substack{m,n>0 \\ 0<k<m \\ 0<l<n}} [\mathcal{P}^*_{m-k}q^*_k q_m, q^*_n \eta_l \pi_{n-l}](n+l/2)$$

$$= \text{N.O.} - 2 \sum_{m>0} \left(\sum_{0<k<m} (m+k/2) \right) q^*_m q_m$$

$$= \text{N.O.} - \sum_{m>0} \tfrac{5}{2} m(m-1) q^*_m q_m \qquad (15.2.3.4)$$

where N.O. stands for the nonanomalous (normal-ordered) expression. The next anomalous term is [③, ④]

$$2i \sum_{\substack{m,n>0 \\ 0<k<m \\ 0<l<n}} [q^*_m \mathcal{P}_{m-n} q_n, \pi^*_{n-m} \eta^*_m q_n]$$

$$= \text{N.O.} - \tfrac{5}{2} \sum_{m>0} m(m-1) q^*_m q_m \qquad (15.2.3.5)$$

It can easily be checked that there is no anomalous term in [①, ⑤] or [③, ⑥], so one is left to consider—besides the zero-mode contributions treated separately—the term [⑤, ⑥], which yields

$$\sum_{\substack{m>0 \ 0<k<m \\ n>0 \ 0<l<n}} \left(k + \frac{m}{2} \right) \left(l + \frac{n}{2} \right) [\eta^*_m \pi_{m-k} q_k, \pi^*_{n-l} q^*_l \eta_n]$$

$$= \text{N.O.} + \sum_{m>0} \eta^*_m \eta_m \left(\sum_{0<k<m} \left(k + \frac{m}{2} \right) \left(\frac{3m}{2} - k \right) \right)$$

$$= \text{N.O.} + \sum_{m>0} \eta^*_m \eta_m \left(\frac{11}{12} m^3 - \frac{3}{4} m^2 - \frac{m}{6} \right) \qquad (15.2.3.6)$$

We now consider the zero-mode anomalous terms, which appear in [⑦, ⑧], [⑨, ⑩], and [⑩, ⑪]:

$$[⑦, ⑧] \to 2i \sum_{m,n>0} n[q^*_m q_m \mathcal{P}_0, (\pi^*_n q_n - q^*_n \pi_n) \eta^0]$$

$$= \text{N.O.} - 2 \sum_{n>0} n q^*_n q_n \qquad (15.2.3.7)$$

[⑨, ⑩] → 3i $\sum_{n,m>0} n[(\mathcal{P}_m^* q_m + q_m^* \mathcal{P}_m) q^0, (\eta_n^* q_n - q_n^* \eta_n) \pi_0]$

$$= -3 \sum_{n>0} n q_n^* q_n + \text{N.O.} \qquad (15.2.3.8)$$

[⑩, ⑪] → $\frac{3}{4} \sum_{n,m>0} mn[(\eta_m^* q_m - q_m^* \eta_m) \pi_0, (\pi_n^* \eta_n - \eta_n^* \pi_n) q_0]$

$$= \text{N.O.} + \frac{3}{4} \sum_{n>0} n^2 \eta_n^* \eta_n \qquad (15.2.3.9)$$

By combining relations (15.2.3.1)-(15.2.3.8), we obtain

$$\Omega^2 = \sum_{n>0} \left[n^3 \left(\frac{d}{8} - \frac{5}{4} \right) + 2\alpha_0 n \right] \eta_n^* \eta_n$$

$$+ \sum_{n>0} \left[n^2 \left(\frac{d}{2} - 5 \right) + 2\alpha_0 \right] q_n^* q_n + \alpha_0 (q^0)^2 \qquad (15.2.3.10)$$

The condition $\Omega^2 = 0$ implies

$$d = 10 \qquad (15.2.3.11a)$$

as in the Neveu-Schwarz model, but also

$$\alpha_0 = 0 \qquad (15.2.3.11b)$$

The intercept parameter is now zero and the tachyon is avoided.

15.2.4. Structure of the Physical Subspace

The BRST operator can again be expanded in powers of η^0 and $\mathcal{P}_0$ to yield

$$\Omega = (\alpha' p^2 + L - \alpha_0) \eta^0 - M \mathcal{P}_0 + \bar{\Omega} \qquad (15.2.4.1)$$

where L, M, and $\bar{\Omega}$ are independent of η^0 and $\mathcal{P}_0$ and will not be written explicitly. The same arguments as in the Neveu-Schwarz case enable one to eliminate the η^0 part of physical states, i.e., we assume $|b\rangle = 0$ in $|\psi\rangle = |a\rangle + |b\rangle \eta^0$. For such states, the mass-shell condition must clearly be statisfied.

One can then further decompose $\bar{\Omega}$ by making explicit the commuting ghost zero-mode dependence:

$$\bar{\Omega} = Fq^0 + G\pi_0 + \bar{\bar{\Omega}} \qquad (15.2.4.2)$$

where F, G, and $\bar{\bar{\Omega}}$ are operators independent of all ghost zero modes. Their form can easily be read from expression (15.2.2.1). Corresponding to the operator decomposition (15.2.4.2), the states can be expressed as

$$|a\rangle = \sum_{n\geq 0} |a_n\rangle q_0^n \qquad (15.2.4.3)$$

We assume throughout that only a finite number of terms appear in equation (15.2.4.3) (forms of finite degree). With this assumption, the higher-power coefficient $|a_N\rangle$ must obey, from $\bar{\Omega}|a\rangle = 0$, the condition

$$F|a_N\rangle = 0 \qquad (15.2.4.4)$$

But this implies $|a_N\rangle = F|a'_{N-1}\rangle$ for some $|a'_{N-1}\rangle$. Hence one can remove the higher power $|a_N\rangle$ from expression (15.2.4.3), and then successively all the powers $n \leq N$, except the zero-order power. Once this is done, we are left with a physical state $|a\rangle$ which involves neither η^0 nor q^0. For such a state, the BRST condition $\Omega|a_0\rangle = 0$ becomes

$$(\alpha'p^2 + L - \alpha_0)|a_0\rangle = 0 \qquad (15.2.4.5a)$$

$$F|a_0\rangle = 0 \qquad (15.1.4.5b)$$

$$\bar{\bar{\Omega}}|a_0\rangle = 0 \qquad (15.2.4.5c)$$

The operator $\bar{\bar{\Omega}}$ is nilpotent on the subspace (15.2.4.5a) and (15.2.4.5b) and, as mentioned earlier, is independent of all ghost zero modes. Furthermore, one can show that it is a "perturbation" (in the sense of Kato and Ogawa) of a linear nilpotent operator to which the Kugo–Ojima quartet mechanism clearly applies. Hence one can remove all remaining ghosts from $|a_0\rangle$ [53].

We will not give any further details on this question here. We will also omit the problem of the regularization of the scalar product, which involves a second ill-defined product "$\delta(0) \cdot 0$," because the ghost zero-mode integral $\int dq^0$ yields infinity, while one collects a vanishing factor from the scalar product associated with the zero modes Γ_0^A. We refer the reader elsewhere [53] for more information on these questions.

The sketchy analysis of this section can be summarized by the following theorem:

Theorem. Any physical state $|\psi\rangle$ solution of $\Omega|\psi\rangle = 0$ can be written as

$$|\psi\rangle = |P\rangle|0\rangle_{\text{ghost}} + \Omega|\chi\rangle \qquad (15.2.4.6)$$

where $|P\rangle$ is a physical state of the covariant approach obeying

$$L_0|P\rangle = L_n|P\rangle = 0 \qquad (n > 0) \qquad (15.2.4.7a)$$

$$F_0|P\rangle = F_n|P\rangle = 0 \qquad (n > 0) \qquad (15.2.4.7b)$$

Proof [53]. Of course, the theorem does not say that $|\chi\rangle$ is normalizable.

Remarks. (1) If one restricts $|\chi\rangle$ by asymptotic conditions, then a doubling associated with the anticommuting ghost zero mode is obtained. One could also get another multiplicity due to the commuting ghost q^0 if conditions are imposed as $q^0 \to \pm\infty$. (2) The ghost number of the states (15.2.4.6) (with $|\chi\rangle = 0$) is 0.

Exercise. Show that any solution of $F|a_N\rangle = 0$ [equation (15.2.4.4)] can be written as $|a_N\rangle = F|a'_{N-1}\rangle$.

Hints. As usual, the momentum and the mass level can be treated as c numbers. Compute $[F(p), F(k)]$, where k^A is an arbitrary d vector. Adjust k^A such that $[F(p), F(k)]|\psi\rangle = |\psi\rangle$. Conclude that if $F(p)|\psi\rangle = 0$, then $|\psi\rangle = F(p)|\psi'\rangle$, with $-|\psi'\rangle = F(k)|\psi\rangle$. Here,

$$F(k) = \frac{1}{\sqrt{2}}\left[\sqrt{2\alpha'}\Gamma_0^A k_A + i\sum_{k>0}\sqrt{k}(a_k^{*A}\Gamma_{Ak} - \Gamma_{Ak}^* a_k^A)\right]$$

$$- 2\sum_{n>0}(\mathcal{P}_n^* q_n + q_n^*\mathcal{P}_n) + \frac{i}{2}\sum_n n(\pi_n^*\eta_n - \eta_n^*\pi_n)$$

with $F(p) \equiv F$.

15.2.5. Remarks on the Closed String

The application of BRST methods to the closed string is straightforward. The closed string is indeed just the direct product of one left-moving sector and one right-moving sector, which are isomorphic to either the (open) Neveu-Schwarz or Ramond model.

Accordingly, one finds that the BRST generator is nilpotent only in ten dimensions. Besides, the intercept parameter is either unity (left- and right-moving sectors isomorphic to Neveu–Schwarz), $\frac{1}{2}$ (left Neveu–Schwarz, right Ramond), or 0 (left and right Ramond).

15.3. LIGHT-CONE GAUGE QUANTIZATION OF THE NEVEU–SCHWARZ MODEL

15.3.1. Poincaré Invariance

The nontrivial question in the light-cone gauge quantum theory is whether the quantum Poincaré algebra remains anomaly-free.

The Lorentz generator that raises difficulties is M^{i-}, which is no longer quadratic in the oscillator variables. One finds that M^{i-} is given by

$$M^{i-} = \frac{1}{2}\left(p^i u_0^- - \frac{L_0^{tr} - \alpha_0}{2\alpha' p^+} X_0^i\right) - \frac{1}{2}\sum_{n>2} \frac{a_n^{*i} L_n^{tr} + L_{-n}^{tr} a_n^i}{\sqrt{2\alpha'} n p^+}$$

$$+ \frac{i}{2}\sum_{r>0}\left(b_r^{i*}\frac{G_r^{tr}}{\sqrt{2\alpha'} p^+} - \frac{G_{-r}^{tr}}{\sqrt{2\alpha'} p^+} b_r^i\right) \quad (15.3.1.1)$$

on application of the same methods as for the bosonic string. Quantity G_r^{tr} in this equation differs from the covariant G_r by a summation over transverse degrees of freedom only. The other generators will not be written here, since they are not needed explicitly.

The Poincaré algebra is fulfilled trivially with the exception of $[M^{i-}, M^{j-}]$, which fails to be zero in general. Lengthy but otherwise straightforward computations indicate that $[M^{i-}, M^{j-}]$ picks up not only an α_0 term, but also anomalous contributions resulting from normal ordering of $L_0[X_0^i, \sum a_n^* L_n + L_{-n} a_n]$ and of the commutators of the cubic terms. The quantum mechanical $[M^{i-}, M^{j-}]$ is explicitly given by

$$[M^{i-}, M^{j-}] = \frac{1}{4\alpha'(p^+)^2}\sum_{n>0}\left[n^2\left(\frac{d-2}{16} - \frac{1}{2}\right) + \alpha_0 - \frac{d-2}{16}\right]$$

$$\times (a_n^{*i} a_n^j - a_n^{*j} a_n^i)$$

$$+ \frac{1}{4\alpha'(p^+)^2}\sum_{r>0}\left[r^2\left(\frac{d-2}{4} - 2\right) + \alpha_0 - \frac{d-2}{16}\right]$$

$$\times (b_r^{*i} b_r^j - b_r^{*j} b_r^i) \quad (15.3.1.2)$$

and vanishes if and only if

$$d = 10 \quad \text{and} \quad \alpha_0 = \tfrac{1}{2} \qquad (15.3.1.3)$$

These critical values coincide with those found by BRST methods.

15.3.2. Neveu–Schwarz Spectrum

The discussion of the Neveu-Schwarz spectrum follows that of the bosonic model. States at a given mass level belong to representations of the little group, which is SO(8) when the states are massless, or SO(9) when they are massive. In the latter case, various distinct SO(8) representations combine to form SO(9) ones.

Since $\alpha_0 = \tfrac{1}{2}$, the ground state is a tachyon, with mass-squared equal to $-(2\alpha')^{-1}$:

$$|0\rangle, \quad a_n^i|0\rangle = b_r^i|0\rangle = 0, \quad \alpha'M^2 = -\tfrac{1}{2} \qquad (15.3.2.1)$$

The mass of a general state is given by

$$\alpha'M^2 = \sum_{n>0} n a_n^{*i} a_{ni} + \sum_{s>0} s b_s^{*i} b_{si} - \alpha_0$$

$$= N - \tfrac{1}{2} \qquad (15.3.2.2)$$

The b^* oscillators increase the mass by half-integer numbers, and the a^* oscillators increase it by integer units. States with an odd number of b^* oscillators will accordingly have integer masses. One says that they have G parity equal to $+1$, where the operator G is defined as

$$G = (-)^{(\sum b_s^{*i} b_{si} - 1)} \qquad (15.3.2.3)$$

States with an even number of b^* oscillators possess half-integer mass and have G parity equal to -1. The ground state tachyon has $G = -1$.

The next excited states are massless, SO(8) vectors:

$$b_{1/2}^{*i}|0\rangle, \quad \alpha'M^2 = 0, \quad G = 1, \quad 8 \text{ states (spin 1)} \quad (15.3.2.4)$$

Again, they must be massless in order to form a representation of the Poincaré group (otherwise, one would not have enough states).

At the next level, with $G = -1$, one finds:

$$(b_{1/2}^{*i} b_{1/2}^{*j} - b_{1/2}^{*j} b_{1/2}^{*i})|0\rangle, \quad \alpha'M^2 = \tfrac{1}{2}, \quad G = -1, \quad 28 \text{ states}$$

$$\alpha_1^{*i}|0\rangle, \quad \alpha'M^2 = \tfrac{1}{2}, \quad G = -1, \quad 8 \text{ states} \qquad (15.3.2.5)$$

Then come massive states with $\alpha'M^2 = 1$:

$$b^{*i}_{1/2}b^{*j}_{1/2}b^{*k}_{1/2}|0\rangle, \qquad \alpha'M^2 = 1, \qquad G = 1, \qquad 56 \text{ states}$$

$$b^{*i}_{1/2}a^{j*}_1|0\rangle, \qquad \alpha'M^2 = 1, \qquad G = 1, \qquad 64 \text{ states} \qquad (15.3.2.6)$$

$$b^{i*}_{3/2}|0\rangle, \qquad \alpha'M^2 = 1, \qquad G = 1, \qquad 8 \text{ states}$$

and so on. All states in the spectrum possess integer spin.

Finally, we recall that the states are also characterized by their d-momentum components p^+, p^i (which determine p^-), although we have not written this dependence explicitly in relations (15.3.2.1)-(15.3.2.6).†

15.3.3. The Closed Neveu–Schwarz Spectrum

When both right and left movers are described by antiperiodic fields, the spectrum of the closed string easily follows from that discussed above. The basic equations are the mass-shell condition

$$\tfrac{1}{2}\alpha'M^2 = N_R + N_L - 1 \qquad (15.3.3.1)$$

and the "right–left equality"

$$N_R = N_L \qquad (15.3.3.2)$$

The first state is the ground state, which is a tachyon:

$$|0\rangle, \qquad \alpha'M^2 = -2, \qquad \text{spin } 0 \qquad (15.3.3.3)$$

It has G_p parity -1, with

$$G_p = (-)^{\Sigma_s b^{*i}_s b_{si} - 1} = (-)^{\Sigma_s \bar{b}^{*i}_s \bar{b}_{si} - 1} \qquad (15.3.3.4)$$

[The last equality is valid for on-shell states obeying relation (15.3.3.2).]
The next states are massless:

$$b^{*i}_1 \bar{b}^{*j}_1|0\rangle, \qquad G_p = 1, \qquad \alpha'M^2 = 0 \qquad (15.3.3.5)$$

† Also to be stressed here is the fact that excited states of the string are single-particle states, as can be seen, for instance, from the fact that the mass spectrum is discrete. Accordingly, there is no contradiction between spin and statistics even though we have fermionic oscillators with a vector index. (Statistics deals with many particles simultaneously.)

These states can be decomposed into a traceless symmetric tensor ("graviton," spin 2), an antisymmetric second-rank tensor, and a spin-0 scalar.

The rest of the spectrum is constructed along the same lines. It only contains integer-spin states.

15.4. LIGHT-CONE GAUGE QUANTIZATION OF THE RAMOND MODEL

15.4.1. Poincaré Invariance

The Lorentz generator M^{i-} is given by

$$M^{i-} = \frac{1}{2}\left(p^i u_0^- - \frac{L_0^{tr} - \alpha_0}{2\alpha' p^+} X_0^i\right) - \frac{1}{2}\sum_{n>0}\frac{a_n^{i*}L_n^{tr} + L_{-n}^{tr}a_n^i}{\sqrt{2\alpha'n}\,p^+}$$

$$+ \frac{i}{4}\sum_{m>0}\sqrt{2}\,\frac{\Gamma_m^{i*}F_m - F_{-m}\Gamma_m^i}{\sqrt{2\alpha'}\,p^+} + \frac{i}{8}\sqrt{2}\,\frac{\Gamma_0^i F_0 - F_0\Gamma_0^i}{\sqrt{2\alpha'}\,p^+} \qquad (15.4.1.1)$$

In the last term of expression (15.4.1.1), we have adopted the antisymmetric ordering for the Γ_0^i zero modes. Any other real order would coincide with equation (15.4.1.1).

The commutator $[M^i, M^j]$ fails to vanish in general. One obtains

$$[M^{i-}, M^{j-}] = \frac{1}{4\alpha'(p^+)^2}\sum_{n>0}\left[n^2\left(\frac{d-2}{16} - \frac{1}{2}\right) + \alpha_0\right](a_n^{*i}a_n^j - a_n^{*j}a_n^i)$$

$$+ \frac{1}{8\alpha'(p^+)^2}\sum_{n>0}\left[n^2\left(\frac{d-2}{4} - 2\right) + \alpha_0\right](\Gamma_n^{*i}\Gamma_n^j - \Gamma_n^{*j}\Gamma_n^i)$$

$$+ \frac{1}{16\alpha'(p^+)^2}\alpha_0(\Gamma_0^i\Gamma_0^j - \Gamma_0^j\Gamma_0^i) \qquad (15.4.1.2)$$

By demanding that expression (15.4.1.2) be zero, one again finds critical values for d and α_0 given by

$$d = 10 \quad \text{and} \quad \alpha_0 = 0 \qquad (15.4.1.3)$$

15.4.2. Ramond Spectrum

The novel feature of the Ramond model is the presence of the fermionic zero modes Γ_0^A. Quantum mechanically, these form a Clifford algebra.

15 • Fermionic String: Quantum Analysis

In the light-cone gauge approach, only the transverse matrices Γ_0^i are independent and one has

$$\Gamma_0^i \Gamma_0^j + \Gamma_0^j \Gamma_0^i = 2\delta^{ij} \tag{15.4.2.1}$$

The irreducible representation space for this relation is 16-dimensional (8 independent Γ_0^i matrices).† Elements of this representation space transform as spinors under $SO(8)$ rotations, which are generated by appropriate products $\Gamma_0^{[i}\Gamma_0^{j]}$. That is why the states of the Ramond spectrum all have half-integer spin.

It is well known that in a space-time of even dimensionality, the Dirac representation of the rotation group is reducible. It reduces to two irreducible representations obtained by imposing definite handedness (Weyl condition). In our case, the 16-dimensional space reduces to two 8-dimensional subspaces.

The first state in the spectrum is the ground state $|0\rangle$ and it is massless. It possesses spin $\frac{1}{2}$. The tachyon is avoided because $\alpha_0 = 0$!

$$|0\rangle u(p^i, p^+), \quad 16 \text{ states}, \quad 2 \text{ spin-}\tfrac{1}{2}, \quad \alpha' M^2 = 0 \tag{15.4.2.2}$$

We have made explicit the dependence of the ground state on all the variables in the problem. The state $|0\rangle u(p^i, p^+)$ is the direct product of the Fock-space vacuum $|0\rangle$:

$$a_n^i |0\rangle = \Gamma_n^i |0\rangle = 0 \tag{15.4.2.3}$$

by a spinor u belonging to the 16-dimensional representation space of relation (15.4.2.1). The state (15.4.2.2) is also characterized by its d-momentum p^i, p^+.

In view of our remarks on the reducibility of the spinor representation, the ground state (15.4.2.2), without any truncation, contains actually two spin-$\frac{1}{2}$ representations of the Poincaré group.‡

---

† In the light-cone gauge, the lightlike Γ matrices $\Gamma_0^\pm$ are functions of the other variables. They cannot be treated independently. We note that the scalar product u^*u [where u is an element of the representation space of relation (15.4.2.1)] is evidently positive definite and makes all Γ_0^i Hermitian. There are no negative-norm states in the light-cone gauge.

‡ Because p^+ and p^i are unrestricted, one also finds that p^0 can be positive or negative, as for the previous models. Negative p^0 states can be interpreted as antiparticles. These can be identified with positive p^0 states (ground-state particles = ground-state antiparticles), which amounts to taking $p^0 > 0$ only (in the free spectrum). This has been done here. In principle, however, one could choose at this stage not to make this identification, which would double the spectrum.

The next states are obtained by acting on the vacuum with the a_1^{*i} or Γ_1^{*i} oscillators, which carry an SO(8) vector index:

$$a_1^{*i}|0\rangle u(p^+, p^-), \qquad \Gamma_1^{*i}|0\rangle u(p^+, p^-) \qquad (256 \text{ states}) \qquad (15.4.2.4)$$

They are massive, $\alpha'M^2 = 1$, and they contain spin $\frac{3}{2}$ and $\frac{1}{2}$. The mass is again given for all states by the mass formula

$$\alpha'M^2 = N - \alpha_0 = N \qquad (15.4.2.5)$$

Higher states are discussed in the same way.

15.4.3. Closed String

We discuss here the interesting combination of the right-moving Neveu-Schwarz sector with the left-moving Ramond sector [50]. The product of two Ramond models will not be treated.

The zero-mode constraints are given by

$$L_0 - \tfrac{1}{2} = 0 \quad \text{and} \quad \bar{L}_0 = 0 \qquad (15.4.3.1)$$

Hence, the mass-shell equation and the "right-left equality" condition are, respectively,

$$\tfrac{1}{2}\alpha'M^2 = (N_R - \tfrac{1}{2}) + N_L \qquad (15.4.3.2)$$

and

$$N_R - \tfrac{1}{2} = N_L \qquad (15.4.3.3)$$

Equation (15.4.3.3) can be used to transform equation (15.4.3.2) to the form

$$\tfrac{1}{2}\alpha'M^2 = 2N_L \qquad (15.4.3.4)$$

There is no tachyon in the spectrum because $N_L \geq 0$.

To understand this result, one notes that the ground state $|0\rangle$ is not a solution of condition (15.4.3.3), since it has $N_R = 0 = N_L$. Hence, it is excluded from the physical subspace defined by the closed-string constraint (15.4.3.3).

The lesson is that the tachyon of one string sector can disappear in closed-string models when this sector is coupled, as in constraint (15.4.3.3), to another tachyon-free sector. A similar mechanism occurs in the heterotic string.

15 • Fermionic String: Quantum Analysis

It follows from condition (15.4.3.3) that N_R must be a half-integer. Only Neveu-Schwarz states with G parity +1 appear.

The first state is

$$b^{i*}_{1/2}|0\rangle u(p^+, p^-) \tag{15.4.3.5}$$

and describes a spin-$\frac{3}{2}$ particle and a spin-$\frac{1}{2}$ particle which are both massless [50].† We leave the reader to construct the higher excited states, which all possess half-integer spin.

15.5. SUPERSYMMETRY IN TEN DIMENSIONS

15.5.1. Open String

The remarkable discovery was made [50] that appropriate truncations of the Neveu-Schwarz and Ramond models yield, respectively, bosonic and fermionic spectra which, when combined, form at each mass level ten-dimensional supersymmetric multiplets.

The truncation of the Neveu-Schwarz sector consists in retaining only the $G = 1$ part of the spectrum. This excludes the tachyon. All states have integer mass. The truncation of the Ramond sector consists in taking a spinor vacuum with definite handedness.

With these truncations, the values of the mass are equal in both sectors. Furthermore, Gliozzi *et al.* [50] have shown that the numbers of states at each mass level are identical, a necessary condition for supersymmetry (equal number of bosons and fermions).

The proof will not be reproduced here, but we shall simply check the assertion at the first two levels.

The massless states in the Neveu-Schwarz model form an SO(8) vector, with eight helicities. The massless spinor ground state of the Ramond model possesses also eight "polarizations" of definite handedness.

There are then 128 bosonic states at $\alpha'M^2 = 1$ [see equation (15.3.2.6)]. There are also 128 fermionic states when the spinor ground state is restricted by the Weyl condition [see equation (15.4.2.4)]. The equality in the number of states therefore holds for $\alpha'M^2 = 0, 1$ and is easily seen to be true for $\alpha'M^2 = 2$ also.

A picture of the combined supersymmetric Neveu-Schwarz-Ramond spectrum is given in Figure 15.1 [50].

The ground state of the supersymmetric Neveu-Schwarz-Ramond open string contains massless spin-$\frac{1}{2}$ and spin-1 particles. These are the same

---

† If the spinor u does not obey the Weyl condition, one has twice as many such states.

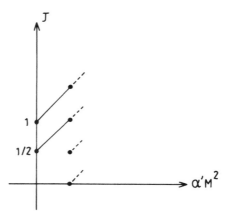

Figure 15.1. The ground state of the supersymmetric truncation of the combined Neveu–Schwarz–Ramond model is the massless gauge multiplet of super-Yang–Mills theory.

particles as those of the gauge supermultiplet of the super-Yang–Mills theory (more precisely, superelectromagnetism since we have not included internal quantum numbers here).

We recall that the gauge field of superelectromagnetism consists of a real vector field A_A and a real (i.e., Majorana) spinor ψ. The reality condition expresses the fact that particles in the gauge multiplet are their own antiparticles. In addition, ψ is a Weyl spinor. This is equivalent to the definite handedness of the string ground state. The Majorana and Weyl conditions are not compatible in any number of dimensions. It turns out that Majorana–Weyl spinors exist only in dimensions 2 mod 8 and hence, in particular, in 10 dimensions.

To complete the proof of supersymmetry, one needs to determine the supersymmetry transformation laws within the supermultiplets. This was not done by Gliozzi *et al.* [50]. The supersymmetry generators were given by Green and Schwarz [55] using their new formalism.

15.5.2. Closed String

Gliozzi *et al.* [50] also showed equality of both mass and number of states for the closed-string bosonic spectrum of Section 15.3.3 (truncated to $G_p = 1$) and the closed-string fermionic spectrum of Section 15.4.3 (truncated by the Weyl condition).

The ground state of this combined Neveu–Schwarz–Ramond closed model contains 64 fermionic and 64 bosonic states. The corresponding particles can be described by the following ten-dimensional local fields: ϕ, $B_{AB} = -B_{BA}$, $g_{AB} = g_{AB}$ (bosonic fields) and χ, ψ_A (spin $\frac{1}{2}$ and $\frac{3}{2}$ Majorana–Weyl fields). This is the gauge multiplet of $N = 1$, $d = 10$ supergravity.

Chapter 16

The Superstring

16.1. COVARIANT ACTION

The ten-dimensional supersymmetry of the Neveu–Schwarz–Ramond spectrum is quite unexpected and obscure in the formalism developed so far. Green and Schwarz have discovered a new action principle which is invariant under ten-dimensional global supersymmetry and for which one can construct supersymmetry generators as Noether charges. Upon quantization, this action yields the supersymmetric part of the Neveu–Schwarz–Ramond spectrum without the undesirable tachyon. This is the "superstring."

The superstring action was first discovered in the light-cone gauge [44, 55]. Its covariant form was found more recently [56].

Following recent work [57], we derive the superstring covariant action as a σ-model action with a Wess–Zumino term.

16.1.1. SUSY(N)/SO($d-1, 1$) as Target Space

The σ models inherit their internal symmetries from the target space. For instance, the Nambu–Goto string is globally Poincaré invariant, because the target space is in that case ordinary Minkowski space.

To construct a σ model which is invariant under the graded extension of the Poincaré group, one needs a target space on which this supergroup acts. The appropriate space turns out to be the quotient space of the SUSY "supergroup" with N supersymmetries, by the Lorentz subgroup $SO(d-1,1)$. This manifold is isomorphic to the "subgroup" of $SUSY(N)$ obtained by taking the translations and the N supersymmetry transformations.

We leave N and d arbitrary at this stage. Restrictions on these parameters will soon arise, already at the classical level.

The supermanifold $SUSY(N)/SO(d-1,1)$ can be parametrized by d commuting coordinates X^A ($A = 0, \ldots, d-1$) and N anticommuting spinors θ^α ($\alpha = 1, 2, \ldots, N$), which are assumed for simplicity to obey the Majorana condition. The coordinates X^A are just Minkowski-space coordinates.

The action of the $SUSY(N)$ supergroup on $SUSY(N)/SO(d-1,1)$ is given by

$$X^A \to X^A + \varepsilon^A$$
$$\theta^\alpha \to \theta^\alpha$$
(16.1.1.1)

(translations) and

$$X^A \to X^A + i \sum_{\alpha=1}^{N} \bar{\varepsilon}^\alpha \gamma^A \theta^\alpha$$
$$\theta^\alpha \to \theta^\alpha + \varepsilon^\alpha$$
(16.1.1.2)

(supersymmetries), where ε^α are N arbitrary real anticommuting spinor parameters and γ^A are the γ matrices in d dimensions (our conventions are summarized in Appendix B). In addition, the variables X^A and θ^α transform, respectively, as a vector and as spinors under Lorentz rotations.

One can easily construct invariant forms for the transformations (16.1.1.1) and (16.1.1.2). These are

$$\omega^A = dX^A - i\bar{\theta}^\alpha \gamma^A d\theta^\alpha$$
(16.1.1.3a)

and

$$d\theta^\alpha$$
(16.1.1.3b)

From now on, summation over the various supersymmetries in expressions like (16.1.1.3a) will always be understood.

Explicit computations of $\delta \omega^A$ show that ω^A is indeed invariant under supersymmetries,

$$\delta \omega^A = \delta(dX^A) - i\delta\bar{\theta}^\alpha \gamma^A d\theta^\alpha - i\bar{\theta}^\alpha \gamma^A \delta d\theta^\alpha$$
$$= d\delta X^A - i\delta\bar{\theta}^\alpha \gamma^A d\theta^\alpha - i\bar{\theta}^\alpha \gamma^A d\delta\theta^\alpha$$

(using $[\delta, d] = 0$)

$$= i\bar{\varepsilon}^\alpha \gamma^A \, d\theta^\alpha - i\bar{\varepsilon}^\alpha \gamma^A \, d\theta^\alpha$$

(since $d\varepsilon = 0$)

$$= 0$$

Similarly, $\delta \, d\theta^\alpha = d \, \delta\theta^\alpha = 0$. The invariance of $d\theta^\alpha$ and ω^A under translations is as easy to verify.

In the sequel, we will need a few properties of differential forms on supermanifolds, which we briefly review here.

The differential operator d obeys relations very similar to the usual exterior derivative operator on an ordinary manifold. One noticeable property is

$$dy \wedge dz = (-)^{\varepsilon_y \varepsilon_z + 1} \, dz \wedge dy \qquad (16.1.1.4)$$

where ε_y and ε_z are the Grassmann parities of y and z, respectively ($\varepsilon_y = 0$ if y is even, $\varepsilon_y = 1$ if y is odd). When y and z are both anticommuting, property (16.1.1.4) becomes

$$dy \wedge dz = dz \wedge dy \qquad (\varepsilon_y = \varepsilon_z = 1)$$

One also has

$$dy \, z = (-)^{\varepsilon_y \varepsilon_z} z \, dy \qquad (16.1.1.5)$$

and

$$d(\omega \wedge \nu) = d\omega \wedge \nu + (-)^{k_\omega} \omega \wedge d\nu \qquad (16.1.1.6)$$

where k_ω is the degree of ω as a form. These rules are such that d is nilpotent,

$$d^2 = 0 \qquad (16.1.1.7)$$

Finally, a mapping from the two-dimensional space $x^\lambda \equiv (\tau, \sigma)$ ($\lambda = 0, 1$) into the supermanifold $SUSY(N)/SO(d - 1, 1)$ is defined by giving X^A and θ^α as functions of x^λ,

$$X^A = X^A(\tau, \sigma), \qquad \theta^\alpha = \theta^\alpha(\tau, \sigma) \qquad (16.1.1.8)$$

The "pullback" of a $SUSY(N)/SO(d - 1, 1)$ differential form on the two-dimensional space time (x^λ) is obtained by replacing everywhere dX^A by

$X^A_{,\lambda} dx^\lambda$ and $d\theta^\alpha$ by $\theta^\alpha_{,\lambda} dx^\lambda$. The pullback commutes with d. For more about differential forms on supermanifolds see, e.g., the book by Wess and Bagger [58].

16.1.2. Invariant Actions

In order to construct an invariant action one first considers the kinetic term, which is quadratic in the derivatives,

$$L_1 = -\frac{1}{4\pi\alpha'}\sqrt{-^{(2)}g}\, g^{\lambda\mu}\omega^A_\lambda \omega_{A\mu} \qquad (16.1.2.1)$$

where we have set

$$\omega^A_\lambda = X^A_{,\lambda} - i\bar{\theta}^\alpha \gamma^A \theta^\alpha_{,\lambda} \qquad (16.1.2.2)$$

($\omega^A = \omega^A_\lambda dx^\lambda$). The form ω^A is invariant under translations and supersymmetry transformations, and rotates furthermore as a vector under Lorentz transformations, so the Lagrangian L_1 is clearly SUSY-invariant.

The Lagrangian (16.1.2.1) was thought for some time to be the desired supersymmetric Lagrangian. However, it does not possess enough gauge freedom for it to be linearized—it is only invariant under two-dimensional reparametrizations. Owing to its nonlinearities, the relevant quantum theory has not been developed. All that can be said is that it does not correspond to the free superstring. This means that relation (16.1.2.1) is only one piece of the full Lagrangian.

One could add to expression (16.1.2.1) terms quadratic in $\theta^\alpha_{,\lambda}$, such as $b_{\alpha\beta}\bar{\theta}^\alpha_{,\lambda}\theta^\beta_{,\mu}\sqrt{-g}g^{\lambda\mu}$. However, such terms lead to second-order spinor equations and seem to imply the existence of negative-norm states in the spectrum. Hence they will be ruled out.

This leaves us with the so-called "Wess–Zumino term" [59], which should always be considered in σ models. This term is SUSY-invariant up to a total divergence, and, hence, the equations of motion are invariant. Furthermore, it is also quadratic in the derivatives of the fields and thus "competes" with the kinetic term L_1.

The Wess–Zumino Lagrangian L_2 is constructed by first determining the closed three-forms Ω_3 in the quotient space $SUSY(N)/SO(d-1,1)$ that are invariant under SUSY transformations. These invariant three-forms are just given by exterior products of the invariant forms (16.1.1.3),

$$\Omega_3 = f_{\Gamma\Delta\Lambda}\omega^\Gamma \wedge \omega^\Delta \wedge \omega^\Lambda \qquad (16.1.2.3)$$

where the capital Greek indices run over the one-forms (16.1.1.3) and $f_{\Gamma\Delta\Lambda}$ are constant.

16 • The Superstring

The three-form (16.1.2.3) is Lorentz invariant if the constants $f_{\Gamma\Delta\Lambda}$ are the components of an appropriate Lorentz-covariant quantity. The only such quantity[†] with matching indices is $(\gamma_A C^{-1})_{ab} a_{\alpha\beta}$, where C is the charge-conjugation matrix and $a_{\alpha\beta}$ is a pure imaginary, symmetric matrix in internal space.[‡] Here, a and b are spinor indices.

The matrix $a_{\alpha\beta}$ can be diagonalized by a rotation in the internal space of the θ coordinates. Hence, one can assume without loss of generality that $a_{\alpha\beta} = 0$ for $\alpha \neq \beta$.

The next step in the construction consists of imposing that Ω_3 be closed, $d\Omega_3 = 0$. One finds explicitly that $d\Omega_3$ is given by

$$d\Omega_3 = a_{11}(d\bar{\theta}^1 \wedge \gamma^A d\theta^1) \wedge (d\bar{\theta}^1 \wedge \gamma_A d\theta^1)$$

$$+ (a_{11} + a_{22})(d\bar{\theta}^1 \wedge \gamma^A d\theta^1) \wedge (d\bar{\theta}^2 \wedge \gamma_A d\theta^2)$$

$$+ \text{similar terms with } (\alpha, \beta) = (2,2), (3,3), (1,3), \text{ etc.} \quad (16.1.2.4)$$

The "diagonal terms" in this equation disappear as a result of the identity

$$\gamma_A \psi_1 \bar{\psi}_2 \gamma^A \psi_3 + \gamma_A \psi_2 \bar{\psi}_3 \gamma^A \psi_1 + \gamma_A \psi_3 \bar{\psi}_1 \gamma^A \psi_2 = 0 \quad (16.1.2.5)$$

which holds for Majorana-Weyl spinors in $d = 10$, and Majorana spinors in $d = 4$ and 3 (and also Weyl spinors in $d = 6$ for which a similar analysis is valid).

Without the identity (16.1.2.5), the term

$$(d\bar{\theta}^1 \gamma^A \wedge d\theta^1) \wedge (d\bar{\theta}^1 \wedge \gamma_A d\theta^1) \quad (16.1.2.6)$$

would not vanish. The condition $d\Omega_3 = 0$ would then require $a_{11} = 0$ (and $a_{22} = a_{33} = \cdots = 0$), which would kill Ω_3. Hence, the possibility of constructing nontrivial Wess-Zumino actions only exists in particular spacetime dimensions, with particular spinors. From now on, we only consider cases when identity (16.1.2.5) holds.

If there is only one supersymmetry ($N = 1$), then identity (16.1.2.5) suffices to guarantee $d\Omega_3 = 0$ and there is no condition on $a_{\alpha\beta}$. For $N \geq 2$, the mixed terms in expression (16.1.2.4) cancel if and only if

$$a_{\alpha\alpha} + a_{\beta\beta} = 0, \quad \alpha \neq \beta \quad (16.1.2.7)$$

[†] When $d = 3$, another candidate is $\varepsilon_{ABC} \omega^A \wedge \omega^B \wedge \omega^C$, which is however not closed in superspace. Furthermore, the case $f \sim \gamma_A \gamma_5 C^{-1} b_{\alpha\beta}$ can be absorbed in a redefinition of θ^α.

[‡] $a_{\alpha\beta}$ is symmetric because $(\gamma_A C^{-1})_{ab}$ is also symmetric and $d\theta^{\alpha a} \wedge d\theta^{\beta b} = d\theta^{\beta b} \wedge d\theta^{\alpha A}$.

This rules out $N > 2$, for then one finds from condition (16.1.2.7) that $a_{\alpha\beta}$ must vanish. The only possibility left is $N = 2$, with

$$a_{11} = -a_{22} \quad \Leftrightarrow \quad \text{tr } a_{\alpha\beta} = 0 \tag{16.1.2.8}$$

The search for invariant closed three-forms has thus led us to the following two cases:

$$N = 1, \quad \Omega_3 = ia\omega^A \wedge d\bar{\theta}\gamma_A \wedge d\theta \tag{16.1.2.9}$$

and

$$N = 2, \quad \Omega_3 = ia(\omega^A \wedge d\bar{\theta}^1 \wedge \gamma_A d\theta^1 - \omega^A \wedge d\bar{\theta}^2 \wedge \gamma_A d\theta^2) \tag{16.1.2.10}$$

together with the particular values of the space-time dimensions mentioned above.

The form Ω_3 is not only closed, but also exact,

$$\Omega_3 = d\Omega_2 \tag{16.1.2.11a}$$

with

$$a^{-1}\Omega_2 = -i\, dX^A \wedge (\bar{\theta}^1 \gamma_A d\theta^1 - \bar{\theta}^2 \gamma_A d\theta^2) + \bar{\theta}^1 \gamma^A d\theta^1 \wedge \bar{\theta}^2 \gamma_A d\theta^2 \tag{16.1.2.11b}$$

for $N = 2$. [For $N = 1$, we simply omit the θ^2 terms in the latter equation.]

Although Ω_3 is SUSY-invariant, the two-form Ω_2, which contains the noninvariant one-forms dX^A, is only invariant up to a total derivative $[\delta\Omega_3 = 0$ implies $\delta\Omega_2 = d(\text{something})]$. It is the Wess–Zumino term. Its pullback on the $(\tau\text{-}\sigma)$ space yields the Lagrangian L_2, where

$$a^{-1}L_2 = -i\varepsilon^{\alpha\beta}\partial_\alpha X^A(\bar{\theta}^1\gamma_A\partial_\beta\theta^1 - \bar{\theta}^2\gamma_A\partial_\beta\theta^2)$$

$$+ \varepsilon^{\alpha\beta}\bar{\theta}^1\gamma^A\partial_\alpha\theta_1\bar{\theta}^2\gamma_A\partial_\beta\theta^2 \tag{16.1.2.12}$$

(for $N = 2$), which is SUSY-invariant up to a total divergence.

Since the $N = 1$ theory can be thought of as a truncation of the $N = 2$ theory we focus here only on the more complex $N = 2$ case.

We note that in ten space-time dimensions, one can take Majorana–Weyl spinors θ^1 and θ^2 of equal or opposite chiralities, since one obtains $d\Omega_3 = 0$ in both cases.

16.1.3. Local Supersymmetry

The Lagrangian $L_1 + L_2$ possesses, for $a = \pm\frac{1}{2}(\pi\alpha')^{-1}$, remarkable properties. It is not only invariant under two-dimensional changes of coordinates, but also under a new type of local symmetry [56]. This local symmetry involves arbitrary functions which are simultaneously anticommuting d-dimensional spinors and 2-dimensional vectors.

In order to describe it, it is useful to introduce projection operators

$$P_\pm^{\lambda\mu} = \frac{1}{2}\left(g^{\lambda\mu} \pm \frac{\varepsilon^{\lambda\mu}}{\sqrt{-g}}\right) \qquad (16.1.3.1)$$

which project a two-dimensional vector v^λ onto its lightlike components $v_\pm^\lambda$,

$$v_\pm^\lambda = P_{\pm\,\mu}^\lambda v^\mu \qquad (16.1.3.2)$$

where the $\pm$ refers to two dimensions. The operators $P_{\pm\,\mu}^\lambda$ obey

$$P_{+\,\mu}^\lambda P_{+\,\nu}^\mu = P_{+\,\nu}^\lambda \qquad (16.1.3.3a)$$

$$P_{-\,\mu}^\lambda P_{-\,\nu}^\mu = P_{-\,\nu}^\lambda \qquad (16.1.3.3b)$$

$$P_{+\,\mu}^\lambda P_{-\,\nu}^\mu = P_{-\,\mu}^\lambda P_{+\,\nu}^\mu = 0 \qquad (16.1.3.3c)$$

$$(P_{+\,\mu}^\lambda)^T = P_{-\,\mu}^\lambda \qquad (16.1.3.3d)$$

as well as the less obvious identity

$$P_\pm^{\lambda\rho} P_\pm^{\mu\sigma} = P_\pm^{\mu\rho} P_\pm^{\lambda\sigma} \qquad (16.1.3.4)$$

If a two-dimensional vector satisfies $v^\lambda = v_\pm^\lambda$, then it is tangent to one of the lightlike directions.†

The local "supersymmetry" transformations under which the action is invariant are given explicitly by

$$\delta_\kappa \theta^1 = 2i\omega_\lambda^A \gamma_A P_-^{\lambda\mu} \kappa_\mu^1 \qquad (16.1.3.5a)$$

---

† If k^λ and n^λ are two nonzero null vectors obeying $k^\lambda = k_+^\lambda$ and $n^\lambda = n_-^\lambda$ ($\Rightarrow k^\lambda k_\lambda = n^\lambda n_\lambda = 0$ and $k^\lambda n_\lambda \neq 0$, usually normalized to minus one), then it is easy to see that $v^\lambda = v_+^\lambda + v_-^\lambda$, $v_+^\lambda = v_+ k^\lambda$, and $v_-^\lambda = v_- n^\lambda$, where v_+ and v_- are the components of v^λ in the null frame $\{k^\lambda, n^\lambda\}$.

$$\delta_\kappa \theta^2 = 2i\omega_\lambda^A \gamma_A P_+^{\lambda\mu} \kappa_\mu^2 \tag{16.1.3.5b}$$

$$\delta_\kappa X^A = i(\bar{\theta}^1 \gamma^A \delta_\kappa \theta^1 + \bar{\theta}^2 \gamma^A \delta_\kappa \theta^2) \tag{16.1.3.5c}$$

$$\delta_\kappa(\sqrt{-g} g^{\lambda\mu}) = -16\sqrt{-g}(P_-^{\lambda\rho} P_-^{\mu\sigma} \bar{\kappa}_\sigma^1 \partial_\rho \theta^1 + P_+^{\lambda\rho} P_+^{\mu\sigma} \bar{\kappa}_\sigma^2 \partial_\rho \theta^2) \tag{16.1.3.5d}$$

where κ_μ^1 and κ_μ^2 are infinitesimal parameters.† As pointed out above, these parameters are two-dimensional vectors and ten-dimensional, pure imaginary spinors of chirality opposite to that of θ^1 or θ^2 (so that $\delta_\kappa \theta^\alpha$ possesses the same chirality as θ^α).

It can be seen from system (16.1.3.5) that only the projected components $\kappa_-^{1\lambda}$ and $\kappa_+^{2\lambda}$ appear in the transformation laws. Furthermore, because the two-dimensional lightlike components of ω_λ^A define ten-dimensional null vectors

$$\eta_{AB} \omega_-^{A\mu} \omega_-^{B\nu} = \eta_{AB} \omega_+^{A\mu} \omega_+^{B\nu} = 0 \tag{16.1.3.6}$$

as a result of the field equations for the metric (see the next section), the matrices $\omega_\pm^{A\mu} \gamma_A$ are noninvertible (they are nilpotent). There is thus a further projection of the space-time spinors $\kappa_-^{1\lambda}$ and $\kappa_+^{2\lambda}$ onto one of their light-cone spinor components in relations (16.1.3.5a)–(16.1.3.5c). The number of effective gauge parameters in these relations is accordingly $2 \times 8 = 16$ (8 is the number of components of a light-cone chiral spinor). This is also true for transformation (16.1.3.5d) when one takes into account the spinor equations of motion.

In order to check the invariance of the action $S_1 + S_2$ [with $a = \frac{1}{2}(\pi\alpha')^{-1}$] subject to transformation (16.1.3.5), one notes that with $\delta_\kappa X^A$ related to $\delta_\kappa \theta^1$ and $\delta_\kappa \theta^2$ as in equation (16.1.3.5c), the variation of S is given by

$$\delta S = -\frac{1}{4\pi\alpha'} \int \delta(\sqrt{-g} g^{\lambda\mu}) \omega_\lambda^A \omega_{A\mu}$$

$$- \frac{i}{\pi\alpha'} \int \sqrt{-g} g^{\lambda\mu} \omega_{A\mu} (\bar{\theta}_{,\lambda}^1 \gamma^A \delta\theta^1 + \bar{\theta}_{,\lambda}^2 \gamma^A \delta\theta^2)$$

$$- \frac{1}{2\pi\alpha'} \int \varepsilon^{\lambda\mu} \bar{\theta}^1 \gamma^A \delta\theta^1 \partial_\lambda \bar{\theta}^1 \gamma_A \partial_\mu \theta^1$$

$$+ \frac{1}{2\pi\alpha'} \int \varepsilon^{\lambda\mu} \bar{\theta}^2 \gamma^A \delta\theta^2 \partial_\lambda \bar{\theta}^2 \gamma_A \partial_\mu \theta^2$$

$$+ \frac{i}{\pi\alpha'} \int \varepsilon^{\lambda\mu} \partial_\lambda X_A (\bar{\theta}_{,\mu}^1 \gamma^A \delta\theta^1 - \bar{\theta}_{,\mu}^2 \gamma^A \delta\theta^2)$$

† There is no need to give the variation of $g_{\lambda\mu}$ itself, since only the unimodular combination $\sqrt{-g} g^{\lambda\mu}$ appears in the action.

16 • The Superstring

$$-\frac{1}{\pi\alpha'}\int \varepsilon^{\lambda\mu}\partial_\lambda\bar{\theta}^1\gamma^A\delta\theta^1\bar{\theta}^2\gamma_A\partial_\mu\theta^2$$

$$-\frac{1}{\pi\alpha'}\int \varepsilon^{\lambda\mu}\bar{\theta}^1\gamma^A\partial_\lambda\theta^1\partial_\mu\bar{\theta}^2\gamma_A\delta\theta^2 \tag{16.1.3.7}$$

where we integrated by parts to eliminate the derivatives of $\delta\theta^1$ and $\delta\theta^2$ (we note that these cancel from $\delta\omega_\lambda^A$).

If one now Fierz-rearranges the $\partial_\lambda\bar{\theta}^1\gamma_A\partial_\mu\theta^1$ and $\partial_\lambda\bar{\theta}^2\gamma_A\partial_\mu\theta^2$ terms with the help of identity (16.1.2.5), there results the expression

$$\delta S = -\frac{1}{4\pi\alpha'}\int \delta(\sqrt{-g}g^{\lambda\mu})\omega_\lambda^A\omega_{A\mu}$$

$$-\frac{2i}{\pi\alpha'}\int \sqrt{-g}(P_-^{\mu\lambda}\bar{\theta}_{,\lambda}^1\gamma^A\delta\theta^1 + P_+^{\mu\lambda}\bar{\theta}_{,\lambda}^2\gamma^A\delta\theta^2)\omega_{A\mu} \tag{16.1.3.8}$$

Further use of equations (16.1.3.5) and (16.1.3.4) then easily leads to the sought result, $\delta S = 0$.

The fact that the parameters of the gauge transformations are anticommuting suggests that one should call these transformations "local supersymmetries." It should be kept in mind, however, that these parameters are two-dimensional vectors. The new gauge transformations are not completely understood (more on this later) and their relationship with two-dimensional supergravity is unclear.† This is probably one of the reasons why the covariant quantization of the superstring has not been carried through so far.

The additional gauge invariance gives one enough gauge freedom to fix a gauge in which the theory can be completely solved at both the classical and quantum levels. Without this gauge invariance it seems that nothing can be said about the quantum theory, so that we reject the cases $a \neq (2\pi\alpha')^{-1}$‡ on grounds of simplicity.

The final covariant superstring action is therefore

$$L = -\frac{1}{4\pi\alpha'}\sqrt{-g}g^{\lambda\mu}\omega_\lambda^A\omega_{A\mu} - \frac{i}{2\pi\alpha'}\varepsilon^{\lambda\mu}\partial_\lambda X^A(\bar{\theta}^1\gamma_A\partial_\mu\theta^1 - \bar{\theta}^2\gamma_A\partial_\mu\theta^2)$$

$$+\frac{\varepsilon^{\lambda\mu}}{2\pi\alpha'}\bar{\theta}^1\gamma^A\partial_\lambda\theta^1\bar{\theta}^2\gamma_A\partial_\mu\theta^2 \tag{16.1.3.9}$$

This is the action given by Green and Schwarz [56].

† For this reason, it took some time to realize that these gauge transformations were present even in the simplest case of the superparticle; see Siegel [61].
‡ $a = -1/2\pi\alpha'$ is obtained from $a = 1/2\pi\alpha'$ by exchanging θ^1 and θ^2.

16.1.4. Equations of Motion and Boundary Conditions

The equations of motion which derive from the Lagrangian (16.1.3.9) are

$$\omega_\lambda^A \omega_{A\mu} - \tfrac{1}{2} g_{\lambda\mu} g^{\rho\sigma} \omega_\rho^A \omega_{A\sigma} = 0 \qquad (16.1.4.1)$$

$$\partial_\lambda [\sqrt{-g}(g^{\lambda\mu}\partial_\mu X^A - 2\mathrm{i} P_-^{\lambda\mu}\bar\theta^1 \gamma^A \partial_\mu \theta^1 - 2\mathrm{i} P_+^{\lambda\mu}\bar\theta^2 \gamma^A \partial_\mu \theta^2)] = 0 \qquad (16.1.4.2)$$

and

$$\gamma_A \omega_\lambda^A P_-^{\lambda\mu} \partial_\mu \theta^1 = 0 \qquad (16.1.4.3a)$$

$$\gamma_A \omega_\mu^A P_+^{\lambda\mu} \partial_\mu \theta^2 = 0 \qquad (16.1.4.3b)$$

The first equation is obtained by varying the action with respect to the metric components $g_{\lambda\mu}$. These only appear in L_1 through the unimodular combinations $\sqrt{-g}\,g^{\lambda\mu}$ (the action is Weyl invariant), so that among equations (16.1.4.1) only two are independent: the trace of equations (16.1.4.1) vanishes identically.

The meaning of equations (16.1.4.1) is again that the metric on the string world sheet (in superspace) can be Weyl-rescaled to the appropriate induced metric $\omega_\lambda^A \omega_{A\mu}$. As a result, the lightlike directions of $g_{\lambda\mu}$ coincide with those of $\omega_\lambda^A \omega_{A\mu}$, as stated earlier in equation (16.1.3.6). [An explicit proof of equation (16.1.3.6) from equation (16.1.4.1) runs as follows: Introduce two null vectors k^λ, n^λ such that $g_{\lambda\mu} = -k_\lambda n_\mu - k_\mu n_\lambda$. Write $\omega_\lambda^A = \omega_+^A k_\lambda + \omega_-^A n_\lambda$. Show that equations (16.1.4.1) imply $\omega_+^A \omega_{A+} = 0$, $\omega_-^A \omega_{A-} = 0$, i.e., $\omega_{+\lambda}^A \omega_{A+\mu} = \omega_+^A \omega_{A+} k_\lambda k_\mu = 0$, $\omega_{-\lambda}^A \omega_{A-\mu} = 0$.]

Equation (16.1.4.2) is obtained by variation of X^A. As to equations (16.1.4.3a) and (16.1.4.3b), they are equivalent to the θ-field equations and simply differ from them by use of the X^A-field equation (16.1.4.2) (and some Fierz rearrangement).

In the closed-string case, the fields are naturally taken to be periodic in σ since they are all two-dimensional scalars. In the open-string case, the equations of motion (16.1.4.1)–(16.1.4.3) must be supplemented by boundary conditions, which we now determine.

The boundary conditions for the open superstring must be such that no unwanted surface term arises in the variation of the action. This will be the case if one imposes at the ends [56]

$$\theta^1 = \theta^2 \qquad (16.1.4.4)$$

together with

$$\omega^{A1} = 0 \qquad (16.1.4.5)$$

When the latter condition holds, the variation of $\int L_1 \, d^2x$ leads to no surface term at $\sigma = 0, \pi$. Hence only $\int L_2 \, d^2x$ need be examined.

It is easy to see that condition (16.1.4.4) guarantees the absence of boundary terms in the variation of $\int L_2 \, d^2x$ with respect to X^A. Verification of this fact for variations with respect to θ^1 and θ^2 is immediate and will not be reproduced here.

The open-string boundary condition (16.1.4.4) implies that θ^1 and θ^2 possess the same chirality. Furthermore, the two global supersymmetries must be related so as to preserve condition (16.1.4.4). There is then only one global supersymmetry left, which means that the open superstring is an $N = 1$ chiral theory.

In the closed case there are two possibilities, according to whether θ^1 and θ^2 possess identical (theory B) or opposite (theory A) chiralities. Without any further restriction, these theories have two supersymmetries ($N = 2$).

16.1.5. Structure of Gauge Symmetries

As indicated above, the superstring possesses three different types of gauge invariance: reparametrization invariance, Weyl invariance, and local fermionic symmetry. While the first two define true groups of transformations, the last possesses a more complicated structure. Computation of the anticommutator of two "supergauge" transformations leads to a transformation apparently of a new kind (plus terms which obviously vanish on-shell). This "new" transformation reads [56]

$$\delta_\Lambda \theta^1 = \sqrt{-g} P_-^{\lambda\mu} \partial_\mu \theta^1 \Lambda_\lambda \tag{16.1.5.1a}$$

$$\delta_\Lambda \theta^2 = \sqrt{-g} P_+^{\lambda\mu} \partial_\mu \theta^2 \Lambda_\lambda \tag{16.1.5.1b}$$

$$\delta_\Lambda X^A = i\bar{\theta}^1 \gamma^A \delta_\Lambda \theta^1 + i\bar{\theta}^2 \gamma^A \delta_\Lambda \theta^2 \tag{16.1.5.1c}$$

$$\delta_\Lambda (\sqrt{-g} g^{\lambda\mu}) = 0 \tag{16.1.5.1d}$$

Is this gauge transformation really a new gauge invariance, implying further degeneracy of the kinetic term in the action (and needing its own Fadde'ev–Popov ghosts, and so on), or is it a gauge transformation of a trivial kind?

It is easy to see that any action $S[q^i]$ always possesses the invariance

$$\delta q^i = \frac{\delta S}{\delta q^j} \varepsilon^{ij} \tag{16.1.5.2}$$

where $\varepsilon^{ij} = (-)^{\varepsilon_i \varepsilon_j + 1} \varepsilon^{ji}$ is completely arbitrary. Indeed, one finds

$$\delta S = \frac{\delta S}{\delta q^i} \delta q^i = \frac{\delta S}{\delta q^i} \frac{\delta S}{\delta q^j} \varepsilon^{ij} = 0 \qquad (16.1.5.3)$$

where ε_i is the Grassmann parity of q^i.

However, this transformation is trivial because it vanishes on-shell and does not imply by itself any degeneracy of the action or ambiguity in the Cauchy problem. Hence it is important to establish whether system (16.1.5.1) is a true gauge transformation or whether it is of the trivial type (16.1.5.2).

Our purpose is to show that transformation (16.1.5.1) reduces to an appropriate "supergauge transformation" (16.1.3.5) when the equations of motion are used. Accordingly, nothing new is introduced.

Exercise. Consider an action $S[q^i]$ invariant under a transformation which vanishes on-shell. Show that this invariance must necessarily be of the form (16.1.5.2) with $\varepsilon^{ij} = (-)^{\varepsilon_i \varepsilon_j + 1} \varepsilon^{ji}$ when the equations of motion are independent. [When they are not independent, it can still be written as in equation (16.1.5.2) by using the dependence of the equations.]

The equations of motion in θ can be rewritten as

$$\gamma_A \omega_+^A \partial_- \theta^1 = 0 \qquad (16.1.5.4a)$$

$$\gamma_A \omega_-^A \partial_+ \theta^2 = 0 \qquad (16.1.5.4b)$$

Together with $(\omega_+^A)^2 = (\omega_-^A)^2 = 0$ (equations for the metric), relations (16.1.5.4) imply

$$\partial_- \theta^1 = \gamma_A \omega_+^A S^1 \qquad (16.1.5.5a)$$

$$\partial_+ \theta^2 = \gamma_A \omega_-^A S^2 \qquad (16.1.5.5b)$$

Exercise. Prove equations (16.1.5.5).

Now, equations (16.1.5.5) enable transformation (16.1.5.1a) and (16.1.5.1b) to be expressed in the form

$$\delta_\Lambda \theta^1 = \sqrt{-g} \partial_- \theta^1 \Lambda_+$$

$$= \sqrt{-g} \gamma_A \omega_+^A S^1 \Lambda_+$$

and

$$\delta_\Lambda \theta^2 = \sqrt{-g} \gamma_A \omega_-^A S^2 \Lambda_-$$

16 • The Superstring

which is accordingly of the form (16.1.3.5a) and (16.1.3.5b) with $2i\kappa^1_- = \sqrt{-g}S^1\Lambda_+$ and $2i\kappa^2 = \sqrt{-g}S^2\Lambda_-$. Once this is established, the equivalence of relations (16.1.5.1c) and (16.1.3.5c) is immediate. It remains to be checked that transformation (16.1.3.5d) vanishes on-shell when κ^1 and κ^2 are given by the above expressions. This is done as follows:

$$\delta_\kappa(\sqrt{-g}g^{\rho\sigma}) = -8ig(n^\rho n^\sigma \Lambda_+ \bar{S}^1 \gamma_A \omega_-^A S^1 + k^\rho k^\sigma \Lambda_- \bar{S}^2 \gamma_A \omega_+^A S^2)$$

$$= 0$$

since $C\gamma_A$ is symmetric; n^ρ and k^ρ are the null vectors introduced previously.

The conclusion is that transformation (16.1.5.1) is not a further local invariance but rather reduces to the form (16.1.3.5) on-shell. Modulo the equations of motion, the system of coordinate changes and of supergauge transformations is accordingly closed.

16.1.6. Super-Poincaré Charges

The super-Poincaré charges are computed by applying the Noether theorem. The only subtlety comes from the supercharge. Indeed, the action is invariant under supersymmetry transformations only up to a total divergence, and the conserved supersymmetry current therefore picks an additional contribution from this divergence.

The translation current is given by

$$j^\mu_A = -\frac{1}{2\pi\alpha'}\sqrt{-g}g^{\mu\nu}X_{A,\nu} + \frac{i}{\pi\alpha'}\sqrt{-g}(P_-^{\mu\nu}\bar{\theta}^1\gamma_A\partial_\nu\theta^1 + P_+^{\mu\nu}\bar{\theta}^2\gamma_A\partial_\nu\theta^2)$$

(16.1.6.1)

while the Lorentz currents are given by

$$j^\mu_{AB} = j^\mu_{[A}X_{B]} - \frac{i}{4\pi\alpha'}\bar{\theta}^1\gamma_{ABC}\theta^1\left(P_+^{\mu\nu}\omega_\nu^C + \frac{i}{2}\varepsilon^{\mu\nu}\bar{\theta}^1\gamma^C\partial_\nu\theta^1\right)$$

$$-\frac{i}{4\pi\alpha'}\bar{\theta}^2\gamma_{ABC}\theta^2\left(P_-^{\mu\nu}\omega_\nu^C - \frac{i}{2}\varepsilon^{\mu\nu}\bar{\theta}^2\gamma^C\partial_\nu\theta^2\right) \quad (16.1.6.2)$$

The expression for the supersymmetry currents can be obtained by using the equations

$$Q^{1\mu} = i\gamma^A\theta^1\frac{\partial L}{\partial X^A_{,\mu}} + \frac{\partial L}{\partial \bar{\theta}^1_{,\mu}} - Y^{1\mu} \quad (16.1.6.3a)$$

$$Q^{2\mu} = i\gamma^A \theta^2 \frac{\partial L}{\partial X^A_{,\mu}} + \frac{\partial L}{\partial \bar{\theta}^2_{,\mu}} - Y^{2\mu} \quad (16.1.6.3b)$$

where $Y^{1\mu}$ and $Y^{2\mu}$ define the total divergences into which the action transforms under supersymmetries. The Fierz-rearrangement property (16.1.2.5) enables one to obtain

$$Y^{1\mu} = \frac{1}{2\pi\alpha'} \gamma_A \theta^1 \varepsilon^{\mu\nu}(i\partial_\nu X^A + \tfrac{1}{3}\bar{\theta}^1 \gamma^A \partial_\nu \theta^1) \quad (16.1.6.4a)$$

$$Y^{2\mu} = \frac{1}{2\pi\alpha'} \gamma_A \theta^2 \varepsilon^{\mu\nu}(-i\partial_\nu X^A - \tfrac{1}{3}\bar{\theta}^2 \gamma^A \partial_\nu \theta^2) \quad (16.1.6.4b)$$

and thus

$$Q^{1\mu} = -\frac{2i}{\pi\alpha'} \gamma_A \theta^1 (\sqrt{-g} P_+^{\mu\nu} \omega_\nu^A + \tfrac{2}{3} i \varepsilon^{\mu\nu} \bar{\theta}^1 \gamma^A \partial_\nu \theta^1) \quad (16.1.6.5a)$$

$$Q^{2\mu} = -\frac{2i}{\pi\alpha'} \gamma_A \theta^2 (\sqrt{-g} P_-^{\mu\nu} \omega_\nu^A - \tfrac{2}{3} i \varepsilon^{\mu\nu} \bar{\theta}^2 \gamma^A \partial_\nu \theta^2) \quad (16.1.6.5b)$$

In the closed-string case the corresponding charges, given by spatial integrals of the current temporal components, are automatically conserved as a result of the continuity equation.

For the open string, there is no flux of energy momentum or angular momentum through the ends thanks to the boundary conditions. This is not true for the supercharges. One finds that only the sum $Q^1 + Q^2$ is conserved. One of the two global supersymmetries is broken by the boundary conditions.

16.1.7. Hamiltonian Formalism

A full, satisfactory Hamiltonian formulation of the superstring has not yet been carried out owing to the following difficulties:

1. The action (16.1.3.9) leads to constraints in the canonical formalism. Some of these constraints are "first class" and generate the gauge invariances of the theory. There are, however, also some "second-class" constraints which originate from the absence of an independent kinetic term for the spinors. In order to go to the quantum theory, one must eliminate all the second-class constraints and replace the Poisson bracket by the Dirac bracket. The main problem

is that this cannot be done in a nice way: it is difficult to disentangle the first- and second-class constraints without at the same time breaking manifest super-Poincaré invariance [62, 63]. The manifest preservation of the global invariances was the main motivation for the constrained approach.
2. The first-class constraint "algebra" is not the super-Virasoro algebra, corresponding to the fact that the theory is not $2d$ supergravity. Actually, the constraint algebra is an open algebra, with structure functions which involve the fields. This corresponds to the fact that the gauge transformations only close on-shell, and implies that one needs the full machinery developed by the Fradkin school [30, 31] in order to construct the BRST charge. It is furthermore unclear how to give a quantum meaning to the nonlinear terms which will arise in Ω.
3. The supergauge spinor parameters κ_-^1 and κ_+^2 are redundant on-shell. If $\omega_+^A \gamma_A \kappa_-^1 = 0$ and $\omega_-^A \gamma_A \kappa_+^2 = 0$, the right-hand sides of transformations (16.1.3.5a)-(16.1.3.5d) vanish. This leads to further complication in the BRST formalism.

This is all we intend to say about the Hamiltonian formulation since, at the time of writing, there is not much more to report. In order to illustrate the difficulties, we will treat in detail below the superparticle, which is the zero-mode truncation of the superstring.

16.1.8. Light-Cone Gauge

The light-cone gauge quantization of the superstring is comparatively much more straightforward. By employing the coordinate freedom, one can pass to a "conformal gauge,"

$$\sqrt{-g}g^{\lambda\mu} = \eta^{\lambda\mu} \tag{16.1.8.1}$$

The global validity of the conformal gauge is submitted to the same conditions as in the pure bosonic case. Contrary to what happens to the bosonic string model, however, the conformal coordinate choice does not linearize the equations of motion. In order to achieve this goal, one must also fix the local fermionic invariance.

The basic difference with the spinning-string model of Neveu–Schwarz–Ramond is that this second step cannot be done covariantly, at least in a simple manner. The reason can be understood heuristically as follows. The spinor parameters κ_-^1 and κ_+^2 are redundant on-shell, i.e., they do not all define independent gauge transformations. The true number of gauge invariances contained in κ_-^1 is not equal to 16, as one might naively think (number of independent components of a Majorana–Weyl spinor), but rather only half that number, namely 8: if $\kappa_-^1 = \omega_+^A \gamma_A \varepsilon$, the associated variations of the fields vanish. This number 8 does not correspond to a representation of the Lorentz group SO(9, 1), and it is therefore difficult to find 8 independent Lorentz-covariant conditions which fix the fermionic gauge symmetry described by κ_-^1.

In view of this problem, one rather imposes, together with the conformal gauge (16.1.8.1), simple, noncovariant conditions on the spinors, namely,

$$\gamma^+ \theta^1 = 0 = \gamma^+ \theta^2 \qquad (16.1.8.2)$$

where $\gamma^+ = (\gamma^0 + \gamma^9)$; the + refers now to ten dimensions.† The conditions (16.1.8.2) contain only 8 + 8 independent equations, as required for freezing an (8 + 8)-fold gauge freedom.

These conditions completely fix the nonredundant freedom contained in $\kappa_-^1 \equiv \kappa^1$ and $\kappa_+^2 \equiv \kappa^2$. Indeed, one can always write‡

$$\kappa^1 = \varepsilon + \omega_+^A \gamma_A \eta \qquad (16.1.8.3)$$

where $\gamma^+ \varepsilon = \gamma^+ \eta = 0$. To see this, we multiply (16.1.8.3) by γ^+ and obtain

$$\gamma^+ \kappa^1 = \gamma^+ \omega_+^A \gamma_A \eta$$
$$= \tfrac{1}{2}\omega_+^A (\gamma^+ \gamma_A + \gamma_A \gamma^+) \eta$$
$$= \tfrac{1}{2}\omega^+{}_+ \eta \qquad (16.1.8.4)$$

where the upper + in $\omega^+{}_+$ refers to ten dimensions and the lower one to two dimensions. If $\omega^+{}_+ \neq 0$, which we assume,§ this equation determines η and then ε. Hence

$$\kappa^1 = \left(\kappa^1 - \frac{2\omega_+^A \gamma_A \gamma^+ \kappa^1}{\omega^+{}_+}\right) + 2 \frac{\omega_+^A \gamma_A \gamma^+}{\omega^+{}_+} \kappa^1 \qquad (16.1.8.5)$$

which proves equation (16.1.8.3).

The κ^1-gauge freedom is completely fixed by the gauge condition (16.1.8.2) if and only if the κ^1-gauge transformations which leave $\gamma^+ \theta^1$

---

† One could take instead, in order to preserve Lorentz invariance, the condition $\omega_-^A \gamma_A \theta^1 = \omega_+^A \gamma_A \theta^2 = 0$, where $\omega_\pm^A$ are the lightlike directions along the string world sheet. This leads, however, to complicated equations. Furthermore, this construction would break down for the ground state (superparticle).
‡ Some light-cone gauge techniques are reviewed in Appendix C.
§ The case $\omega^+{}_+ = 0$ is the usual infrared singularity of the light-cone gauge.

16 • The Superstring

equal to zero are characterized by $\varepsilon = 0$, i.e., have no on-shell effect on the fields. This is the case since $\delta\theta^1$ is given by

$$\delta\theta^1 = 2i\omega_+^A \gamma_A \kappa^1$$

Therefore, the variation of the gauge conditions (16.1.8.2) yields

$$0 = \delta(\gamma^+ \theta^1) = 2i\gamma^+ \omega_+^A \gamma_A \kappa^1$$
$$= 2i\gamma^+ \omega_+^A \gamma_A \varepsilon \qquad (16.1.8.6)$$

On using the anticommutation relation of the γ matrices and $\gamma^+ \varepsilon = 0$, this relation implies $\varepsilon = 0$, as desired.

One proves in a similar way that the condition $\gamma^+ \theta^2 = 0$ fixes the nonredundant part of the κ^2-gauge freedom.

When condition (16.1.8.2) holds, the equations of motion for the fields simplify substantially. The bilinear terms $\bar{\theta}^1 \gamma^+ \partial_\mu \theta^1$, $\bar{\theta}^2 \gamma^+ \partial_\mu \theta^2$, $\bar{\theta}^1 \gamma^i \partial_\mu \theta^1$, and $\bar{\theta}^2 \gamma^i \partial_\mu \theta^2$ vanish,

$$\bar{\theta}^1 \gamma^+ \partial_\mu \theta^1 = \bar{\theta}^2 \gamma^+ \partial_\mu \theta^2 = 0 \qquad (16.1.8.7a)$$

$$\bar{\theta}^1 \gamma^i \partial_\mu \theta^1 = -\tfrac{1}{2} \bar{\theta}^1 (\gamma^+ \gamma^- + \gamma^- \gamma^+) \gamma^i \partial_\mu \theta^1$$
$$= -\tfrac{1}{2} \bar{\theta}^1 \gamma^- \gamma^+ \gamma^i \partial_\mu \theta^1$$
$$= \tfrac{1}{2} \bar{\theta}^1 \gamma^- \gamma^i \gamma^+ \partial_\mu \theta^1$$
$$= 0 \qquad (16.1.8.7b)$$

and

$$\bar{\theta}^2 \gamma^i \partial_\mu \theta^2 = 0 \qquad (16.1.8.7c)$$

so that the equations for X^+ and X^i become ordinary d'Alembertian equations in two dimensions,

$$\Box X^+ = \Box X^i = 0 \qquad (16.1.8.8)$$

At the same time, one can replace the spinor equations $\gamma_A \omega_+^A \partial_- \theta^1 = 0$ and $\gamma_A \omega_-^A \partial_+ \theta^2 = 0$ by

$$\partial_- \theta^1 = 0 \qquad (16.1.8.9a)$$

and

$$\partial_+\theta^2 = 0 \tag{16.1.8.9b}$$

as can easily be seen upon multiplication by γ^+.

Advantage can be taken of the harmonicity (16.1.8.8) of X^+ to perform a conformal coordinate transformation in which the new time is given by X^+,

$$X^+ = 2p^+\alpha'\tau \quad \text{(open string)} \tag{16.1.8.10}$$

where p^+ is the total space-time momentum in the + direction.

This is just as in the bosonic case, and one finds also that the momentum density in the + direction is constant,

$$j^{+0}(\sigma) = \frac{\partial L}{\partial \dot{X}_+} = \frac{p^+}{\pi} \quad \text{(open string)} \tag{16.1.8.11}$$

The conditions (16.1.8.10), (16.1.8.11), and (16.1.8.2) define the light-cone gauge for the open superstring. No residual gauge invariance remains when they hold. The light-cone gauge for the closed superstring is defined by similar equations, with 2π in place of π due to our conventions. Again, the conditions (16.1.8.10) and (16.1.8.11) do not completely fix the coordinate system in this latter case. The residual gauge invariance is given by σ-independent translations along the σ direction. The corresponding constraint relates the number of right and left movers.

In order to describe the dynamics in the light-cone gauge, it is convenient to introduce the light-cone gauge action, which is simply obtained by inserting the gauge conditions into the covariant action. This yields

$$S^{\text{L.C.}} = \int d\tau\, d\sigma \left[-(4\pi\alpha')^{-1} \sum_{i=1}^{8} \eta^{\lambda\mu}\partial_\lambda X^i \partial_\mu X^i + i\frac{p^+}{\pi} \bar{\theta}\gamma^-\rho^\lambda \partial_\lambda \theta \right] \tag{16.1.8.12}$$

where

$$\theta = \begin{pmatrix} \theta^1 \\ \theta^2 \end{pmatrix} \quad \text{and} \quad \bar{\bar{\theta}} = \begin{pmatrix} \bar{\theta}^1 \\ \bar{\theta}^2 \end{pmatrix} \rho^0 \tag{16.1.8.13}$$

and ρ^λ are two-dimensional Dirac matrices,

$$\rho^0 = \begin{pmatrix} 0 & 1 \\ -1 & 0 \end{pmatrix} \quad \text{and} \quad \rho^1 = \begin{pmatrix} 0 & 1 \\ 1 & 0 \end{pmatrix} \tag{16.1.8.14}$$

In deriving expression (16.1.8.12), we have used $\dot{p}^+ = 0$ (which is a consequence of the equation of motion for X^-) and omitted appropriate total time derivatives.†

One can readily check that the Euler–Lagrange equations following from equation (16.1.8.12) are the correct light-cone gauge equations (16.1.8.8) and (16.1.8.9) for the physical transverse modes X^i, θ^1, and θ^2. One sees in particular that equations (16.1.8.9) comprise just Dirac equations in two dimensions.

The light-cone gauge action completely describes the open superstring in the light-cone gauge. In the closed case, the action must be supplemented by the constraint on the equality of the level numbers in the right-moving and left-moving sectors.

Finally, in order to complete the light-cone gauge analysis of the superstring, one needs the form of the super-Poincaré generators. This is obtained as in the previous models, by expressing the dependent in terms of the independent variables. We leave it to the reader to actually write down the light-cone gauge form of the super-Poincaré charges [55]. All that is needed is the expression for $\dot{X}^-$ and X'^-, which follows from the constraints (16.1.4.1), $\eta_{AB}\omega^A_\lambda \omega^B_\mu = \phi g_{\lambda\mu}$:

$$\dot{X}^- = \frac{1}{4\alpha'p^+}(\dot{X}^{i2} + X'^{i2}) + i\bar{\theta}^1\gamma^-\dot{\theta}^1 + i\bar{\theta}^2\gamma^-\dot{\theta}^2 \qquad (16.1.8.15a)$$

$$X'^- = \frac{1}{2\alpha'p^+}\dot{X}^i X'^i + i\bar{\theta}^1\gamma^-\theta^{1'} + i\bar{\theta}^2\gamma^-\theta^{2'} \qquad (16.1.8.15b)$$

The quantity X^- is quadratic in the other variables, so the super-Poincaré generators acquire cubic contributions.

16.2. QUANTUM THEORY

Only the light-cone gauge quantization of the superstring has been developed so far for the reasons explained above. The reader should now be able to work out the light-cone gauge quantum analysis of the superstring, which proceeds along lines similar to the bosonic case [36]. Accordingly, we only reproduce the results.

One finds that the quantum super-Poincaré algebra develops no anomaly only in 10 dimensions [36, 44, 55]. The $d = 10$ open-superstring

† We are a little bit cavalier in our treatment of the zero modes p^+ and X_0^-, since we use their equations of motion inside the action. For a more careful analysis in which these modes are consistently treated as dynamical variables, the reader is referred to the sections on the bosonic string and on the superparticle.

spectrum reproduces the supersymmetric truncation of the open Neveu-Schwarz-Ramond fermionic string, which is $N = 1$ supersymmetric. The ground-state particles yield a super-Yang-Mills gauge multiplet.

The closed-superstring spectrum can again be truncated by retaining only states which are symmetric under the interchange $\sigma \to -\sigma$ (nonoriented superstring $\equiv$ type I closed superstring). This symmetric truncation is not invariant under both supersymmetries, but rather only possesses $N = 1$ supersymmetry. The ground state is the $d = 10$, $N = 1$ supergravity gauge multiplet. The spectrum coincides with the supersymmetric truncation of the closed Neveu-Schwarz-Ramond fermionic string. One can show that interactions of open superstrings can produce type I closed superstrings, so that they combine to form a single "type I theory."

The full closed-superstring spectrum is noninvariant under $\sigma \to -\sigma$ and hence corresponds to an oriented string ("type II"). It possesses $N = 2$ supersymmetry. Depending on the relative chiralities of θ^1 and θ^2, the theory is called type II_A or type II_B. The ground states of these theories are the gauge multiplets of two different $d = 10$, $N = 2$ supergravity models (type II_A corresponds to dimensional reduction of $d = 11$, $N = 1$ supergravity; type II_B to $d = 10$, $N = 2$ chiral supergravity). For more on this, see Schwarz [36].

16.3. THE SUPERPARTICLE

16.3.1. Action–Gauge Symmetries

The action which describes the massless superparticle is given by [64]

$$S[V, X^A, \theta] = -\tfrac{1}{2} \int V^{-1} \omega^A \omega_A \, d\tau \qquad (16.3.1.1)$$

with

$$\omega^A = \dot{X}^A - i\bar{\theta}\gamma^A \dot{\theta} \qquad (16.3.1.2)$$

It is the obvious supersymmetric extension of the action for a massless relativistic particle without spin. Here, V is the "einbein," X^A ($A = 0, \ldots, 9$) are the particle coordinates, and θ is an internal degree of freedom which obeys the Majorana-Weyl condition for simplicity.

If one truncates the superstring to its zero-mode sector by discarding the higher modes, the result is action (16.3.1.1). Hence, this action represents the superstring ground state and it is clearly important to understand its dynamics.

16 • The Superstring

An interesting feature of the superparticle is that the major obstacles to a satisfactory covariant quantization of the superstring are already illustrated in that simpler model.

The action (16.3.1.1) is invariant under reparametrizations,

$$\delta X^A = \varepsilon \dot{X}^A, \qquad \delta\theta = \varepsilon \dot{\theta} \tag{16.3.1.3a}$$

$$\delta V = \varepsilon \dot{V} + \dot{\varepsilon} V = (\varepsilon V)^{\cdot} \tag{16.3.1.3b}$$

as well as under a fermionic gauge symmetry analogous to (16.1.3.5) [61],

$$\delta_\kappa \theta = i\omega\!\!\!/\,\kappa \tag{16.3.1.4a}$$

$$\delta_\kappa X^A = i\bar{\theta}\gamma^A \delta_\kappa \theta \tag{16.3.1.4b}$$

$$\delta_\kappa V = 4V\dot{\bar{\theta}}\kappa \tag{16.3.1.4c}$$

where we have set $\omega\!\!\!/ = \omega_A \gamma^A$.

The Euler-Lagrange equations are

$$\omega^A \omega_A = 0 \tag{16.3.1.5a}$$

$$(V^{-1}\omega^A)^{\cdot} = 0 \tag{16.3.1.5b}$$

$$\omega\!\!\!/\,\dot{\theta} = 0 \tag{16.3.1.5c}$$

It is seen from equation (16.3.1.5a) that ω^A is a null vector. Therefore the matrix $\omega\!\!\!/$ is nilpotent and cannot be inverted, in complete similarity with the superstring.

Equation (16.3.1.5c) implies

$$\dot{\theta} = \omega\!\!\!/\, s \tag{16.3.1.6}$$

because the kernel of $\omega\!\!\!/$ coincides with its image when ω^A is null, a property we have already used. Integration of equation (16.3.1.6) yields

$$\theta(\tau) = \theta_0 + \omega\!\!\!/\,\phi(\tau) \tag{16.3.1.7}$$

where $\dot{\phi} + (\dot{V}/V)\phi = s$. The spinor $\phi(\tau)$ is not determined by the equations of motion, in agreement with the gauge invariance (16.3.1.4).

It follows from equations (16.3.1.6) and (16.3.1.5a) that the fermionic gauge transformations (16.3.1.4) are redundant on-shell. If $\kappa = \omega\!\!\!/\,\varepsilon$, then

$\delta_\kappa \theta$ and $\delta_\kappa X^A$ evidently vanish while $\delta_\kappa V$ becomes

$$\delta_\kappa V = -4 V \bar{s} \not{\omega} \kappa = 0$$

It was pointed out by Green and Schwarz [56] that the action possesses also the following bosonic invariance:

$$\delta_\lambda \theta = \lambda \dot{\theta} \tag{16.3.1.8a}$$

$$\delta_\lambda X^A = i\bar{\theta}\gamma^A \delta_\lambda \theta \tag{16.3.1.8b}$$

$$\delta_\lambda V = 0 \tag{16.3.1.8c}$$

which arises when one tries to close the algebra of systems (16.3.1.3) and (16.3.1.4). However, invariance (16.3.1.8) differs from a transformation of type (16.3.1.4) only by terms which vanish with the field equations. This can be seen by the same reasoning as for the superstring. Accordingly, system (16.3.1.8) adds nothing new.

16.3.2. Super-Poincaré Charges

The action is invariant under the global super-Poincaré transformations

$$\delta\theta = \tfrac{1}{4}\omega_{AB}\gamma^{AB}\theta + \varepsilon \tag{16.3.2.1a}$$

$$\delta X^A = \omega^A{}_B X^B + i\bar{\varepsilon}\gamma^A \theta + a^A \tag{16.3.2.1b}$$

Straightforward application of the Noether theorem yields the super-Poincaré charges,

$$P^A = -V^{-1}\omega^A \tag{16.3.2.2a}$$

$$M^{AB} = P^{[A} X^{B]} - \frac{i}{4} P_C \bar{\theta}\gamma^{ABC}\theta \tag{16.3.2.2b}$$

$$Q = 2iP_A \gamma^A \theta \tag{16.3.2.2c}$$

16.3.3. Hamiltonian Formalism

The theory possesses gauge invariances; therefore one expects constraints in the Hamiltonian formalism.

The momenta conjugate to V, X^A, and θ are

$$p_V = 0 \tag{16.3.3.1}$$

16 • The Superstring

$$p_A = \frac{\partial L}{\partial \dot{X}^A} = -V^{-1}\omega_A = P_A \qquad (16.3.3.2)$$

$$\bar{p}_\theta = \frac{\partial L}{\partial \dot{\theta}} = -i\bar{\theta}\slashed{p} \qquad (16.3.3.3)$$

Expressions (16.3.3.1) and (16.3.3.3) yield the primary constraints

$$p_V = 0 \qquad (16.3.3.4a)$$

$$\bar{\chi} \equiv \bar{p}_\theta + i\bar{\theta}\slashed{p} = 0 \qquad (16.3.3.4b)$$

The Hamiltonian is given by

$$H = \dot{V}p_V + \dot{X}^A p_A + \bar{p}_\theta \dot{\theta} - L + \lambda p_V + (\bar{p}_\theta + i\bar{\theta}\slashed{p})\mu$$

$$= -\tfrac{1}{2}V p_A p^A + \lambda p_V + (\bar{p}_\theta + i\bar{\theta}\slashed{p})\mu \qquad (16.3.3.5)$$

where λ and μ are Lagrange multipliers (λ is a scalar and μ is a Majorana–Weyl spinor).

The preservation in time of the primary constraint $p_V = 0$ leads to the secondary constraint

$$\mathcal{H} \equiv \tfrac{1}{2}p_A p^A = 0 \qquad (16.3.3.6)$$

(mass-shell condition).

The preservation in time of the constraint (16.3.3.4b) implies conditions on the multiplier μ,

$$C\slashed{p}\mu = 0 \to \slashed{p}\mu = 0 \to \mu = \slashed{p}\nu \qquad (16.3.3.7)$$

The secondary constraint (16.3.3.6) is maintained in time by the Hamiltonian, so that the "consistency algorithm" ends.

One can eliminate the momentum p_V since the pair (p_V, V) is pure gauge, and treat V as a Lagrange multiplier for the constraint (16.3.3.6). The superparticle is accordingly described by the canonical pairs (X^A, p_A), $(\theta, \bar{p}_\theta)$ obeying

$$[X^A, p_B] = \delta^A{}_B \qquad (16.3.3.8a)$$

$$[\theta^\alpha, \bar{p}_{\theta\beta}] = [\bar{p}_{\theta\beta}, \theta^\alpha] = \delta^\alpha_\beta \qquad (16.3.3.8b)$$

(where α is now a spinor index), by the constraints (16.3.3.6) and (16.3.3.4b), and by the Hamiltonian

$$H = -\tfrac{1}{2} V p_A p^A + \bar{p}_\theta \not{p} \nu \qquad (16.3.3.9)$$

In the latter expression, the multiplier V has been conveniently redefined in order to attain a simpler form for H.

The Hamiltonian vanishes weakly, as is appropriate for generally covariant systems.

16.3.4. Meaning of the Constraints

Among the constraints, some are related to the gauge invariances of the theory while some result from the absence of an independent kinetic term quadratic in the fermions.

The constraints related to the gauge invariances are called "first class" and are characterized by the property that they commute weakly with the other constraints. In order to identify them, we compute the constraint algebra:

$$[\mathcal{H}, \bar{\chi}_\alpha] = 0 \qquad (16.3.4.1a)$$

$$[\bar{\chi}_\alpha, \bar{\chi}_\beta] = 2i(\gamma_0 \not{p})_{\alpha\beta} \qquad (16.3.4.1b)$$

The constraint $\mathcal{H}$ is clearly first class. It is just the generator of local reparametrizations accompanied by an appropriate transformation (16.3.1.8), so that $\delta\theta = 0$.

The matrix on the right-hand side of relation (16.3.4.1b) does not vanish, however, hence the constraints $\bar{\chi}$ are not all first class. As a consequence of the mass-shell condition, the rank of the matrix $(\gamma_0 \not{p})_{\alpha\beta}$ is 8 (in the subspace of Weyl spinors), so that among the 16 Majorana–Weyl constraints $\bar{\chi}_\alpha = 0$, 8 are second class and 8 first class.

The first-class constraints can be isolated covariantly, in a redundant way, by acting with $\not{p}$ on $\bar{\chi}$,

$$\bar{\chi}\not{p} = \bar{p}_\theta \not{p} + i\bar{\theta} p^2 \approx \bar{\phi} \equiv \bar{p}_\theta \not{p} \qquad (16.3.4.2)$$

One has

$$[\bar{\phi}_\alpha, \mathcal{H}] = 0 \qquad (16.3.4.3a)$$

$$[\bar{\phi}_\alpha, \bar{\chi}_\beta] = 2i(\gamma_0)_{\alpha\beta}\mathcal{H} \approx 0 \qquad (16.3.4.3b)$$

and also

$$[\bar{\phi}_\alpha, \bar{\phi}_\beta] = 0 \qquad (16.3.4.4)$$

It is this last relation which motivates our replacement of $\bar\chi p\!\!\!/$ by $\bar\phi$, since it makes abelian the algebra of the first-class constraints.

One can easily check that $\bar\phi_\alpha$ (weakly) generates the fermionic gauge symmetry (16.3.1.4) (plus a suitable reparametrization),

$$[\theta, \bar\phi\nu] = p\!\!\!/\nu \tag{16.3.4.5a}$$

$$[X^A, \bar\phi\nu] = \bar p_\theta \gamma^A \nu \approx -i\bar\theta p\!\!\!/ \gamma^A \nu = i\bar\theta\gamma^A p\!\!\!/\nu - 2i\bar\theta\nu p^A \tag{16.3.4.5b}$$

The last term on the right-hand side of relation (16.3.4.5b) is a particular reparametrization generated by $\mathcal{H}$, corresponding to the above redefinition $\bar\chi p\!\!\!/ \to \bar\phi$.

Two things should be stressed at this point from our analysis: the matching between the gauge transformations in Lagrangian and Hamiltonian forms only holds on-shell, as is generically the case for noninternal gauge symmetries; furthermore, there is no independent Hamiltonian first-class constraint associated with the "fake" gauge invariance (16.3.1.8). Actually, one of the interests of the Hamiltonian formalism is that only "true" gauge invariances lead to first-class constraints, and this enables one to identify and count them in a straightforward manner.

Although the first-class constraints contained in $\bar\chi \approx 0$ have been isolated covariantly, the price paid for maintaining manifest covariance is the redundancy of the constraints $\bar\phi$. One indeed finds the following strong relation between $\bar\phi$ and $\mathcal{H}$:

$$\bar\phi p\!\!\!/ - 2\bar p_\theta \mathcal{H} = 0 \tag{16.3.4.6}$$

This reflects the redundancy of the gauge parameters of transformations (16.3.1.4) and leads to complications in the BRST construction. We note that there appears to be no simple covariant way to isolate $(8 + 1)$ independent constraints among the $(16 + 1)$ redundant first-class constraints $\bar\phi, \mathcal{H}$.

In the same way, there is no simple covariant manner to identify the remaining 8 second-class constraints contained in $\bar\chi$. It will be shown below that these second-class constraints play an important role in providing the correct anticommutation relation for the spinors, so that they should not be taken too lightly. It is indispensable to keep them, and one must accordingly face the manifest covariance problem just mentioned.

16.3.5. Siegel's Model

We have shown that the first-class constraints contained in $\bar\chi$ can be identified covariantly, although in a redundant way.

Siegel [65] has suggested a new theory in which one retains only the first constraints $\bar\phi$ and $\mathcal{H}$, and forgets about the second-class constraints contained in $\bar{\chi}$. The canonical action for this modified theory is given by

$$S = \int (p_A \dot{X}^A + \bar{p}_\theta \dot{\theta} - H) \, d\tau \qquad (16.3.5.1)$$

and the sole constraints are

$$\mathcal{H} = 0 \quad \text{and} \quad \bar{\phi} = 0 \qquad (16.3.5.2)$$

The problem with this new model is that it contains negative-norm states, because such unwanted states are necessary to represent the canonical anticommutation relations for the fermions which follow from expression (16.3.5.1), without the second-class constraints.

To see this, we consider the light-cone gauge

$$\gamma^+ \theta = 0 \quad \text{and} \quad X^+ \sim \tau \qquad (16.3.5.3)$$

The kinetic term for the fermions becomes

$$\int \bar{p}_\theta^a \dot{\theta}^a \, d\tau \qquad (16.3.5.4)$$

where a is an SO(8) spinor index. Also, θ^a is an SO(8) light-cone gauge spinor obeying relations (16.3.5.3). It contains 8 independent real components taking into account the Majorana–Weyl condition; see Appendix C.

It follows from the term (16.3.5.4) that θ^a and $\bar{p}_\theta^a$, which are unconstrained in Siegel's model, obey the Poisson-bracket relations

$$[\theta^a, \theta^b] = 0, \quad [\theta^a, \bar{p}_\theta^b] = \delta^{ab}, \quad [p_\theta^a, p_\theta^b] = 0 \qquad (16.3.5.5)$$

The corresponding anticommutators are

$$\theta^a \theta^b + \theta^b \theta^a = 0, \quad \theta^a \bar{p}_\theta^b + \bar{p}_\theta^b \theta^a = i \delta^{ab} \qquad (16.3.5.6a)$$

$$\bar{p}_\theta^a \bar{p}_\theta^b + \bar{p}_\theta^b \bar{p}_\theta^a = 0 \qquad (16.3.5.6b)$$

Because θ^a is real, the only way to implement $(\theta^a)^2 = 0$ is to use a Hilbert space with indefinite metric. [In a true Hilbert space, $(\theta^a)^2 = 0$ and $\theta^a = (\theta^a)^*$ would imply $\theta^a = 0$, in contradiction to equations (16.3.5.6a).] The purpose of the second-class constraints of the original superparticle model is precisely to avoid this problem, by giving to the fermions the correct kinetic term $\theta^a \dot{\theta}^a$.

Although Siegel's model suffers from the presence of negative-norm states, it is nevertheless of interest to go ahead with its quantization by BRST methods. One can truncate, at this end, the theory to the positive-definite sector. This procedure should presumably be equivalent to the superparticle model.

The noticeable property of Siegel's model within the BRST context is that it is the first example of a theory which needs an infinite tower of compensating ghosts for ghosts if one wishes to preserve manifest covariance. The theory is a reducible theory "of infinite stage" [66]. This feature is shared by the original model (16.3.1.1).

We have stressed already many times that the constraints $\bar{\phi}$ and $\mathcal{H}$ are not independent but obey

$$\bar{\phi}\slashed{p} - 2\mathcal{H}\bar{p}_\theta = 0 \qquad (16.3.5.7)$$

(identically).

Now, in the usual BRST formalism, only one canonically conjugate pair of ghosts should be introduced per independent constraint. However, because there is no simple covariant way to single out the independent constraints among $\bar{\phi}$ and $\mathcal{H}$, it would seem that this orthodox application of BRST methods would encounter the same difficulties of nonmanifest covariance that we wanted to avoid by omitting the second-class constraints.

The way to deal with this problem is well known [66] and reviewed in Ref. 31. One associates a canonically conjugate pair of ghosts to each constraint and then introduces further "ghosts for ghosts," which compensate for the wrong counting. Therefore, one would first have "primary" ghosts (η, π) and $(\overset{(1)}{C}, \bar{\mathcal{P}}^{(1)})$ obeying

$$[\eta, \pi] = 1 = [\pi, \eta] \qquad (16.3.5.8a)$$

$$[\overset{(1)}{C}{}^\alpha, \bar{\mathcal{P}}^{(1)}_\beta] = \delta^\alpha_\beta = -[\bar{\mathcal{P}}^{(1)}_\beta, \overset{(1)}{C}{}^\alpha] \qquad (16.3.5.8b)$$

The ghosts (η, π) are anticommuting scalars, while $(\overset{(1)}{C}, \bar{\mathcal{P}}^{(1)})$ are commuting Majorana–Weyl spinors of appropriate chirality. They possess the following ghost numbers:

$$\text{gh}(\eta) = \text{gh}(\overset{(1)}{C}) = -\text{gh}(\pi) = -\text{gh}(\bar{\mathcal{P}}^{(1)}) = 1 \qquad (16.3.5.9)$$

At this first stage, the BRST generator is given by [66, 31]

$$\overset{(1)}{\Omega} = \tfrac{1}{2}p^2 \eta + \bar{p}_\theta \slashed{p} \overset{(1)}{C} \qquad (16.3.5.10)$$

and is clearly nilpotent.

Next, one introduces secondary ghosts which compensate for the overcounting of constraints. It has been explained elsewhere [66, 31] that one canonically conjugate pair of ghosts must be introduced for each independent relation among the constraints.

What is the number of independent relations (16.3.5.7) obeyed by the constraints? Not 16, of course, as the fact that condition (16.3.5.7) is a Majorana-Weyl spinor might superficially indicate, but only 8. The coefficients $\overset{(1)}{X}$ of the constraints in relation (16.3.5.7) indeed satisfy

$$\overset{(1)}{X}{}^\alpha{}_\beta(\not{p})^\beta{}_\rho \approx 0 \quad \text{and} \quad X^{(1)}{}_\beta(\not{p})^\beta{}_\rho \approx 0 \qquad (16.3.5.11)$$

with $\overset{(1)}{X}{}^\alpha{}_\beta$ and $\overset{(1)}{X}_\beta$ defined by

$$\bar{\phi}_\alpha \overset{(1)}{X}{}^\alpha{}_\beta + \mathcal{H} \overset{(1)}{X}_\beta = 0 \iff X^{(1)\alpha}{}_\beta = (\not{p})^\alpha_\beta, \quad \overset{(1)}{X}_\beta = -2(\not{p}_\theta)_\beta$$

$$(16.3.5.12)$$

Hence they cannot be independent.

Can one isolate covariantly 8 independent relations among the equations (16.3.5.7)? One runs here into the same type of difficulties as above [equation (16.3.5.7) defines an irreducible representation of SO(9, 1)]. Accordingly, it is better not to separate relations (16.3.5.7) into independent and dependent components, but rather to introduce as many canonically conjugate pairs of secondary ghosts as there are relations in equations (16.3.5.7) and compensate later by adding "ghosts for ghosts for ghosts." The secondary ghosts, denoted by $\overset{(2)}{C}$ and $\bar{\mathcal{P}}^{(2)}$, are anticommuting spinors and obey

$$[\overset{(2)}{C}{}^\alpha, \bar{\mathcal{P}}^{(2)}_\beta] = \delta^\alpha_\beta = [\bar{\mathcal{P}}^{(2)}_\beta, \overset{(2)}{C}{}^\alpha], \quad \text{gh}(\overset{(2)}{C}) = -\text{gh}(\bar{\mathcal{P}}^{(2)}) = 2 \qquad (16.3.5.13)$$

At this second stage, the BRST generator is [66, 31]

$$\Omega = \overset{(1)}{\Omega} + \overset{(2)}{\Omega} \qquad (16.3.5.14a)$$

$$\overset{(2)}{\Omega} = \bar{\mathcal{P}}^{(1)} \not{p} \overset{(2)}{C} - 2\pi \bar{p}_\theta \overset{(2)}{C} \qquad (16.3.5.14b)$$

and is still clearly nilpotent.

The next step is to introduce tertiary ghosts, which take care of the fact that equations (16.3.5.11) involving $\overset{(1)}{X}{}^\alpha_\beta$ and $\overset{(1)}{X}_\beta$ are not all independent, as we have emphasized. Instead, one has

$$\overset{(2)}{X}{}^{\alpha}{}_{\beta}\overset{}{p}{}^{\beta}{}_{\rho} \approx 0 \qquad (16.3.5.15)$$

with $\overset{(2)}{X}{}^{\alpha}{}_{\beta}$ defined as the coefficients of $\overset{(1)}{X}{}^{\alpha}{}_{\beta}$ and $\overset{(1)}{X}{}_{\beta}$ in equations (16.3.5.11),

$$\overset{(1)}{X}{}^{\alpha}{}_{\beta}\overset{(2)}{X}{}^{\beta}{}_{\rho} \approx 0, \qquad \overset{(1)}{X}{}_{\beta}\overset{(2)}{X}{}^{\beta}{}_{\rho} \approx 0 \quad \Leftrightarrow \quad \overset{(2)}{X}{}^{\beta}{}_{\rho} = \overset{}{p}{}^{\beta}{}_{\rho} \qquad (16.3.5.16)$$

Equations (16.3.5.15) can be rewritten as

$$\overset{(2)}{X}{}^{\alpha}{}_{\beta}\overset{(3)}{X}{}^{\beta}{}_{\rho} \approx 0 \quad \text{and} \quad \overset{(3)}{X}{}^{\beta}{}_{\rho} = \overset{}{p}{}^{\beta}{}_{\rho} \qquad (16.3.5.17)$$

The recipe [66, 31] is to associate as many pairs of ghosts as there are independent equations on $\overset{(2)}{X}{}^{\alpha}{}_{\beta}$. What is this number? Again, it is equal to 8 since the coefficients $\overset{(3)}{X}{}^{\beta}{}_{\rho}$ are not independent,

$$\overset{(3)}{X}{}^{\alpha}{}_{\beta}\overset{(4)}{X}{}^{\beta}{}_{\rho} \approx 0 \quad \text{with} \quad \overset{(4)}{X}{}^{\beta}{}_{\rho} = \overset{}{p}{}^{\beta}{}_{\rho} \qquad (16.3.5.18)$$

Furthermore, one cannot easily separate $\overset{(3)}{X}{}^{\beta}{}_{\rho}$ covariantly into independent and dependent components. Therefore, what one does is to introduce more tertiary ghosts than really needed, and take care of relations (16.3.5.18) later by adding "ghosts for ghosts for ghosts for ghosts."

At the tertiary stage, the BRST generator is [66, 31]

$$\Omega = \overset{(1)}{\Omega} + \overset{(2)}{\Omega} + \overset{(3)}{\Omega} \qquad (16.3.5.19a)$$

$$\overset{(3)}{\Omega} = \bar{\mathcal{P}}^{(2)} p \overset{(3)}{C} - 2\pi \bar{\mathcal{P}}^{(1)} \overset{(3)}{C} \qquad (16.3.5.19b)$$

and, again, is clearly nilpotent. Here, $\overset{(3)}{C}$ and $\bar{\mathcal{P}}^{(3)}$ are commuting Majorana-Weyl spinors obeying

$$[\overset{(3)}{C}{}^{\alpha}, \bar{\mathcal{P}}^{(3)}_{\beta}] = -[\bar{\mathcal{P}}^{(3)}_{\beta}, \overset{(3)}{C}{}^{\alpha}] = \delta^{\alpha}_{\beta} \qquad (16.3.5.20a)$$

$$\text{gh}(C^{(3)}) = -\text{gh}(\bar{\mathcal{P}}^{(3)}) = 3 \qquad (16.3.5.20b)$$

The procedure proceeds forever along the same pattern if one does not wish to break manifest covariance at some stage. An infinite tower of ghosts for ghosts is required. The BRST generator is given by

$$\Omega = \tfrac{1}{2}p^2\eta + \bar{p}_\theta \not{p} C^{(1)} + (\bar{\mathcal{P}}^{(1)}\not{p} - 2\pi\bar{p}_\theta)\overset{(2)}{C}$$

$$+ \sum_{n\geq 2}(\bar{\mathcal{P}}^{(n)}\not{p} - 2\pi\bar{\mathcal{P}}^{(n-1)})\overset{(n+1)}{C} \qquad (16.3.5.21)$$

It is unclear to the author, however, that one can make sense of this infinite series. The number of true fermionic degrees of freedom, which should be 8 (16 minus 8 independent first-class constraints) is superficially given by the infinite sum $16 - 16 + 16 - 16 + \cdots$ (commuting and anticommuting degrees of freedom contribute with a relative minus sign), which does not possess an obvious meaning. It would be interesting to investigate in detail the cohomology of the BRST operator (16.3.5.21) and see whether one can define the function space on which Ω acts in such a way that the BRST cohomology reproduces the desired results.

This question falls outside the scope of our discussion, the main purpose of which was to exhibit the first explicit known example of a theory which is infinitely reducible.

Exercise. Consider a theory with a single commuting degree of freedom (q, p), characterized by the redundant constraints $\phi_1 = p$, $\phi_2 = p$.

   a. Discuss the Dirac quantization.

   b. Compute the BRST cohomology associated with the wrong BRST operator $\Omega^w = p\eta^1 + p\eta^2$.

   c. Write the correct BRST operator and compute its cohomology. Show that it coincides, in the appropriate function space, with the physical space of (a). (In particular, if C is the "ghost for ghost" of ghost number 2, one should work with formal series of positive degree in C.)

16.3.6. Light-Cone Gauge

Since Siegel's model suffers from difficulties, we revert to the discussion of the original superparticle action.

The second-class constraints can be taken to be $\bar{\chi}\gamma^+$ because

$$[(\bar{\chi}\gamma^+)_\alpha, (\bar{\chi}\gamma^+)_\beta] = 2\sqrt{2}ip^+(\gamma^-\gamma^+)_{\alpha\beta} \qquad (16.3.6.1)$$

We assume $p^+ \neq 0$, as in the usual light-cone gauge treatment (although, so far, the gauge has not been fixed: we simply isolate noncovariantly the second-class constraints).

In the representation of the γ matrices of Appendix C, $(\bar{\chi}\gamma^+)_\alpha$ has nonvanishing components only for $16 < \alpha \leq 32$. Furthermore, the determinant of the 16 × 16 matrix $(\gamma^-\gamma^+)_{\alpha\beta}$, $16 < \alpha, \beta \leq 32$ is $(-2)^{16}$ and hence does not vanish (of course, 16 and 32 are effectively halved by the Weyl condition).

One could follow at this stage the standard method for computing the Dirac brackets corresponding to the second-class constraints (16.3.6.1) and then try to redefine new variables which give them the canonical form. It is more effective, however, to work at the level of the action. The Dirac brackets are obtained by injecting the second-class constraints in the action. One "diagonalizes" them by redefining new variables such that the kinetic term takes the standard form "$\int p\dot{q} + \frac{i}{2}\theta\dot{\theta}$."

We will follow this second route and adopt slightly different notation. Instead of referring to the $+$ and $-$ null directions, we will use two arbitrary lightlike vectors obeying [61]

$$n^2 = r^2 = 0, \qquad n \cdot r = -1 \qquad (16.3.6.2)$$

The superparticle action is equivalent to the first-order action

$$S[X, p, \theta, V] = \int d\tau [p_A(\dot{X}^A - i\bar{\theta}\gamma^A\dot{\theta}) + \tfrac{1}{2}Vp^A p_A] \qquad (16.3.6.3)$$

This equation is obtained by applying the Legendre transformation to the variables $(\dot{X}^A, p_A)$ only.

We now introduce the following change of variables in equation (16.3.6.3) [61, 64]:

$$X^A = q^A - \frac{i}{2}(n \cdot p)^{-1} n_B p_C \bar{\theta}\gamma^{ABC}\theta \qquad (16.3.6.4a)$$

$$\theta = \eta + \not{p}\zeta \qquad (16.3.6.4b)$$

where the light-cone Majorana–Weyl spinors η and ζ possess only 8 independent components in view of the conditions

$$\not{n}\eta = 0 = \not{n}\zeta \qquad (16.3.6.4c)$$

We note that ζ and θ have opposite chiralities.

Appendix C includes useful relations pertaining to decomposition (16.3.6.4b) and (16.3.6.4c). When these are taken into account, one finds for the action

$$S[q, p, \eta, \zeta, V] = \int d\tau [p_A\dot{q}^A + i(n \cdot p)\bar{\eta}\not{r}\dot{\eta} - i(n \cdot p)p^2\bar{\zeta}\not{r}\dot{\zeta} + \tfrac{1}{2}Vp^A p_A] \qquad (16.3.6.5)$$

This equation can be simplified by an elementary rescaling of η and ζ that absorbs the factor $(n \cdot p)^{-1}$. In terms of the new η and ζ (which we denote by the same letters), the action becomes ($\bar{\eta}\gamma\eta = \bar{\zeta}\gamma\zeta = 0$)

$$S[q, p, \eta, \zeta, V] = \int d\tau [p_A \dot{q}^A + i\bar{\eta}\gamma\dot{\eta} - ip^2\bar{\zeta}\gamma\dot{\zeta} + \tfrac{1}{2}Vp^2] \qquad (16.3.6.6)$$

A further change of variables

$$V \to V - 2i\bar{\zeta}\gamma\dot{\zeta} \qquad (16.3.6.7)$$

leads to

$$S[q, p, \eta, \zeta, V] = \int d\tau [p_A \dot{q}^A + i\bar{\eta}\gamma\dot{\eta} + \tfrac{1}{2}Vp^2] \qquad (16.3.6.8)$$

The action is manifestly independent of ζ, which is just a consequence of the fermionic gauge invariance of the theory (ζ is arbitrary and is not governed by any equation of motion).

In order to attain the full canonical form for S, one must reintroduce the momenta $\bar{p}_\zeta$ conjugate to ζ, which obey the primary first-class constraints

$$\bar{p}_\zeta \equiv \partial L/\partial \dot{\zeta} = 0 \qquad (16.3.6.9)$$

and add equation (16.3.6.9) to S with Lagrange multiplier ν:

$$S = \int d\tau [p_A \dot{q}^A + i\bar{\eta}\gamma\dot{\eta} + \bar{p}_\zeta \dot{\zeta} + \tfrac{1}{2}Vp^2 + \bar{p}_\zeta \nu] \qquad (16.3.6.10)$$

The constraints (16.3.6.9) are, of course, the generators of the fermionic gauge invariance and are just equivalent to $\bar{\phi} = 0$ ($\bar{p}_\zeta = \bar{p}_\theta \rlap{/}p$). One of the virtues of the above changes of variables is that q^A and η are gauge invariant for the local fermionic symmetry in Hamiltonian form, i.e., they commute with $\bar{p}_\zeta$.

The action (16.3.6.10) is what we would have obtained had we followed the standard Dirac procedure for eliminating the second-class constraints by means of the Dirac bracket (only the first-class constraints $\bar{p}_\zeta = 0$ are left in the absence of gauge fixing). One can read the Dirac brackets of p, q, η, ζ, and $\bar{p}_\zeta$ from the kinetic term of equation (16.3.6.10). One finds explicitly

$$[q^A, p_B] = \delta^A_B \qquad (16.3.6.11a)$$

16 • The Superstring

$$[\eta, \bar{\eta}] = \frac{i}{4} \not{p} \tag{16.3.6.11b}$$

$$[\zeta^a, \bar{p}_{\zeta b}] = \delta^a_b \tag{16.3.6.11c}$$

while all other Dirac brackets vanish. The second virtue of the above redefinitions of X^A and θ is that the Dirac brackets possess canonical form in terms of the new variables. [In relation (16.3.6.11c), a and b are SO(8) indices corresponding to light-cone Majorana–Weyl spinors. We note that η also possesses only 8 independent components. Its associated Dirac brackets can be written as in relation (16.3.6.11b), since $\not{p}\eta$ commute identically with everything else.]

We are now ready to impose the light-cone gauge condition and reduce the theory to its true physical degrees of freedom. We revert at this point to the more traditional notation $\not{p} \to \gamma^+$ and $\not{r} \to \gamma^-$.

The light-cone gauge conditions are

$$X^+ = p^+ \tau \quad \text{and} \quad \gamma^+ \theta = 0 \tag{16.3.6.12}$$

The first condition fixes the reparametrizations. The second one freezes the fermionic gauge invariance and is equivalent to

$$\zeta = 0 \tag{16.3.6.13}$$

It is a good gauge condition, since $\det[\zeta, \bar{p}_\zeta] \neq 0$.

The constraint $p^A p_A = 0$ can be solved to yield p^- as a function of p^+ and p^i,

$$p^- = \frac{1}{2p^+} \sum_i (p^i)^2, \quad i = 1, \ldots, 8 \tag{16.3.6.14}$$

The light-cone gauge action is obtained by inserting equations (16.3.6.12), (16.3.6.14), and $\bar{p}_\zeta = 0$ into expression (16.3.6.10) to yield

$$S[q^i, p_i, p^+, u^-, \eta^a] = \int d\tau \left[p_i \dot{q}^i - p^+ \dot{u}^- + i\sqrt{2} \sum_a \eta^a \dot{\eta}^a - H \right] \tag{16.3.6.15a}$$

with

$$H = \tfrac{1}{2} \sum_i (p^i)^2 \tag{16.3.6.15b}$$

and

$$u^- = X^- - p^-\tau \qquad (16.3.6.15c)$$

In equation (16.3.6.15a), a is an SO(8) spinor index associated with the 8_s representation; see Appendix C. For later convenience, we note that in the light-cone gauge η is related to θ as follows:

$$\eta = \sqrt{p^+}\,\theta \qquad (16.3.6.16)$$

The factor $\sqrt{p^+}$ is due to the rescaling of $\eta \to \eta(n \cdot p)^{1/2}$.

The light-cone gauge Dirac brackets follow from the kinetic term of equation (16.3.6.15a) and are given by

$$[q^i, p_j] = \delta^i{}_j$$

$$[u^-, p^+] = -1 \qquad (16.3.6.17)$$

$$[\eta^a, \eta^b] = -\frac{i}{2\sqrt{2}}\delta^{ab}$$

(all other Dirac brackets vanish).

Exercise. Eliminate the momenta p_i from equation (16.3.6.15a) by using their equations of motion. Compare with the cavalier treatment of the second-order light-cone gauge superstring action in Section 16.1.8, in which the degree of freedom (p^+, u^-) was effectively frozen.

The light-cone gauge Poincaré charges are obtained by substituting the light-cone gauge conditions and the constraints into expressions (16.3.2.2a) and (16.3.2.2b) to yield

$$P^+ = p^+, \qquad P^i = p^i, \qquad P^- = p^- \qquad (16.3.6.18a)$$

$$M^{ij} = p^{[i}X^{j]} + \frac{i\sqrt{2}}{4}\eta^T\gamma^{ij}\eta \qquad (16.3.6.18b)$$

$$M^{i-} = \tfrac{1}{2}(p^i u^- - p^- X^i) + \frac{i\sqrt{2}}{4p^+}p_j\eta^T\gamma^{ij}\eta \qquad (16.3.6.18c)$$

$$M^{i+} = -\tfrac{1}{2}p^+ X^i \qquad (16.3.6.18d)$$

$$M^{-+} = -\tfrac{1}{2}p^+ u^- \qquad (16.3.6.18e)$$

with p^- given by expansion (16.3.6.14). Relation (16.3.6.18b) was derived with the aid of the relation

$$\bar{\theta}\gamma^{ijk}\theta = 0 \qquad (16.3.6.19)$$

valid in the light-cone gauge.

The matrices γ^{ij} in expressions (16.3.6.18b) and (16.3.6.18c) are the 8 by 8 generators of the 8_s representation of SO(8), induced by the 32 by 32 matrices γ^{ij} on the 8-dimensional space of light-cone gauge Majorana–Weyl spinors. They are, of course, antisymmetric. More details are given elsewhere [67].

In order to write the supercharge, it is convenient to split it into its irreducible SO(8) components according to

$$Q = Q_+ + Q_- \qquad (16.3.6.20a)$$

with

$$Q_+ = -\tfrac{1}{2}\gamma^-\gamma^+ Q, \qquad Q_- = -\tfrac{1}{2}\gamma^-\gamma^+ Q \qquad (16.3.6.20b)$$

where Q_+ belongs to the spinor representation 8_s of SO(8) and Q_- to 8_c (see Appendix C; we note that Q possesses opposite chirality to θ).

After appropriate renormalizations, we obtain

$$Q_+^a = 2^{5/4}\sqrt{p^+}\,\eta^a \qquad (16.3.6.21a)$$

$$Q_-^{\dot{a}} = \frac{2^{3/4}}{\sqrt{p^+}}\,p_i(\gamma^i\eta)^{\dot{a}} \qquad (16.3.6.21b)$$

where the 8 by 8 matrices $(\gamma^i)^{a\dot{a}}$ are given in Ref. 67. Relation (16.3.6.21a) indicates that the spinorial variable η^a is essentially the supercharge.

The light-cone gauge components of the supercharge obey the classical Poisson-bracket relations of light-cone supersymmetry [67],

$$[Q_+^a, Q_+^b] = -2ip^+\delta^{ab} \qquad (16.3.6.22a)$$

$$[Q_-^{\dot{a}}, Q_-^{\dot{b}}] = -2ip^-\delta^{\dot{a}\dot{b}} \qquad (16.3.6.22b)$$

$$[Q_+^a, Q_-^{\dot{b}}] = -i\sqrt{2}(\gamma^i)^{a\dot{b}}p_i \qquad (16.3.6.22c)$$

One passes to the quantum theory by replacing the variables X^i, p_i, u^-, p^+, and η^a by operators obeying the commutation relations

$$X^i p_j - p_j X^i = i\delta^i_j, \qquad u^- p^+ - p^+ u^- = -i \qquad (16.3.6.23a)$$

$$\eta^a \eta^b + \eta^b \eta^a = \delta^{ab}, \qquad a = 1, \ldots, 8 \qquad (16.3.6.23b)$$

($= i$ Dirac brackets; the other commutators vanish).

These commutation relations are realized in the positive-definite Hilbert space obtained by taking the direct product of the Hilbert space for relations (16.3.6.23a) by the Hilbert space for relations (16.3.6.23b).

System (16.3.6.23b) is a Clifford algebra with eight generators, so its irreducible representation space is 16-dimensional. For given momentum, the superparticle can therefore be in one of 16 different states. These states can be shown to correspond to the helicity, and break into the $8_c + 8_v$ representation of SO(8) [67]. One has eight spinorial states and eight vector states, as required by ten-dimensional supersymmetry [58, 67, 68]. The transformation properties of the states under SO(8) follow, of course, from the expression for the SO(8) generators M^{ij} in terms of the basic canonical variables.

In the case of the superstring, the super-Poincaré generators also acquire a contribution from the excited modes. However, these annihilate the ground state, which is accordingly described by the same quantum space as the superparticle.

Exercise. Write the action of the SO(8) generators on the 16 Clifford states. Decompose the SO(8) representation which they yield into its irreducible components.

Chapter 17

The Heterotic String

For a brief introduction to the heterotic string, we refer the reader to the following published material:

1. Bosonic strings in space-times with compactified dimensions (case of tori):
    General formalism and closed-string winding number: see Cremmer and Scherk [69].
    Special tori and Kac–Moody symmetries; see Kac *et al.* [22, 23], Dolan and Slansky [70], and Englert and Neveu [71].
2. The heterotic string: see Gross *et al.* [72] and Chapter 5.

The remarkable feature of the heterotic string is the intrinsic incorporation of Yang–Mills symmetries in the string spectrum [with gauge group $E_8 \times E_8$ or $SO(32)$].

Appendix A

BRST-Based Demonstration of the No-Ghost Theorem for the Bosonic String

In their paper, Kato and Ogawa [19] have shown that any BRST physical state $|b\rangle$ solution to $\bar{\Omega}|b\rangle = 0$, where $\bar{\Omega}$ is the restriction of the BRST charge to a given zero-mode sector, can be written as

$$|b\rangle = |P\rangle|0\rangle_{\text{ghost}} + \bar{\Omega}|c\rangle \tag{A.1}$$

where $|P\rangle$ is a purely transverse DDF state (obtained from the vacuum by acting with the DDF oscillators). This result was used in Section 13.2.6. We briefly sketch here the idea of their proof.

The main ingredient is that $\bar{\Omega}$ can be regarded as a "perturbation" of the operator

$$\bar{\Omega}' = i \sum_{n>0} (a_n^{+*} \eta_n - \eta_n^* a_n^+) \tag{A.2}$$

for which the absence of ghosts is a mere consequence of the Kugo–Ojima "quartet mechanism." Kato and Ogawa have then proved that the no-ghost theorem extends to the full $\bar{\Omega}$.

In expression (A.2), a_n^+ and a_n^{+*} are destruction and creation operators of appropriately normalized lightlike modes. They commute:

$$[a_n^+, a_n^{+*}] = \eta^{++} = 0 \tag{A.3}$$

The commutator-conjugate operators are a_n^{-*} and a_n^{-}. (We do not fix the gauge here, but work in a given Lorentz frame.)

The main qualitative features of the no-ghost theorem are already present in the simpler model based on operator (A.2), so we prove it in that case only. The analog of the DDF oscillators for expression (A.2) are just the transverse oscillators a_n^i, since these commute with $\bar{\Omega}'$. A transverse state in the theory based on relation (A.2) is accordingly a state containing only transverse excitations.

To show that $\bar{\Omega}'|b\rangle = 0$ implies

$$|b\rangle = |P'\rangle|0\rangle_{\text{ghost}} + \bar{\Omega}'|c\rangle \tag{A.4}$$

where $|P'\rangle$ is a transverse state involving only the transverse operators a_n^{i*}, it suffices to proceed mode by mode. We assume that

$$\eta_n|b\rangle = \mathcal{P}_n|b\rangle = 0$$
$$a_n^+|b\rangle = a_n^-|b\rangle = 0 \quad \text{for } n > M \tag{A.5}$$

which must be true for some M since N_f, N_g, and N are bounded by $-\alpha'p^2 + \alpha_0$. The conditions (A.5) are clearly compatible with $\bar{\Omega}'|b\rangle = 0$.

We wish to remove the Mth mode from $|b\rangle$, i.e., to transform equation (A.5) into the same one with $M - 1$ in place of M. For that purpose, we write

$$|b\rangle = |\alpha\rangle + |\beta\rangle\eta_M^* + |\gamma\rangle\mathcal{P}_M^* + |\delta\rangle\eta_M^*\mathcal{P}_M^* \tag{A.6}$$

where $|\alpha\rangle$, $|\beta\rangle$, $|\gamma\rangle$, and $|\delta\rangle$ are states which do not involve the nth ghost modes ($n \geq M$) and are also destroyed by a_n^+ and a_n^- ($n > M$). It is easy to compute $\bar{\Omega}'|b\rangle$:

$$\bar{\Omega}'|b\rangle = \bar{\Omega}'_{M-1}|b\rangle + ia_M^{+*}|\gamma\rangle - ia_M^{+*}|\delta\rangle\eta_M^*$$
$$- ia_M^+|\alpha\rangle\eta_M^* - ia_M^+|\gamma\rangle\eta_M^*\mathcal{P}_M^* \tag{A.7}$$

with $\bar{\Omega}'_{M-1} = i\sum_{0<n<M}(a_n^{+*}\eta_n - \eta_n^* a_n^+)$.

It is seen from equation (A.7) that one can set $|\delta\rangle = 0$, and then $|\beta\rangle = 0$, by adding to $|b\rangle$ an appropriate state of the form $\Omega|\chi\rangle$ (take $ia_M^+|\gamma'\rangle = |\delta\rangle$, etc.). Once this is done, we find from $\bar{\Omega}'|b\rangle = 0$ that $a_M^+|\alpha\rangle = a_M^+|\gamma\rangle = 0$, so that it can actually even also be assumed that $|\alpha\rangle$ is independent of a_M^{+*}, i.e., $a_M^-|\alpha\rangle = 0$, again by adding to $|b\rangle$ an appropriate state of the form $\Omega|\chi\rangle$.† When this condition holds, the remaining implications of the BRST

---

† We note that a_M^{+*} is commutator-conjugate to a_M^-. To implement $a_M^-|\alpha\rangle = 0$, one adjusts $|\tilde{\gamma}\rangle$ in $|\chi\rangle = |\tilde{\gamma}\rangle\mathcal{P}_M^*$ appropriately.

Appendix A: BRST Demonstration of No-Ghost Theorem

annihilation condition are

$$\bar{\Omega}'_{M-1}|\gamma\rangle = 0 \quad \text{and} \quad \bar{\Omega}'_{M-1}|\alpha\rangle + a^{+*}_M|\gamma\rangle = 0$$

from which it follows in particular that $|\gamma\rangle = 0$ (there is no a^{+*}_M in $|\alpha\rangle$). This proves relation (A.4) and the "no-ghost theorem" (i.e., the absence of negative-norm states) for the simple quadratic operator (A.2), since transverse states form a positive-definite subspace.

The mechanism by which the ghosts can be removed (for a quadratic Fock-space BRST charge) is called the Kugo–Ojima quartet mechanism [32]. It should be compared with a similar mechanism described elsewhere [31] for systems defined in the (q, p) representation.

The quartet mechanism not only removes the ghosts η_n and $\mathcal{P}_n$, but also the lightlike oscillators a^+_n and a^-_n. It is sometimes convenient not to completely fix the BRST gauge as was done above, but rather to keep the freedom to add null states created by a^{+*}_n. This is done by imposing only the ghost vacuum condition on $|b\rangle$,

$$\eta_n|b\rangle = \mathcal{P}_n|b\rangle = 0 \quad (n > 0) \tag{A.8a}$$

This condition is consistent with $\bar{\Omega}'|b\rangle = 0$ if

$$a^+_n|b\rangle = 0 \quad (n > 0) \tag{A.8b}$$

The purely transverse states clearly obey system (A.8) as well as the "gauge condition"

$$a^-_n|b\rangle = 0 \quad \text{for } n > 0 \tag{A.9}$$

Finally, in order to compare with the light-cone conventions, we note that the operators a^+_n and a^-_n used in this appendix should be exchanged with a^-_n and a^+_n, respectively.

Exercise. Which transformations $|b\rangle \to |b\rangle + \bar{\Omega}'|c\rangle$ do preserve conditions (A.8)?

Appendix B

γ Matrices in Ten Dimensions

We briefly summarize here our conventions on γ matrices in the case of ten dimensions, of particular interest in the present context. γ matrices obey

$$\gamma_A \gamma_B + \gamma_B \gamma_A = 2\eta_{AB} \tag{B.1}$$

and are taken to be real (Majorana representation). The nine spatial γ_k are symmetric, while $\gamma_0 = C$ (charge conjugation matrix) is antisymmetric. The matrices $C\gamma_A$ are symmetric.

The matrix $\gamma_{11} = \gamma_0 \cdots \gamma_9$ is real and symmetric, so that it is compatible to impose the Weyl chirality constraint

$$(1 \pm \gamma_{11})\theta = 0 \tag{B.2}$$

together with the reality (= Majorana) condition. These conditions define Majorana–Weyl spinors.

B.1. SYMMETRY PROPERTIES

A basis of 32×32 matrices is given by the $(32)^2$ matrices B_Λ

$$\{I, \gamma_A, \gamma_{AB}, \gamma_{ABC}, \gamma_{ABCD}, \gamma_{ABCDE},$$
$$\gamma_{ABCD}\gamma_{11}, \gamma_{ABC}\gamma_{11}, \gamma_{AB}\gamma_{11}, \gamma_A\gamma_{11}, \gamma_{11}\}, \qquad A < B < C \ldots \tag{B.3}$$

which obey

$$\operatorname{tr} B_\Lambda B^\Delta = 32 \delta_\Lambda^\Delta (-)^{\varepsilon_\Lambda} \tag{B.4}$$

We have defined

$$\gamma_{A_1 \cdots A_k} = \gamma_{[A_1 \cdots A_k]} = \frac{1}{k!} (\gamma_{A_1} \cdots \gamma_{A_k} - \gamma_{A_2} \gamma_{A_1} \cdots \gamma_{A_k} + \cdots) \tag{B.5}$$

and

$$B^\Lambda = \{I, \gamma^A, \gamma^{AB}, \ldots, \gamma^{ABCD}\gamma_{11}, \ldots, \gamma_{11}\} \tag{B.6}$$

In equation (B.4), the phase $(-)^{\varepsilon_\Lambda}$ is given by

$$\varepsilon_\Lambda = 0 \quad \text{for } I, \gamma_A, \gamma_{ABCD}, \gamma_{ABCDE}, \gamma_{ABCD}\gamma_{11}, \gamma_{ABC}\gamma_{11}, \gamma_{11}$$
$$\varepsilon_\Lambda = 1 \quad \text{for } \gamma_{AB}, \gamma_{ABC}, \gamma_{AB}\gamma_{11}, \gamma_A\gamma_{11} \tag{B.7}$$

and is such that $B_\Lambda B^\Lambda = (-)^{\varepsilon_\Lambda} I$ (without summation over Λ).

The matrices CB_Λ have definite symmetry properties, which one easily derives from their definition. These symmetry properties imply the following identities:

$$\bar\psi \chi = \bar\chi \psi, \qquad \bar\psi \gamma_A \chi = -\bar\chi \gamma_A \psi$$

$$\bar\psi \gamma_{AB} \chi = -\bar\chi \gamma_{AB} \psi, \qquad \bar\psi \gamma_{ABC} \chi = \bar\chi \gamma_{ABC} \psi$$

$$\bar\psi \gamma_{ABCD} \chi = \bar\chi \gamma_{ABCD} \psi, \qquad \bar\psi \gamma_{ABCDE} \chi = -\bar\chi \gamma_{ABCDE} \psi$$

$$\bar\psi \gamma_{ABCD}\gamma_{11} \chi = -\bar\chi \gamma_{ABCD}\gamma_{11} \psi, \qquad \bar\psi \gamma_{ABC}\gamma_{11} \chi = \bar\chi \gamma_{ABC}\gamma_{11} \psi$$

$$\bar\psi \gamma_{AB}\gamma_{11} \chi = \bar\chi \gamma_{AB}\gamma_{11} \psi, \qquad \bar\psi \gamma_A \gamma_{11} \chi = -\bar\chi \gamma_A \gamma_{11} \psi, \qquad \bar\psi \gamma_{11} \chi = -\bar\chi \gamma_{11} \psi$$
$$\tag{B.8}$$

where ψ and χ are (anticommuting) Majorana spinors, and where $\bar\psi$ stands for $\psi^T \gamma_0$.

If ψ and χ are also Weyl spinors of the same chirality, the only nonvanishing bilinear quantities are easily seen from relations (B.8) to be $\bar\psi \gamma_A \chi$, $\bar\psi \gamma_{ABC} \chi$, and $\bar\psi \gamma_{ABCDE} \chi$. On the other hand, if ψ and χ are of opposite handedness, only the even bilinear invariants $\bar\psi \chi$, $\bar\psi \gamma_{AB} \chi$, and $\bar\psi \gamma_{ABCD} \chi$ can differ from zero.

Appendix B: γ Matrices in Ten Dimensions

B.2. FIERZ REARRANGEMENTS

The matrices B_Λ form a basis; therefore any 32×32 matrix M can be expanded as

$$M = a^\Lambda B_\Lambda \tag{B.9}$$

The coefficients a^Λ are easily evaluated with the aid of condition (B.4):

$$a^\Lambda = \tfrac{1}{32}(-)^{\varepsilon_\Lambda} \operatorname{tr}(B^\Lambda M) \tag{B.10}$$

We now consider the four spinors λ, χ, ψ, and ϕ. The Fierz identities enable the expression

$$\bar\lambda M\chi \cdot \bar\psi N\phi \equiv \bar\lambda_i M^i{}_j \chi^j \cdot \bar\psi_k N^k{}_l \phi^l$$

where M and N are 32×32 matrices, to be rearranged in terms of the 32^2 "fundamental" bilinear functions $\bar\psi B_\Lambda \chi$. This is done by simply noting that the numbers $M^i{}_j N^k{}_l$, for fixed i and l, can be viewed as the components of a 32×32 matrix, and hence

$$M^i{}_j N^k{}_l = \sum_\Lambda a^{\Lambda i}{}_l (B_\Lambda)^k{}_j \tag{B.11a}$$

with $a^{\Lambda i}{}_j$ given by

$$a^{\Lambda i}{}_j = \tfrac{1}{32}(-)^{\varepsilon_\Lambda}(MB^\Lambda N)^i{}_j \tag{B.11b}$$

This leads to

$$\bar\lambda M\chi \cdot \bar\psi N\phi = -\tfrac{1}{32} \sum_\Lambda (-)^{\varepsilon_\Lambda} \bar\lambda MB^\Lambda N\phi \cdot \bar\psi B_\Lambda \chi \tag{B.12}$$

where the minus sign arises because $\bar\psi_k$ and χ^j anticommute.

Of great importance for deriving the covariant form of the superstring action is the Fierz rearrangement (B.12) for Majorana-Weyl spinors with the same chirality. One obtains

$$\bar\lambda M\chi \cdot \bar\psi N\phi = -\tfrac{1}{32}(\bar\lambda M\gamma^A N\phi - \bar\lambda M\gamma^A \gamma_{11} N\phi)\bar\psi \gamma_A \chi$$
$$+\tfrac{1}{32}(\bar\lambda M\gamma^{ABC} N\phi - \bar\lambda M\gamma^{ABC}\gamma_{11} N\phi)\bar\psi \gamma_{ABC}\chi$$
$$-\tfrac{1}{32}\bar\lambda M\gamma^{ABCDE} N\phi \bar\psi \gamma_{ABCDE}\chi \tag{B.13}$$

where it has been assumed that $(1 - \gamma_{11})\lambda = 0 = (1 - \gamma_{11})\chi = \cdots$, and where the summation is again restricted by $A < B < C \ldots$.

If one sets $\bar{\lambda} = \bar{\psi}_1$, $M = \gamma^A$, $\chi = \psi_2$, $\psi = \alpha$, $N = \gamma_A$, and $\phi = \psi_3$, then equation (B.13) reduces to the form

$$\bar{\psi}_1 \gamma^A \psi_2 \cdot \bar{\alpha} \gamma_A \psi_3 = -\tfrac{1}{16}\bar{\psi}_1 \gamma^C \gamma^A \gamma_C \psi_3 \bar{\alpha}\gamma_A \psi_2 + \tfrac{1}{16}\bar{\psi}_1 \gamma^C \gamma^{ABD} \gamma_C \psi_3 \bar{\alpha}\gamma_{ABD}\psi_2$$

$$-\tfrac{1}{32}\bar{\psi}_1 \gamma^C \gamma^{ABDEF} \gamma_C \psi_3 \bar{\alpha}\gamma_{ABDEF}\psi_2 \qquad (B.14)$$

This expression can be transformed using the identities

$$\gamma^C \gamma^A \gamma_C = -8\gamma^A \qquad (B.15a)$$

$$\gamma^C \gamma^{ABD} \gamma_C = -4\gamma^{ABD} \qquad (B.15b)$$

$$\gamma^C \gamma^{ABDEF} \gamma_C = 0 \qquad (B.15c)$$

which are easily inferred from the γ matrices anticommutation relations. (The analogous identities for even $\gamma^{A_i \cdots A_k}$ read $\gamma^C \gamma^{AB} \gamma_C = 6\gamma^{AB}$ and $\gamma^C \gamma^{ABDE} \gamma_C = 2\gamma^{ABDE} \cdots$.) Identities (B.15) allow equation (B.14) to be reduced to

$$\bar{\psi}_1 \gamma^A \psi_2 \cdot \bar{\alpha}\gamma_A \psi_3 = \tfrac{1}{2}\bar{\psi}_1 \gamma^A \psi_3 \cdot \bar{\alpha}\gamma_A \psi_2 - \tfrac{1}{4}\bar{\psi}_1 \gamma^{ABD} \psi_3 \cdot \bar{\alpha}\gamma_{ABD}\psi_2 \qquad (B.16)$$

As a final step, one completely antisymmetrizes the latter relation in ψ_1, ψ_2, and ψ_3. Since α is arbitrary, this yields

$$\gamma_A \psi_1 \bar{\psi}_2 \gamma^A \psi_3 + \gamma_A \psi_2 \bar{\psi}_3 \gamma^A \psi_1 + \gamma_A \psi_3 \bar{\psi}_1 \gamma^A \psi_2 = 0 \qquad (B.17)$$

which is the identity used in the text for establishing the closedness of the 3-form Ω_3 (Section 16.1.2). The term (16.1.2.6) found there, namely

$$(d\bar{\theta}\gamma^A \wedge d\theta^1) \wedge (d\bar{\theta}^1 \wedge \gamma_A d\theta^1)$$

$$= \bar{\theta}^1_{,\lambda} \gamma^A \theta^1_{,\mu} \bar{\theta}^1_{,\nu} \gamma_A \theta^1_{,\rho}\, dx^\lambda \wedge dx^\mu \wedge dx^\nu \wedge dx^\rho$$

vanishes, because the effect of the exterior product $dx^\lambda \wedge dx^\mu \wedge dx^\nu \wedge dx^\rho$ is to antisymmetrize the spinors, as in equation (B.17).

The identity (B.17) plays a key role in the study of the super-Yang–Mills theory and was established in Ref. 73.

Appendix C

Light-Cone Gauge Decomposition of Ten-Dimensional Spinors

The light-cone γ matrices γ^+ and γ^- obey

$$(\gamma^+)^2 = (\gamma^-)^2 = 0 \quad \text{and} \quad \gamma^+\gamma^- + \gamma^-\gamma^+ = 2\eta^{+-} = -2 \quad \text{(C.1)}$$

It is seen from these relations that P^+ and P^- are projectors,

$$P^- = -\tfrac{1}{2}\gamma^+\gamma^-, \qquad P^+ = -\tfrac{1}{2}\gamma^-\gamma^+ \quad \text{(C.2a)}$$

$$(P^+)^2 = P^+, \quad (P^-)^2 = P^-, \quad P^+ + P^- = 1 \quad \text{(C.2b)}$$

In addition

$$(P^+)^T = P^+ \quad \text{and} \quad (P^-)^T = P^- \quad \text{(C.3)}$$

since $(\gamma^+)^T = -\gamma^-$ and $(\gamma^-)^T = -\gamma^+$.

Any spinor can be decomposed into its "+" and "−" components,

$$\psi = \psi^+ + \psi^- \quad \text{(C.4a)}$$

$$\psi^+ = P^+\psi, \qquad \psi^- = P^-\psi \quad \text{(C.4b)}$$

By construction, one finds

$$\gamma^-\psi^+ = \gamma^+\psi^- = 0 \quad \text{(C.5)}$$

The decomposition (C.4) is not covariant under Lorentz transformations, which in general mix the + and − parts of ψ. However, system (C.4) is covariant under the SO(8) subgroup of SO(9, 1) containing the rotations in the transverse directions. In other words, the spinor representation of SO(9, 1) is reducible for SO(8).

It is convenient to choose γ matrices in ten dimensions which clearly display this property. We set

$$\gamma^i = \begin{pmatrix} \tilde{\gamma}^i & 0 \\ 0 & -\tilde{\gamma}^i \end{pmatrix} \tag{C.6}$$

and

$$\gamma^+ = \begin{pmatrix} 0 & \sqrt{2} \\ 0 & 0 \end{pmatrix}, \quad \gamma^- = \begin{pmatrix} 0 & 0 \\ -\sqrt{2} & 0 \end{pmatrix} \tag{C.7a}$$

$$\gamma^0 = \begin{pmatrix} 0 & 1 \\ -1 & 0 \end{pmatrix}, \quad \gamma^9 = \begin{pmatrix} 0 & 1 \\ 1 & 0 \end{pmatrix} \tag{C.7b}$$

Quantities $\tilde{\gamma}^i$ are transverse γ matrices in eight spatial dimensions. They are 16-dimensional, while the γ matrices in ten dimensions are 32 by 32. In expression (C.7b), 1 denotes the 16-dimensional unit matrix.

The projectors P^+ and P^- are given by

$$P^- = \begin{pmatrix} 1 & 0 \\ 0 & 0 \end{pmatrix} \quad \text{and} \quad P^+ = \begin{pmatrix} 0 & 0 \\ 0 & 1 \end{pmatrix} \tag{C.8}$$

while γ^{11} becomes

$$\gamma^{11} = \begin{pmatrix} \tilde{\gamma}^9 & 0 \\ 0 & -\tilde{\gamma}^9 \end{pmatrix} \tag{C.9}$$

where $\tilde{\gamma}^9 = \tilde{\gamma}^1 \cdots \tilde{\gamma}^8$.

A spinor obeying $\gamma^+ \psi = 0$, like ψ^-, possesses only an upper component, while a spinor obeying $\gamma^- \psi = 0$, like ψ^+, possesses only a lower component:

$$\psi = \begin{pmatrix} u \\ v \end{pmatrix}, \quad \psi^- = \begin{pmatrix} u \\ 0 \end{pmatrix}, \quad \psi^+ = \begin{pmatrix} 0 \\ v \end{pmatrix} \tag{C.10}$$

If ψ is a Weyl spinor of definite chirality, e.g.,

$$\gamma_{11} \psi = \psi \tag{C.11a}$$

its SO(8) components u and v possess also definite 8-dimensional chirality (γ_{11} commutes with P^+ and P^-). More precisely, one finds that

$$\tilde{\gamma}^9 u = u \quad \text{and} \quad \tilde{\gamma}^9 v = -v \tag{C.11b}$$

One can show that u and v yield inequivalent, irreducible, 8-dimensional representations of SO(8) (of a spinorial character). They are denoted by 8_s and 8_c, respectively. The other 8-dimensional representation of SO(8) is the vector one, 8_v.

It follows from our discussion that a Weyl spinor obeying the light-cone gauge condition $\gamma^+ \theta = 0$ belongs to the 8_s representation of SO(8) (with our conventions).

It is useful to use SO(8) notations when dealing with light-cone gauge spinors, and to write everything in terms of the SO(8)-invariant scalar product

$$\sum_{a=1}^{8} \theta^{*a} \psi^a \tag{C.12}$$

($= \sum_{a=1}^{8} \theta^a \psi^a$ for Majorana spinors).

The correspondence rules are

$$\theta^\alpha, \quad \alpha = 1, \ldots, 32 \to \theta^a, \quad a = 1, \ldots, 8 \tag{C.13a}$$

$$\bar{\theta}\psi \to \sum_a \theta^a \psi^a, \quad a = 1, \ldots, 8 \tag{C.13b}$$

$$\bar{\eta}_1 \gamma^- \eta_2 \to \sqrt{2} \sum_a \eta_1^a \eta_2^a \tag{C.13c}$$

and so on. [In rule (C.13b), ψ is a Weyl spinor of chirality opposite to θ and obeys $\gamma^- \psi = 0$; in rule (C.13c), η_1 and η_2 are like θ.]

The above light-cone gauge decomposition can, of course, be carried out with respect to any pair of null vectors n^A and r^A satisfying

$$n^2 = r^2 = 0 \quad \text{and} \quad n^A r_A = -1 \tag{C.14}$$

One must simply replace γ^+ and γ^- by $\slashed{n}$ and $\slashed{r}$, respectively, where $\slashed{r} = r^A \gamma_A$ and $\slashed{n} = n^A \gamma_A$.

We suppose θ is an arbitrary Majorana-Weyl spinor. Instead of decomposing it in terms of θ^+ and θ^-, it is also useful to express it in the form

$$\theta = \eta + \slashed{p}\zeta \tag{C.15a}$$

where η and ζ both satisfy

$$\gamma^+\eta = \gamma^+\zeta = 0 \tag{C.15b}$$

(η possesses the same chirality as θ, where ζ is of opposite chirality). In equation (C.15a), p^A is a null vector (e.g., it can be ω_+^A for the superstring, or p^A for the superparticle).

We now derive some equations related to the decomposition (C.15) in terms of the notation employed in relations (C.14), i.e.,

$$\theta = \eta + p\!\!\!/\zeta \tag{C.16a}$$

$$n\!\!\!/\eta = n\!\!\!/\zeta = 0 \tag{C.16b}$$

in place of system (C.15a) and (C.15b). First, one easily checks that the transformation (C.16) is nonsingular and invertible provided $p^A n_A \neq 0$, which we assume. In fact

$$\zeta = \frac{1}{2(n \cdot p)} n\!\!\!/\theta \tag{C.17a}$$

$$\eta = \theta - \frac{1}{2(n \cdot p)} p\!\!\!/n\!\!\!/\theta \tag{C.17b}$$

Then, one establishes the following identities, valid for arbitrary Majorana spinors obeying $n\!\!\!/\xi_1 = n\!\!\!/\xi_2 = 0$,

$$\bar{\xi}_1 \xi_2 = 0 \tag{C.18}$$

$$\bar{\xi}_1 \gamma_A \xi_2 = -n_A \bar{\xi}_1 n\!\!\!/\xi_2 \tag{C.19}$$

and

$$n_A \bar{\xi}_1 \gamma^{ABC} \xi_1 = 0 \tag{C.20}$$

Equations (C.18) and (C.19) are simply obtained by inserting $(\gamma\!\!\!/n\!\!\!/ + n\!\!\!/\gamma\!\!\!/) \times (-\tfrac{1}{2}) = 1$ in $\bar{\xi}_1 \xi_2$ or $\bar{\xi}_1 \gamma_A \xi_2$ and using

$$\overline{K\xi} = -\bar{\xi}K \tag{C.21}$$

($\Rightarrow n\!\!\!/\xi_1 = 0 \Leftrightarrow \bar{\xi}_1 n\!\!\!/ = 0$). The relation (C.20) is a direct consequence of $\bar{\xi}_1 \gamma^A \xi_1 = -\bar{\xi}_1 \gamma^A \xi_1 = 0$ and $n\!\!\!/\xi_1 = 0$.

Appendix C: Decomposition of Ten-Dimensional Spinors

It follows from equations (C.16), (C.20), and (C.21) that

$$n_B \bar{\theta} \gamma^{ABC} \theta = 2 n_B \bar{\eta} \gamma^{ABC} \not{p} \zeta \tag{C.22}$$

while relations (C.18) and (C.19) easily lead to

$$\bar{\theta} \not{p} \dot{\theta} = (n \cdot p)[-\bar{\eta} \not{r} \dot{\eta} + p^2 \bar{\zeta} \not{r} \dot{\zeta}] + \bar{\eta} \not{p} \dot{\not{p}} \zeta \tag{C.23}$$

The last term of this equation can be transformed as follows:

$$\bar{\eta} \not{p} \dot{\not{p}} \zeta = \bar{\eta} \gamma^{ABC} \not{p} \zeta p_A \dot{p}_B n_C (n \cdot p)^{-1} \tag{C.24}$$

by inserting $(\not{n}\not{p} + \not{p}\not{n})(2n \cdot p)^{-1} = 1$ and using the well-known identity

$$\gamma^A \gamma^B \gamma^C = \gamma^{ABC} + \eta^{AB} \gamma^C + \eta^{BC} \gamma^A - \eta^{AC} \gamma^B \tag{C.25}$$

[as well as identity (C.18)]. Substitution of expression (C.24) into equation (C.23) yields

$$\bar{\theta} \not{p} \dot{\theta} = (n \cdot p)[-\bar{\eta} \not{r} \dot{\eta} + p^2 \bar{\zeta} \not{r} \dot{\zeta}] + \bar{\eta} \gamma^{ABC} \not{p} \zeta p_A \dot{p}_B n_C (n \cdot p)^{-1} \tag{C.26}$$

This relation is used in the main text (Section 16.3.6).

References

1. Y. Nambu, Lectures for the Copenhagen Summer Symposium, 1970 (unpublished). Interest in string theory originates from the Veneziano model: G. Veneziano, *Nuovo Cimento* **57A** (1968) 190 (although the string picture was unavailable at that time).
2. T. Goto, *Prog. Theor. Phys.* **46** (1971) 1560; see also O. Hara, *Prog. Theor. Phys.* **46** (1971) 1549.
3. J. Scherk and J. H. Schwarz, *Nucl. Phys.* **B81** (1974) 118; *Phys. Lett.* **52B** (1974) 347.
4. P. A. M. Dirac, *Proc. R. Soc.* (*London*) **A268** (1962) 57; C. Teitelboim and T. Regge, in: *Proceedings of the First Marcel Grossmann Meeting on General Relativity* (R. Ruffini, ed.), North-Holland, Amsterdam (1977).
5. L. P. Eisenhart, *Riemannian Geometry*, Princeton University Press, Princeton (1964).
6. M. Henneaux, *Ann. Phys.* (*N.Y.*) **140** (1982) 45; *J. Phys.* **A17** (1984) 75; M. Henneaux and L. C. Shepley, *J. Math. Phys.* **23** (1982) 2101.
7. a. A. M. Polyakov, *Phys. Lett.* **103B** (1981) 207, 211.
   b. C. W. Misner, *Phys. Rev.* **D18** (1978) 4510.
8. S. Fubini, A. J. Hanson, and R. Jackiw, *Phys. Rev.* **D7** (1973) 1732.
9. S. Mandelstam, *Phys. Rep.* **13** (1974) 259.
10. C. Rebbi, *Phys. Rep.* **12** (1974) 1.
11. P. A. M. Dirac, *Can. J. Math.* **2** (1950) 129; *Proc. R. Soc.* (*London*) **A246** (1958) 326; *Lectures on Quantum Mechanics*, Academic Press, New York (1967).
12. L. N. Chang and F. Mansouri, *Phys. Rev.* **D5** (1972) 2535; F. Mansouri and Y. Nambu, *Phys. Lett.* **39B** (1972) 375.
13. A. J. Hanson, T. Regge, and C. Teitelboim, *Constrained Hamiltonian Systems*, Accademia Nazionale dei Lincei, Rome (1976).
14. T. Regge and C. Teitelboim, *Ann. Phys.* (*N.Y.*) **88** (1974) 286.
15. G. Segal, *Commun. Math. Phys.* **80** (1981) 301.
16. J. D. Brown, M. Henneaux, and C. Teitelboim, *Phys. Rev.* **D33** (1986) 319.

17. P. Goddard, J. Goldstone, C. Rebbi, and C. B. Thorn, *Nucl. Phys.* **B56** (1973) 109.
18. R. C. Brouwer, *Phys. Rev.* **D6** (1972) 1655; P. Goddard and C. B. Thorn, *Phys. Lett.* **40B** (1972) 235.
    A clear explanation of the "no-ghost theorem" can be found in: J. Scherk, *Rev. Mod. Phys.* **47** (1975) 123.
19. K. Fujikawa, *Phys. Rev.* **D25** (1982) 2584; M. Kato and K. Ogawa, *Nucl. Phys.* **B212** (1983) 443; S. Hwang, *Phys. Rev.* **D28** (1983) 2614.
20. a. J. A. Wheeler, in: *Battelle Rencontres* (C. M. De Witt and J. A. Wheeler, eds.), Benjamin, New York (1967); B. S. De Witt, *Phys. Rev.* **160** (1967) 1113.
    b. M. Pilati, Ph.D. Thesis, Princeton (1980); *Phys. Rev.* **D26** (1982) 2645; **D28** (1983) 729; C. J. Isham and A. C. Kakas, *Class. Quant. Grav.* **1** (1984) 621, 633.
21. A. A. Belavin, A. M. Polyakov, and A. B. Zamolodchikov, *Nucl. Phys.* **B241** (1984) 339; D. Friedan, Z. Qiu, and S. Shenker, *Phys. Rev. Lett.* **52** (1984) 1575; Chicago preprint EFI 84-35 (1984); P. Goddard and D. Olive, *Nucl. Phys.* **B257** (1985) 226.
22. The Virasoro conditions, which obey the Virasoro algebra, were found in: M. A. Virasoro, *Phys. Rev.* **D1** (1970) 2933. Representations of the Virasoro algebra are studied in: V. G. Kac, in: *Proceedings of the International Congress of Mathematics*, Helsinki (1978) (O. Lehto, ed.), Vol. 1, p. 299, Academia Scientiarum Fennica, Helsinki (1980); B. L. Feigin and D. B. Fuks, *Funct. Anal. Appl.* **16** (1982) 144; P. Goddard, A. Kent, and D. Olive, *Phys. Lett.* **152B** (1985) 88; J. L. Gervais and A. Neveu, *Commun. Math. Phys.* **100** (1985) 15; as well as in Refs. 21 and 23.
23. V. G. Kac, *Math. USSR Izv.* **2** (1968) 1271; R. Moody, *J. Algebra* **10** (1968) 211; V. Kac, *Infinite Dimensional Lie Algebras*, Birkhaüser, Cambridge, Massachusetts (1983). The use of Kac-Moody algebra in the recent heterotic string model occurs through the Frenkel-Kac construction: I. B. Frenkel and V. G. Kac, *Invent. Math.* **62** (1980) 23; P. Goddard and D. I. Olive, in: *Algebras, Lattices and Strings* (J. Lepowski, ed.), MSRI publications No. 3, p. 419, Springer, Berlin (1984); and Ref. 15.
24. H. Sugawara, *Phys. Rev.* **170** (1968) 1659.
25. D. Nemeschansky and S. Yankielowicz, *Phys. Rev. Lett.* **54** (1985) 620, 1736; see also: D. Altschüler and H. P. Nilles, *Phys. Lett.* **154B** (1985) 135.
26. E. Witten, *Commun. Math. Phys.* **92** (1984) 455.
27. R. Dashen and Y. Frishman, *Phys. Rev.* **D11** (1975) 1781.
28. E. S. Fradkin and A. A. Tseytlin, *Phys. Lett.*, **158B** (1985) 316; **160B** (1985) 69; C. G. Callan, D. Friedan, E. J. Martinec, and M. J. Perry, *Nucl. Phys.* **B262** (1985) 593.
29. W. Siegel, *Phys. Lett.* **149B** (1984) 157, 162; **151B** (1985) 391, 396; W. Siegel and B. Zwiebach, *Nucl. Phys.* **B263** (1986) 105; T. Banks and M. Peskin, SLAC preprint (1985); D. Friedan, S. Shenker, and E. Martinec, *Phys. Lett.* **160B** (1985) 55; E. Witten, Princeton preprint (1985). In this context, see also: A. Neveu, J. Schwarz, and P. West, *Phys. Lett.* **164B** (1985) 51; A. Neveu, H. Nicolai, and P. West, CERN preprint TH4297/85.
30. E. S. Fradkin and G. A. Vilkovisky, *Phys. Lett.* **55B** (1975) 224; CERN report TH2332 (1977); I. A. Batalin and G. A. Vilkovisky, *Phys. Lett.* **69B** (1977) 309; E. S. Fradkin and T. E. Fradkina, *Phys. Lett.* **72B** (1978) 343; I. A. Batalin and E. S. Fradkin, in: *Group Theoretical Methods in Physics*, Vol II, Moscow (1980).
31. M. Henneaux, *Phys. Rev. Lett.* **55** (1985) 769; *Phys. Rep.* **126** (1985) 1; *Bull. Cl. Sci. Acad. R. Belg.* **LXXI** (1985) 198; see also chapter in *Quantum Mechanics of Fundamental Systems* (C. Teitelboim, ed.), Plenum Press, New York (1988).
32. G. Curci and R. Ferrari, *Nuovo Cimento* **35A** (1976) 273; T. Kugo and I. Ojima, *Suppl. Prog. Theor. Phys.* **66** (1979) 1.
33. M. Henneaux and C. Teitelboim, *Ann. Phys. (N.Y.)* **143** (1982) 127.
34. M. Henneaux and C. Teitelboim, in: *Quantum Field Theories and Quantum Statistics*, E. S. Fradkin Festschrift (I. A. Batalin, C. J. Isham, and G. A. Vilkovisky, eds.), Adam Hilger, Bristol (in press).

References

35. J. H. Schwarz, Caltech preprint CALT-68-1304.
36. J. H. Schwarz, *Phys. Rep.* **89** (1982) 223.
37. R. C. Brower and C. B. Thorn, *Nucl. Phys.* **B31** (1971) 163; C. B. Thorn, *Nucl. Phys.* **B248** (1984) 551.
38. J. H. Schwarz, *Phys. Rep.* **8** (1973) 269.
39. S. Fubini and G. Veneziano, *Nuovo Cimento* **67A** (1970) 29; *Ann. Phys.* (*N.Y.*) **63** (1971) 12; J. L. Gervais, *Nucl. Phys.* **B21** (1970) 192.
40. E. Del Giudice, P. Di Vecchia, and S. Fubini, *Ann. Phys.* (*N.Y.*) **70** (1972) 378.
41. L. Brink, *Supersymmetry* (K. Dietz, R. Flume, G. V. Gehlen, and V. Rittenberg, eds.), p. 89, Plenum Press, New York (1984); see also L. Brink and H. B. Nielsen, *Phys. Lett.* **45B** (1973) 332.
42. A. Neveu and J. H. Schwarz, *Nucl. Phys.* **B31** (1971) 86.
43. P. Ramond, *Phys. Rev.* **D3** (1971) 2415. The two-dimensional supersymmetry of these models was analyzed in: J. L. Gervais and B. Sakita, *Nucl. Phys.* **B34** (1971) 477.
44. M. B. Green and J. H. Schwarz, *Phys. Lett.* **109B** (1982) 444.
45. L. Brink, P. de Vecchia, and P. Howe, *Phys. Lett.* **65B** (1976) 471; S. Deser and B. Zumino, *Phys. Lett.* **65B** (1976) 369.
46. L. Brink and J. H. Schwarz, *Nucl. Phys.* **B121** (1977) 285; M. Ademollo, L. Brink, A. D'Adda, R. D'Auria, E. Napolitano, S. Sciuto, E. Del Giudice, P. Di Vecchia, S. Ferrara, F. Gliozzi, R. Musto, and R. Pettorino, *Phys. Lett.* **62B** (1976) 105; *Nucl. Phys.* **B114** (1976) 297; same authors and J. H. Schwarz, *Nucl. Phys.* **B111** (1976) 77; M. Pernici and P. van Nieuwenhuizen, Stony Brook preprint ITP-85-84 (1985).
47. C. Teitelboim (unpublished). The square-root approach was applied to $d = 4$ supergravity in: R. Tabensky and C. Teitelboim, *Phys. Lett.* **69B** (1977) 453; C. Teitelboim, *Phys. Rev. Lett.* **38** (1977) 1106.
48. Y. Iwasaki and K. Kikkawa, *Phys. Rev.* **D8** (1973) 440.
49. M. Pilati, Ph.D. Thesis, Princeton University (1980).
50. F. Gliozzi, J. Scherk, and D. Olive, *Nucl. Phys.* **B122** (1977) 253; see also: *Phys. Lett.* **65B** (1976) 282.
51. J. Fisch, *Quantification BRST des Théories de cordes*, Mémoire de licence, Université Libre de Bruxelles (1985-1986).
52. N. Ohta, *Covariant Quantization of Superstrings Based on BRST Invariance*, University of Texas preprint (1985); H. Terao and S. Uehara, *Covariant Second Quantization of Free Superstring*, Kyoto preprint RRK85-32 (1985).
53. M. Henneaux, *Phys. Lett.* **177B** (1986) 35; **183B** (1987) 59.
54. E. F. Corrigan and P. Goddard, *Nucl. Phys.* **B68** (1974) 189; R. C. Brower and K. A. Friedman, *Phys. Rev.* **D7** (1973) 535; J. H. Schwarz, *Nucl. Phys.* **B46** (1972) 61. See also J. Scherk, *Rev. Mod. Phys.* **47** (1975) 123 and in Ref. 18.
55. M. B. Green and J. H. Schwarz, *Nucl. Phys.* **B181** (1981) 502; **B198** (1982) 252, 441.
56. M. B. Green and J. H. Schwarz, *Phys. Lett.* **136B** (1984) 367.
57. M. Henneaux and L. Mezincescu, *Phys. Lett.* **152B** (1985) 340.
58. J. Wess and J. Bagger, *Supersymmetry and Supergravity*, Princeton University Press, Princeton (1983).
59. J. Wess and B. Zumino, *Phys. Lett.* **37B** (1971) 95; E. Witten, *Nucl. Phys.* **B223** (1983) 422.
60. T. L. Curtright, L. Mezincescu, and C. K. Zachos, *Phys. Lett.* **161B** (1985) 79; M. T. Grisaru, P. Howe, L. Mezincescu, B. E. W. Nilsson, and P. K. Townsend, *Phys. Lett.* **162B** (1985) 116; E. Witten, *Nucl. Phys.* **B266** (1986) 245.
61. W. Siegel, *Phys. Lett.* **128B** (1983) 397.
62. I. Bengtsson and M. Cederwall, Göteborg preprint 84-21 (1984).
63. T. Hori and K. Kamimura, *Prog. Theor. Phys.* **73** (1985) 476.
64. L. Brink and J. H. Schwarz, *Phys. Lett.* **100B** (1981) 310.
65. W. Siegel, *Class. Quant. Grav.* **2** (1985) L95; *Nucl. Phys.* **B263** (1986) 93.

66. I. A. Batalin and E. S. Fradkin, *Phys. Lett.* **122B** (1983) 157.
67. L. Brink, O. Lindgren, and B. E. W. Nilsson, *Nucl. Phys.* **B212** (1983) 401; S. Mandelstam, *Nucl. Phys.* **B213** (1983) 149; L. Brink, in: *Unified String Theories* (M. B. Green and D. J. Gross, eds.), p. 244, World Scientific Publishing Company, Singapore (1986).
68. W. Nahm, *Nucl. Phys.* **B135** (1978) 149.
69. E. Cremmer and J. Scherk, *Nucl. Phys.* **B103** (1976) 399.
70. L. Dolan and R. Slansky, *Phys. Rev. Lett.* **54** (1985) 2075.
71. F. Englert and A. Neveu, *Phys. Lett.* **163B** (1985) 349.
72. D. J. Gross, J. A. Harvey, E. Martinec, and R. Rohm, *Phys. Rev. Lett.* **54** (1985) 502; *Nucl. Phys.* **B256** (1985) 253; **B267** (1986) 75. D. J. Gross, in: *Quantum Mechanics of Fundamental Systems* (C. Teitelboim, ed.), Plenum Press, New York (1988).
73. L. Brink, J. H. Schwarz, and J. Scherk, *Nucl. Phys.* **B121** (1977) 77.

Index

Action
  of bosonic string, 10, 99
  of classical point particle, 9, 97, 103
  of heterotic string, 42
  of spinning string, 22, 24–25, 205
  of superparticle, 256
  of superstring, 31, 245, 254
Algebra: *see* Clifford, Conformal, Kac–Moody, Super-conformal, Super-Poincaré, Super-Virasoro, Virasoro
Amplitudes: *see* Scattering amplitudes
Analytic continuation, 51
Anomaly
  BRST, 168–169, 219, 224–226
  Lorentz–Poincaré, 182, 229
  gauge, 79
  see also Central charge
Antisymmetric tensor fields
  fourth rank, 38
  second rank, 38–39, 186
Area, 10, 98

Baker–Hausdorff formula, 50
Boundary conditions
  for bosonic string
    closed, 12, 15, 99
    open, 12, 15, 110–113, 123–124

Boundary conditions (*Cont.*)
  for spinning string
    closed, 24, 209
    open (Neveu–Schwarz), 22, 26, 207
    open (Ramond), 22, 26, 207
  for superstring, 32, 246
BRST charge
  of bosonic string, 166, 168
  of spinning string, 218, 223–224
BRST invariant extension
  of constraints, 180
  of level operator, 171–220
BRST nilpotency
  classical, 163
  quantum (bosonic string), 169–170
  quantum (spinning string), 219, 226
BRST observable, 164, 179
BRST physical subspace
  of bosonic string, 170–174
  of spinning string (Neveu–Schwarz), 222
  of spinning string (Ramond), 228
BRST proof of no ghost theorem: *see* Quartet mechanism

Calabi–Yau spaces, 84
Cauchy–Riemann conditions, 106, 109

Central charge
  in Virasoro algebra, 14, 151, 155, 162
  in super-Virasoro algebra, 27–28, 215
Changes of coordinates: see Reparametrization invariance
Chan–Paton method
  allowed gauge groups, 39
  factors, 36
Chan variables, 55
Chirality in ten dimensions, 37, 247, 279
Clebsch–Gordon coefficients, 87
Clifford algebra, 35, 87, 214, 232, 272
Coherent states, 46
Cohomology: see BRST physical subspace
Compactification on a hypertorus, 43
Components of string field, 66
Conformal algebra in two dimensions, 107–109, 121; see also Virasoro algebra
Conformal factor, 102
Conformal gauge, 106, 133–136
Conformal invariance, 107, 121, 180–181
Conformal Killing vectors, 107–108
Conformal spin, 54, 192
Conformal transformations, 105–107
Constraints: see Hamiltonian, Primary, Secondary, Super-Virasoro, Virasoro
Cosmological constant, 85
Counterterms, 64, 68
Covariant quantization
  for bosonic string, 13–14, 187–191
  for spinning string, 24–28, 214–215
Critical dimension
  of bosonic string, 17, 151, 169, 187
  of spinning string, 24, 217, 219, 224
  of superstring, 255
Cyclic symmetry, 56

DDF
  operators, 178, 193–196
  states, 193–196
Decoupling of null states, 164, 174, 199
Diffeomorphism group, 107, 121; see also Reparametrization invariance
Differential forms on supermanifolds, 238–239
Dilaton, 186
Dirac bracket, 144–145, 148, 267–269
Doubling, 174–178, 221–222, 228
Duality, 56

E_8, 45
$E_8 \times E_8$, 45, 273
Embedding, 98

End point motion, 112
Energy momentum tensor in two dimensions, 101, 107–110, 117–188, 162
Equations of motion
  for bosonic string
    in conformal gauge, 12, 25
    solutions of, 139–141
  for superstring, 246
Excited levels
  for bosonic string, 184, 186
  for spinning string, 230–231, 234
  for superstring, 36
Exterior derivative operator, 239

Fadde'ev–Popov ghosts, 49, 57; see also Ghosts
Fermionic symmetry
  for superparticle, 257
  for superstring, 243–245
Feynman pole-prescription, 51
Feynman rules, 51
Field theory of strings
  higher-string interactions, 78
  reality constraints, 63, 68
  representation of super Poincaré algebra
    free case, 61–64
    interacting case (type IIB), 69–72
  three-string interaction, 69–75, 78
Fierz formula, 88
Fierz rearrangements, 245, 261–262
Finiteness, 79–81
First class constraints, 12, 118, 204, 260
Fock representation, 152–153
Fock space, 153–214
Fourier modes: see Oscillator modes
Frenkel–Kac construction, 46
Functional measure, 66

Gauss's law, 114
Ghost number, 178–179, 228
Ghosts
  Fock space of
    bosonic string, 166–168
    spinning string (Neveu–Schwarz), 217
    spinning string (Ramond), 222–223
    zero modes of, 166–168, 223
  see also BRST
Ghosts of ghosts, 263–266
G parity, 230
Graviton, 38, 186, 232
GSO projection, 24, 235

Index

Hamiltonian constraints, 11, 125, 205
Harmonic map, 103

Induced metric, 98, 116
Infinite reducibility, 266
Interactions among strings, 69–75, 78; see also Vertex operator
Intercept parameter, 183, 224

Kac–Moody algebra, 159–162
Klein–Gordon scalar product, 175, 177
Koba–Nielsen amplitude, 54

Lagrange multiplier, 43, 117, 119, 206
Lapse, 115, 117, 119
Lattice
  even, 44
  $\Gamma_8 \times \Gamma_8$, 45
  Γ_{16}, 45
  self-dual, 45
Light cone gauge
  for bosonic string, 14, 136–138
  Lorentz invariance in, 17, 181–183, 229, 232
  for spinning string, 25, 209
  for superparticle, 269
  for superstring, 251–255
Light cone spinor, 244, 262, 271
Light cone supersymmetry, 32
Little group, 183, 230
Loop graphs, 57–59
Lorentz generators: see Poincaré charges

Majorana spinor, 236, 241, 279
Majorana–Weyl spinor, 24, 236, 241–242, 279
Mass formula
  for bosonic string
    closed, 18, 186
    open, 17, 183
  for heterotic string, 43
  for spinning string
    open (Neveu–Schwarz), 23, 230
    open (Ramond), 21, 234
    closed, 231, 234
Mean (extrinsic) curvature, 100
Membranes, 99, 103
Möbius
  symmetry $SU(1, 1)$, 28
  transformations, 53

Negative-norm states, 153, 167, 178, 188, 218, 262

Neveu–Schwarz sector
  boundary conditions, 22, 26, 207
  fermionic generators, 27–28, 211
  spectrum, 230–231
No-ghost theorem, 178, 196–199, 222, 227
Nonoriented strings, 39
Normal ordering, 14, 153, 182
Null states, 189, 191

One-loop diagrams, 57
One-loop finiteness, 79
Ordering ambiguity, 153, 215
Oriented strings, 39
Orthonormal gauge, 12; see also Conformal gauge
Oscillator modes
  for bosonic string
    open, 13, 15, 129–130
    closed, 15, 131–132
  for heterotic string, 43
  for spinning string
    open (Neveu–Schwarz), 23, 211
    open (Ramond), 21, 210
  for superstring, 32

Path integral, 49
Perturbation expansion of S-matrix, 49–59
Physical state projection operator, 57
Poincaré charges, 16, 105, 124, 131, 133, 210
Poincaré generators: see Poincaré charges
Poincaré invariance, 10, 104, 181–183, 206, 229, 232
Poles: see Regge poles, Feynman pole prescription
Primary constraints, 11, 115, 117–118, 259
Pullback of differential forms, 239

Quadratic form of string action, 10, 100–102
Quantization
  BRST, 162–165; see also BRST physical subspace
  with constraints, 14, 28, 155
  see also Covariant quantization
Quantum gauge invariance, 163, 165, 180, 199–200
Quartet mechanism, 178, 221, 227, 275–277

Ramond sector
  boundary conditions, 22, 26, 207
  fermionic generators, 28, 215
  spectrum, 232–234
Rarita–Schwinger field, 24

Rarita–Schwinger multiplet, 37
Rarita–Schwinger states, 38–39
Regge poles, 3
Regge slope, 17
Regge trajectory, 3, 183
Reparametrization invariance, 10, 97, 99, 103–104, 119–121, 155–157
Representations of Poincaré group, 21, 183
Representations of super-Poincaré algebra, 61–65; see also Super-Poincaré algebra, Super-Poincaré invariance
Riemann zeta function, 17
Rigidly rotating string, 113–114

Scattering amplitudes
 as correlation functions, 53
 for point particles, 49
 and residues, 53–54
Secondary constraints, 11, 118, 259
Second class constraints, 250, 260
Second fundamental form, 100
Shadow world, 85
Shapiro–Virasoro model
 extended, 186
 restricted, 186
Shift, 115, 119
Sigma model, 103, 161, 237
SO(8)
 covariance, 35
 eight-dimensional representation of, 87
 and spinning string spectrum, 230–234
 triality, 37
SO(24), 183–184
SO(25), 183–184
SO(32), 45, 273
SO(d-2), 31
SO(n), 36, 39, 161–162
Spectrum
 of bosonic string
  closed, 186
  open, 184–185
 of heterotic string, 45
 of spinning string (Neveu–Schwarz), 230–231
 of spinning string (Ramond), 232–234
 of superstring
  type I closed, 38–39
  type I open, 35–37
  type IIA, 38
  type IIB, 37–38
Speed of light renormalization, 18
Spin (32)/Z_2, 45

Spurious states, 199
Square root of bosonic string, 202–207
String field, 63
String field conjugate momemtun, 63
String tension, 10, 113
SU(4) × U(1), 66–67
SU(N), 161–162
Sugawara construction, 161
Supercharge, 32–33, 250, 271
Superconformal algebra, 27–28, 205; see also Super-Virasoro algebra
Superconformal invariance, 26
Superfield formalism, 66
Supergauge transformation, 26, 209
Supergravity
 in ten dimensions, 46, 236, 256
 in two dimensions
  extended, 28
  $N=1$, 24, 205
Super-Poincaré algebra
 second quantized representation of, 62
 in the light cone, 32–34
Super-Poincaré invariance, 32, 41–42, 237, 258
Superspace formulation of vertex operators, 57
Superspace in two dimensions, 25
Supersymmetry
 in ten dimensions, 32, 235, 237
 in two dimensions, 24
Super-Virasoro algebra, 27–28, 215
Super-Virasoro constraints, 27–28, 215–216
Super-Yang–Mills supermultiplet, 36, 236, 256
Super-Yang–Mills theory, 46, 236
Super-Weyl invariance, 206
SUSY(N)/SO($d-1$, 1), 237

Tachyon, 18, 23, 44, 184, 230–231, 233–234
Target space, 237
Transverse gauge: see Light cone gauge
Transverse states: see DDF states
Tree diagrams, 52–56
Trivial gauge transformations, 247
Truncation, 175, 235
Twist operator
 for bosonic string, 57
 for fermionic string, 58
Type I superstrings, 34, 36, 256
Type IIA superstrings, 34, 38, 256
Type IIB superstrings, 34, 37, 256

U(n), 36, 39

Index

Veneziano model, 3, 14
Vertex operator
  for bosonic string
    tachyon emission, 52, 192
    vector emission, 193
  for spinning string
    fermion emission, 59
    tachyon emission, 57
  for type IIB theory (light cone), 73
Virasoro algebra, 14, 107, 125, 128–129, 154–155
Virasoro conditions: *see* Virasoro constraints
Virasoro constraints, 13, 55, 114–115, 127–129, 130, 132, 156, 187

Wess–Zumino term, 240–242
Weyl invariance, 10, 102, 117, 119–120
Wheeler–DeWitt equation, 155–157
Wick rotation, 51
Winding number, 43

Zero-mode translations, 138, 147
Zero-norm states, 180
Zero-point fluctuations
  for bosonic string, 17
  for heterotic string, 44
  for spinning string, 22